SPACES OF GLOBAL KNOWLEDGE

Studies in Historical Geography

Series Editor: Professor Robert Mayhew, University of Bristol, UK

Historical geography has consistently been at the cutting edge of scholarship and research in human geography for the last fifty years. The first generation of its practitioners, led by Clifford Darby, Carl Sauer and Vidal de la Blache presented diligent archival studies of patterns of agriculture, industry and the region through time and space.

Drawing on this work, but transcending it in terms of theoretical scope and substantive concerns, historical geography has long since developed into a highly interdisciplinary field seeking to fuse the study of space and time. In doing so, it provides new perspectives and insights into fundamental issues across both the humanities and social sciences.

Having radically altered and expanded its conception of the theoretical underpinnings, data sources and styles of writing through which it can practice its craft over the past 20 years, historical geography is now a pluralistic, vibrant and interdisciplinary field of scholarship. In particular, two important trends can be discerned. Firstly, there has been a major "cultural turn" in historical geography which has led to a concern with representation as driving historical-geographical consciousness, leading scholars to a concern with text, interpretation and discourse rather than the more materialist concerns of their predecessors. Secondly, there has been a development of interdisciplinary scholarship, leading to fruitful dialogues with historians of science, art historians and literary scholars in particular which has revitalised the history of geographical thought as a realm of inquiry in historical geography.

Studies in Historical Geography aims to provide a forum for the publication of scholarly work which encapsulates and furthers these developments. Aiming to attract an interdisciplinary and international authorship and audience, Studies in Historical Geography will publish theoretical, historiographical and substantive contributions meshing time, space and society.

Spaces of Global Knowledge

Exhibition, Encounter and Exchange in an Age of Empire

Edited by

DIARMID A. FINNEGAN
Queen's University Belfast, UK

JONATHAN JEFFREY WRIGHT
Maynooth University, Ireland

Routledge
Taylor & Francis Group

LONDON AND NEW YORK

First published 2015 by Ashgate Publishing

Published 2016 by Routledge
2 Park Square, Milton Park, Abingdon, Oxon OX14 4RN
711 Third Avenue, New York, NY 10017, USA

First issued in paperback 2018

Routledge is an imprint of the Taylor & Francis Group, an informa business

British Library Cataloguing in Publication Data
A catalogue record for this book is available from the British Library

The Library of Congress has cataloged the printed edition as follows:
Finnegan, Diarmid A.
    Spaces of global knowledge : exhibition, encounter and exchange in an age of empire /
by Diarmid A. Finnegan and Jonathan Jeffrey Wright.
      pages cm. – (Studies in historical geography)
    Includes bibliographical references and index.
    ISBN 978-1-4724-4436-3 (hardback)
   1. Historical geography—History--19th century. 2. Scientific expeditions--History--19th
century. 3. Intellectual life--History--19th century. I. Wright, Jonathan Jeffrey. II. Title.
    G141.F56 2015
    909.81–dc23
                                                                    2015012405

ISBN 13: 978-1-138-54681-3 (pbk)
ISBN 13: 978-1-4724-4436-3 (hbk)

Contents

List of Figures and Tables

Figures

Table

List of Contributors

Angela Byrne teaches in the Department of History, Politics and Social Sciences at the University of Greenwich. She is the author of *Geographies of the Romantic North: Science, Antiquarianism, and Travel, 1790–1830* (2013).

Caroline Cornish is an Honorary Research Fellow of the Department of Geography at Royal Holloway, University of London, where she completed her PhD on the historical geography of the former Museum of Economic Botany at the Royal Botanic Gardens, Kew. Her postdoctoral research has explored the ways in which botany was constructed as a discipline in the nineteenth and twentieth centuries by Kew and the Natural History Museum.

Diarmid A. Finnegan is Senior Lecturer in Human Geography at the School of Geography, Archaeology and Palaeoecology, Queen's University Belfast. He is the author of *Natural History Societies and Civic Culture in Victorian Scotland* (2009).

Louise C. Henderson was awarded her PhD from Royal Holloway, University of London in April 2012. Her research, which encompasses the history of geographical thought, publishing and print histories and nineteenth-century travel and exploration, has featured in a special issue of the *Scottish Geographical Journal* marking the bicentenary of David Livingstone's birth.

Sarah Hunter is a PhD student at Trinity College Dublin. Her research explores the interaction of Irish non-governmental organisations and Indian populations in British Bengal, c.1880–1935.

Nuala C. Johnson is Reader in Human Geography at Queen's University Belfast. She is the editor of *Culture and Society* (2008) and the author of *Ireland, the Great War and the Geography of Remembrance* (2003) and *Nature Displaced, Nature Displayed: Order and Beauty in Botanical Gardens* (2011).

Robert J. Mayhew is Professor of Historical Geography and Intellectual History at the University of Bristol. His most recent book, *Malthus: The Lives and Legacies of an Untimely Prophet* (2014), provides a study of Malthus's ideas and reception. He has also recently edited Malthus's *An Essay on the Principle of Population* for Penguin Classics.

Sarah Louise Millar is a PhD student at the University of Edinburgh. Her doctoral research addresses eighteenth- and nineteenth-century maritime exploration and she has published an article, exploring the process of depth sounding in early nineteenth-century British Polar expeditions, in the *Journal of Historical Geography*.

Polina Nikolaou is a postdoctoral researcher who has recently completed a PhD at the University of Exeter. Her doctoral research explored the historical geographies of Cypriot archaeology in the second half of the nineteenth century.

Karen N. Salt is a Lecturer in the School of Languages and Literature at the University of Aberdeen. Her research focuses on the geographies of race and sovereignty in the nineteenth-century Caribbean and on the interworking of race, power and place in the management and protection of island systems.

Charles W.J. Withers holds the Ogilvie Chair of Geography in the School of Geosciences, Edinburgh University. His publications include *Placing the Enlightenment: Thinking Geographically About the Age of Reason* (2007), and *Geography and Science in Britain, 1831–1939: A Study of the British Association for the Advancement of Science* (2010).

Jonathan Jeffrey Wright is Lecturer in British History at Maynooth University. He is the author of *The 'Natural Leaders' and Their World: Politics, Culture and Society in Belfast, c.1801–1832* (2012).

Acknowledgements

This collection brings together established and early-career scholars from across the UK and Ireland interested in the spatiality of knowledge and knowledge practices in an era marked by the rapid proliferation and intensification of global networks of circulation and exchange. In part responding to recent efforts to construct 'global histories', the essays also take inspiration from work on the spatialities of modern forms of knowledge. This productive multi-disciplinary endeavour would not have been possible without financial support provided by the Arts and Humanities Research Council (research grant AH/J004952/1) and by the Historical Geography Research Group (a study group of the RGS–IBG).

The editors are grateful for the encouragement of Professor Robert Mayhew, editor of the 'Studies in Historical Geography' series and for the support provided by editorial staff at Ashgate. We also wish to thank Charles Withers for advice throughout and to place on record our gratitude for the encouragement we received from the late Christopher Bayly. Finally, we reserve our chief thanks for the contributors to this volume. Despite the differing levels of experience with the world of academic publishing, all have acted with professionalism and patience. It has been a pleasure to work with authors willing to engage with the overall theme of the volume despite the challenges and hesitations occasioned by a concern with a category as contested as 'global knowledge'. Their varying perspectives on the potential and pitfalls involved in tracing the geographies of global knowledge in the nineteenth century have enriched and enlivened our own understanding. We hope that readers of these chapters also experience a deeper appreciation of the varied spaces in which knowledge was encountered, exhibited and exchanged in a global 'age of empire'.

Introduction

Placing Global Knowledge in the Nineteenth Century

Diarmid A. Finnegan and Jonathan Jeffrey Wright

In the late eighteenth century, the world became simultaneously smaller and larger. Smaller, in the sense that European (and particularly British) imperial expansion encompassed ever more remote and unexplored parts of the globe, establishing tentative connections between metropolitan cores and colonial peripheries that would be strengthened in the century to follow; and larger, in the sense that this process of exploration revealed cultures, species and other phenomena that had, hitherto, lain beyond the bounds of 'western' knowledge. In short, as attention was turned to the opportunities that existed in regions unexplored by Europeans and the world became, in geographical terms, *terra cognita*, it became clear that there was much more, in scientific and ethnographic terms, to know.[1]

This process is, of course, illustrated by the well-known Pacific voyages of Captain James Cook. As James M. Hodge notes, these voyages, which combined exploration, imperial expansion and scientific observation, 'have been seen by many as signalling the beginning of a novel phase in European exploration, setting new standards for the scope and accuracy of surveying and empirical observation'.[2] The first of Cook's three voyages provides a case in point.[3] Leaving Britain in 1768, Cook sailed first for Tahiti where his expedition was tasked with recording the Transit of Venus across the Sun.[4] Famously, Sir Joseph Banks joined Cook in the capacity of 'official botanist', bringing with him 'a mass of equipment' and 'an eight-man natural history "suite"', whose number included his close friend,

---

1   Joseph M. Hodge, 'Science and Empire: An Overview of the Historical Scholarship', in *Science and Empire: Knowledge and Networks of Science across the British Empire, 1800–1970*, ed. Brett M. Bennett and Joseph M. Hodge (Basingstoke: Palgrave Macmillan, 2011), 5–7.

2   Hodge, 'Science and Empire', 5.

3   The literature on Cook is extensive, but see, for an overview of his life and changing reputation, Glyn Williams, *The Death of Captain Cook: A Hero Made and Unmade* (London: Profile, 2008). The following account of his first voyage draws primarily on Richard Holmes, *The Age of Wonder: How the Romantic Generation Discovered the Beauty and Terror of Science* (London: Harper Press, 2009), 1–59. But see also Hodge, 'Science and Empire', 6, and Miles Ogborn, *Global Lives: Britain and the World, 1550–1800* (Cambridge: Cambridge University Press, 2008), 295–328.

4   Holmes, *Age of Wonder*, 5.

Dr Daniel Solander.[5] While at sea, Banks occupied himself with fishing, shooting and scientific study, and upon arrival at Tahiti he composed a wide-ranging journal, rich in ethnographic detail, and gathered numerous zoological and botanical specimens.[6] This process of collecting continued as the expedition moved on to Australia and New Zealand, and when they arrived back in Britain in 1771 Banks and Solander brought with them 'over a thousand new plant specimens, over five hundred animal skins and skeletons, and innumerable native artefacts'.[7] But if the *Endeavour* voyage was important in scientific terms, it was important also in imperial terms. That Banks, later portrayed as the 'the staunchest imperialist of the day', was present on the voyage is, in itself, significant, as is the fact that the expedition had an overtly imperial mission.[8] In addition to tracking the Transit of Venus, it was instructed by the British admiralty to investigate claims that New Zealand formed the tip of an unexplored southern continent. If true, Cook was directed to map this continent and claim it for the British.[9]

Cook's first Pacific voyage thus provides a striking example both of European exploration, and of the interrelation of imperial expansion and knowledge construction, but it is just one of many such examples. The *Endeavour* was not, for instance, the first European ship to visit Tahiti: Captain Wallis' *Dolphin* had done so in 1767, and the French arrived in 1768.[10] Moreover, the Pacific was not the only ocean to be explored at this time. The Atlantic and Arctic Oceans were also explored and Cook and Banks had first met on a voyage to Labrador and Newfoundland.[11] Reflective of 'an international movement in the spirit of the Enlightenment', such voyages were not simply concerned with exploring unknown quarters of the world, but with bringing those quarters 'cartographically, anthropologically, botanically, zoologically, and geologically within the bounds of European science'.[12]

* * *

This volume addresses the phenomena – imperial expansion, global connection and constructions of knowledge – so neatly linked in the exploratory voyages of the late eighteenth century. Yet, while taking these staple concerns as a given, the chapters also push beyond such well-known instances of a 'global' quest for knowledge both in empirical and conceptual terms. Although differing in focus,

---

5   Holmes, *Age of Wonder*, 10.

6   Holmes, *Age of Wonder*, 1, 6–7, 20–21, 29–30.

7   Holmes, *Age of Wonder*, 43.

8   David N. Livingstone, *Putting Science in its Place: Geographies of Scientific Knowledge* (Chicago: University of Chicago Press, 2003), 174.

9   Holmes, *Age of Wonder*, 10.

10   Holmes, *Age of Wonder*, 3.

11   Holmes, *Age of Wonder*, 9; N.A.M. Rodger, *The Command of the Ocean: A Naval History of Britain, 1649–1815* (London: Penguin, 2006), 327–8.

12   Rodger, *The Command of the Ocean*, 327; Livingstone, *Putting Science in its Place*, 10.

methodology and disciplinary orientation, the chapters that follow work with and against the composite and contested category of 'global knowledge' and explore the numerous ways and places in which such knowledge was constructed, communicated and contested during the long nineteenth century. This raises an obvious definitional question: what, exactly, might be meant by the phrase 'global knowledge'? But before addressing this question, the volume's chronological focus deserves brief comment.

Broadly speaking, *Spaces of Global Knowledge* focuses on the 'long nineteenth century'. As much recent scholarship in history and historical geography has demonstrated, this era, which stretched from the late eighteenth to the early twentieth century, was one that was characterised by imperialism, exploration and economic exploitation, and that brought with it a huge increase in the creation and crossing of cultural and intellectual boundaries. It was, in C.A. Bayly's influential analysis, an era which witnessed the development of 'global *uniformities* in the state, religion, political ideologies, and economic life' and the emergence of the interconnected, modern world we know today.[13] Of course, it goes without saying that imperial expansion and global interconnection were in no sense unique to the long nineteenth century. The expansion of European empires can be traced back to the renaissance and it is widely accepted that globalisation has a long 'prehistory', even if the nature of that prehistory has been subject to debate.[14] To give just a few examples, Immanuel Wallerstein famously posited the emergence of a 'modern world-system' in the mid sixteenth century, linking it to the development of the Spanish and Portuguese empires, and Miles Ogborn has, more recently, adopted a similar chronology.[15] While alive to the elisions inherent in sweeping arguments relating to the emergence 'of a single capitalist world system whose changing geography of core, periphery and semi-periphery defines the fortunes of different parts of the globe', Ogborn concedes that the world *did* change 'between the sixteenth and eighteenth centuries', and that it was during the sixteenth century that England, under Elizabeth I, 'stepped – or rather, sailed – on to the global stage', if only as 'minor players' in a game still dominated by the Spanish and the Portuguese.[16] Indeed, Bayly concedes the existence of 'older networks and dominances created by geographical expansion of ideas and social forces from the local and regional level to the inter-regional and inter-continental level', characterising these as 'archaic' and 'early modern' globalisations.[17]

---

13   C.A. Bayly, *The Birth of the Modern World, 1780–1914: Global Connections and Comparisons* (Oxford: Blackwell, 2004), 1.

14   Jürgen Osterhammel and Niels P. Petersson, *Globalization: A Short History* (Princeton: Princeton University Press, 2005), 28.

15   Osterhammel and Petersson, *Globalization*, 28, 31; Immanuel Wallerstein, *The Modern World System I: Capitalist Agriculture and the Origins of the European World-Economy in the Sixteenth Century* (New York: Academic Press, 1974).

16   Ogborn, *Global Lives*, 1, 4, 16.

17   Bayly, *The Birth of the Modern World*, 41–2.

As long and complex as its prehistory is, however, it is difficult to escape the conclusion that, in European perspective, from around the late eighteenth century (the era of Cook's voyages) onwards, the process of global interconnection accelerated.[18] War and imperial expansion drew and redrew global political geographies and opportunities for voyaging, discovery and transcultural encounter and exchange were created that, if not unprecedented, were arguably greater than at any time in human history. Notwithstanding his identification of its archaic and early-modern antecedents, globalisation remains, for Bayly, primarily a feature of the long nineteenth century, whether understood in the sense of 'growing interconnectedness' or, more problematically, of 'growing uniformity'.[19] Likewise, while agreeing with Wallerstein that 'the beginning of a basically irreversible process of worldwide integration' can be identified in the early sixteenth century, Jürgen Osterhammel and Niels P. Petersson argue that 'from about 1750 to 1880' there occurred 'an expansion of worldwide integration unprecedented in its intensity and influenced by the new capacities in production, transportation, and communication created by the Industrial Revolution'.[20]

The making, circulation and reception of knowledge – the theme with which this book is concerned – was, of course, central to this process of worldwide integration, whether manifest in the expansion of formal empires, the creation of informal empires and spheres of influence or the intercultural processes of exhibition, encounter and exchange that imperial enterprise occasioned. Hand in hand with the expansion of imperial power and 'western' influence came attempts both to 'know' the wider world, and, indeed, to inform and educate it. Thus, alongside the ubiquitous administrators and soldiers, the global world of the nineteenth century was one populated by missionaries, merchants and explorers (at times one and the same), doctors and cartographers, engineers and botanists. Yet these were not, of course, the only actors. Representatives of non-western populations were not simply the passive recipients of attempts to spread 'civilisation', but historical actors who could – and did – exert agency in their cultural and political encounters with occidental others. Such actors, western and non-western, connected and networked the global world of the nineteenth century, and it is their life-geographies, practices and varied contributions to knowledge – be they medical or botanical, cartographic or cultural – that constitute the focus of this volume. Through the activities of this range of actors, knowledge of the world was constructed, communicated and contested in ways that helped to multiply and reinforce already burgeoning global connections and trans-cultural contacts.

If the long nineteenth century can be usefully characterised as a global age this points to the possibility of identifying in that period a developing global knowledge

---

18   We are aware that we have not escaped from a largely European temporal frame of reference and acknowledge that any periodisation is inescapably perspectival and open to contestation.

19   Bayly, *The Birth of the Modern World*, 2.

20   Osterhammel and Petersson, *Globalization*, 28.

economy. Yet the meaning of the category suggested by this possibility – 'global knowledge' – is not easily determined. Here we take such definitional indeterminacy as an opportunity for critical inquiry rather than analytical paralysis. As a point of departure, global knowledge might be understood in at least two ways. First, it can be defined as a form of knowledge that moves between and beyond multiple and widely dispersed territories or locales and connects individuals or institutions distributed across the world. That definition, rather than confirming the idea of knowledge as universal truth, can be used to raise questions about how knowledge was transported through global space, in what form and in whose interests. Exploring such questions can highlight the ways in which knowledge exceeded as well as expressed the aims of one dominant group or prevailing political arrangements. Global knowledge, in escaping the confines of certain territorial logics, might be taken to name the results of inquiries that were either 'trans-imperial' or, indeed, 'more-than-imperial' in nature.

A second definition takes global knowledge to refer to the investigation or pursuit of patterns, distributions or trends considered trans-continental or worldwide in extent by historical actors. This construal might be usefully deployed to draw attention to motivations and mental constructs more than 'actual' worldwide connections, inter-cultural contacts and long-distance communications and to point to a regulative ideal that was a commonplace characteristic of many knowledge enterprises in the long nineteenth century. The two senses are not, of course, entirely independent but together begin to prise open the range of meanings and research possibilities contained within the capacious and suggestive category of 'global knowledge'.

Taking the global placement of knowledge and knowledge of ostensibly global phenomena during the long nineteenth century as a starting point directly connects this volume with a burgeoning literature on 'global histories'. As has been often pointed out, this recent historiographical concern has emerged from an intellectual and cultural environment that takes the 'global' to be a crucial analytical term, a prime object of scholarly investigation and an urgent social and political issue. The resultant scholarly enterprise might be said to generate as well as scrutinise certain kinds of 'global knowledge'. Not surprisingly, then, the turn to global histories has been regarded as at once promissory and problematic. On the one hand, crafting histories that locate particular episodes in a global context have been offered as a way to circumvent the dangers of parochialism that lurked within the valorisation of 'contextual history'. In the words of one recent introduction, global history works 'against the static geography of place in favour of a geography of linkages' thereby overcoming historiographical myopia.[21] Adopting a global approach has also been presented as a strategy to overcome the privileging of particular global regions and as an opportunity to generate new 'polycentric' narratives of world history that are not bound with what Patrick O'Brien has called the 'fishy glue of

21 Emily Rosenberg, *A World Connecting, 1870–1945* (Cambridge, MA: Harvard University Press, 2012), 8–9.

Eurocentrism'.[22] On the other hand, the enthusiastic rush to embrace the global has been criticised for smoothing over the profoundly unequal effects of political power and for ignoring the irreducible specificities of cultural difference.[23] Mike Hulme's argument that global climate science tends to produce a 'brittle' form of knowledge that cannot accommodate or do justice to genuine diversity can be more generally applied to all attempts to capture 'global' trends or processes of change.[24]

The opportunities and challenges of writing global histories of knowledge enterprises can be usefully elaborated by engaging with recent work in the history of science, a scholarly field particularly relevant to the empirical bearing of this volume. Following wider historiographical trends, historians of science have in recent years queried the discipline's limited geographical ambitions and scope. A widespread penchant for micro-histories of science that located the production or consumption of scientific knowledge in precise and carefully delimited local settings has been challenged by scholars wishing to reconstruct 'generalist' accounts of scientific change that do not presume a positivistic understanding of science as universal by definition, but which move beyond the limits of highly detailed case studies.[25] Alongside this, a concern with understanding how science became a quintessentially mobile form of knowledge that apparently travelled with unique efficiency across various kinds of boundaries has further encouraged a move towards global histories of science.[26] As a result, calls have been made to study the 'circulating practices' of global communication as a crucial context for understanding the spread and influence of scientific knowledge.[27] Others have echoed this and suggested 'scaling up' histories of particular scientific practices or objects of inquiry by examining both how global claims about natural phenomena have been constructed and how scientific projects exploited and expanded global networks and infrastructures in the past.[28]

Those interested in pursuing more globally orientated histories of science have also called for engagement with a widened and more heterogeneous repertoire of primary sources to allow for a 'cross-contextualisation' of very different sorts of

---

22   Patrick O'Brien, 'Historiographical Traditions and Modern Imperatives in the Restoration of Global History', *Journal of Global History* 1(1) (2006): 33.

23   Warwick Anderson, 'From Subjugated Knowledge to Conjugated Subjects: Science and Globalisation, or Postcolonial Studies of Science?' *Postcolonial Studies* 12 (2009): 389–400.

24   Mike Hulme, 'Problems with Making and Governing Global Kinds of Knowledge', *Global Environment Change* 20(4) (2010): 558–64.

25   For example, Robert E. Kohler, 'A Generalist's Vision', *Isis* 96(2) (2005): 224–9.

26   For one example of this, see Steven J. Harris, 'Long-Distance Corporations, Big Science, and the Geography of Knowledge', *Configurations* 6(2) (1998): 269–304.

27   James Secord, 'Knowledge in Transit', *Isis* 95(4) (2004): 654–72.

28   See, for example, Jeremy Vetter, ed., *Knowing Global Environments: New Historical Perspectives on the Field Sciences* (New Brunswick, NJ: Rutgers University Press, 2011).

ideas, practices and knowledge communities.[29] This archival move has been offered as a way to subvert over-reliance on records that reinforce a view of science's global trajectories and worldwide developments from 'the West' or from self-designated centres of imperial power. The conceptual lexicon of this kind of global history includes such terms as contact zone, hybridity, go-betweens, networks and assemblages. If those concepts derive in part from post-colonial theory, they are redeployed to look beyond the structuring effects of colonial relations, to 'fragment' knowledge traditions and bring to view the 'shifts and reinventions of a variety of ways of doing science across the world'.[30] The proposed aim, in other words, is to find ways of writing the history of science outside the 'limitations of a single national or imperial frame' that capture the 'itinerant' and 'extra-imperial' connections that shaped knowledge making activities in a global age.[31]

In this thinking, a global history of science offers a more inclusive account of knowledge construction than histories that, though adopting a critical stance, tend to remain within the terms of reference generated by the epistemic regimes of colonial power. Working against the 'bipolar vision' of postcolonial critique and concentrating instead on the 'fluid crossing of scales' by actors instrumental in the 'making of a planetary modernity' is suggested as a way to recover interactions and mediations that transgressed the standard dualisms of imperial imaginaries.[32] Proponents of this line of inquiry may risk eliding forms of resistance that were anti-global or deliberately disconnective, but argue nevertheless that it brings to view a much wider set of actors that functioned within, and made possible, global economies of knowledge.[33]

A more concrete expression of the global turn in the history of science can be found in one of most lively areas of interest in the history of science, the so-called Darwin industry. A key moment was the appearance of a series of articles published in *Nature* during the anniversary year of 2009 under the rubric of 'global Darwin'.[34] James Secord, using the same designation, further developed the argument that Darwin and Darwinism can be productively and provocatively set within a globalising nineteenth-century world. As Secord observes, Darwin

---

29 Sujit Sivasundaram, 'Science and the Global: On Methods, Questions and Theory', *Isis* 101 (2010): 146–58.

30 Sivasundaram, 'Science and the Global', 155.

31 Neil Safier, 'Global Knowledge on the Move: Itineraries, Amerindian Narratives, and Deep Histories of Science', *Isis* 101(1) (2010): 133–45, 138.

32 Simon Schaffer, Lissa Roberts, Kapil Raj and James Delbourgo, eds, *The Brokered World: Go-Betweens and Global Intelligence, 1780–1820* (Sagamore Beach, MA: Science History Publications, 2009), xv, xix.

33 Schaffer et al., *Brokered World*, xxx.

34 Marwa Elshakry, 'Global Darwin: Eastern Enchantment', *Nature* 461 (29 October 2009): 1200–1201; Daniel Todes, 'Global Darwin: Contempt for Competition', *Nature* 462 (5 November 2009): 36–7; James Pusey, 'Global Darwin: Revolutionary Road', *Nature* 462 (12 November 2009): 162–3; Jürgen Buchenau, 'Global Darwin: Multicultural Mergers', *Nature* 462 (19 November 2009): 284–5.

himself was thoroughly embedded in worldwide networks of exchange not least 'through the nineteenth century's first and most significant global communication system – the post office'. Moreover, Darwin's science – or, more concretely, his books – also crossed multiple political and linguistic boundaries and achieved a global influence. In tracking this movement, and in recording the response of readers in different parts of the world, Secord re-describes Darwinism as 'an outcome of criss-crossing lines of communication between diverse and often scattered localities' and as a 'shared context' for conversations about global cultural transformations.[35] As Secord notes, this move need not be set in straightforward opposition to more geographically circumscribed studies. The global movement of evolutionary texts can understood as contributing to, rather than detracting from, the articulation of local differences. Whether or not this does justice to the conditioning effects of local circumstances remains an open question. As David N. Livingstone has argued, reactions to Darwin could be dramatically different even among scattered but strongly linked communities that shared very similar cultural assumptions.[36] In such cases, it seems to make more sense to bring local rather than global concerns to the fore.

Arguably, then, the danger remains that global history flattens the highly varied political and cultural topography of knowledge-making enterprises. This has certainly been the view of a number of critics of global histories of science. Warwick Anderson, for example, has referred to widespread talk of 'global flows' and 'fluid' movements as a marking an uncritical 'hydraulic turn' among historians.[37] To Anderson, historical imaginations drawn by the lure of global history and forgetful of colonialism tend to evoke a world of relatively unencumbered agents and malleable power structures.[38] This, in turn, can be tied to new 'meta-narratives' that, while eschewing a view from nowhere, somewhat disingenuously offer a view from everywhere. As Claudia Stein and Roger Cooter have argued, even where there is an explicit sensitivity to the heterogeneity of past knowledge cultures, reconstructions made under the sign of the global, 'implicitly reproduce and foster [that] unifying construct'.[39] Global history cannot help but return to a form of universal history.

---

35 James Secord, 'Global Darwin', in *Darwin*, ed. William Brown and Andrew C. Fabian (Cambridge: Cambridge University Press, 2010), 45, 51.

36 For a recent and detailed statement, see David N. Livingstone, *Dealing with Darwin: Place, Politics and Rhetoric in Religious Engagements with Evolution* (Baltimore: Johns Hopkins University Press, 2014).

37 Warwick Anderson, 'Making Global Health History: The Postcolonial Worldliness of Biomedicine', *Social History of Medicine* 27(2) (2014): 372–84.

38 For a related critique of the use of 'circulation' and 'trading zones', see Fa Ti Fan, 'The Global Turn in the History of Science', *East Asian Science Technology and Society* 6(2) (2012): 249–58.

39 Claudia Stein and Roger Cooter, 'Visual Objects and Universal Meanings: AIDS Posters and the Politics of Globalisation and History', *Medical History* 55 (2011): 85–108, 107.

According to this perspective, rather than global histories, what is desired are 'critical studies of globalist ambitions', a move that resonates with an approach to 'global knowledge' that regards it as a description of intent rather than of historical realities.[40] The emphasis should not be on an all-encompassing global context for knowledge making but rather on exposing the genealogies of ostensibly global claims and the conditions that make those claims possible and plausible in different times and places. As a result, the term 'global' shifts from being an analytical category to the naming of an aspiration held by historical actors or, as Samuel Moyn and Andrew Sartori put it, a 'category belonging to the archive'.[41] Figured thus, knowledge has never been, and never can be, properly global even if that impossibility has helped to structure and stimulate intellectual endeavour.[42]

Others have developed a more radical version of this claim. To Bruno Latour, for example, the global does not describe a larger scale or context in which to situate knowledge making enterprises. Instead it simply refers to 'another equally local, equally micro place ... *connected* to many others'.[43] Scale in general, for Latour, is not something given in advance and thus useful for developing an analytical framework but rather 'is the actor's own achievement'.[44] As a consequence, when 'the global' is appealed to Latour suggests raising such questions as 'In which building? In which bureau? Through which corridor is it accessible? Which colleagues has it been read to? How has it been compiled? ... Through which optics is it projected?'[45]

One response to this deconstructive move and methodological reorientation has been to suggest that global history remains, in heuristic terms, an effective way of identifying an approach that 'experiments with [contingent] geographical boundaries'.[46] In this view, there is no reason why 'the global' must denote an account of the 'whole earth' or refer to something that applies to, or is manifest at, every point on the planet. Rather, the global can be usefully employed to refer to relations and movements that are in more modest sense stretched across the world. And while it may be possible to substitute the term global with concepts such as intercultural or inter/trans-national, this arguably comes at a cost. A global approach, for example, may want to unsettle the privileging of national boundaries or focus on the crossing or transcending of other kinds of borders than those of the

40   Anderson, 'Making Global History of Medicine', 381.

41   Samuel Moyn and Andrew Sartori, 'Approaches to Global Intellectual History', in *Global Intellectual History*, ed. Samuel Moyn and Andrew Sartori (New York: Columbia University Press, 2013), p. 17.

42   For a related history of 'witnessing' the globe or earthly sphere over the *longue durée*, see Denis Cosgrove, *Apollo's Eye: A Cartographic Genealogy of the Earth in the Western Imagination* (Baltimore: Johns Hopkins University Press, 2001).

43   Bruno Latour, *Reassembling the Social: An Introduction to Actor-Network-Theory* (Oxford: Oxford University Press, 2005), 176.

44   Latour, *Reassembling the Social*, 185.

45   Latour, *Reassembling the Social*, 183, 187.

46   Moyn and Sartori, 'Approaches', 22.

nation-state. Equally, intercultural exchange can occur at a scale or in a form that hardly merits the description global.

However these methodological – even metaphysical – questions are settled, there are clear benefits to encouraging a greater sensitivity to the *historical geographies* of global knowledge or, as in the title of this volume, the *spaces* of global knowledge. Whether understood in a more radical sense or not, it remains productive to ask in what kinds of spaces were global claims transacted and along which specific long-distance route-ways were objects of knowledge transported. Further, there is value in seeking to ascertain and analyse the sorts of sites or settings produced through the encounters, forms of exchange and exhibitionary practices that attached to efforts to collect, classify and communicate knowledge of distant places or widely-dispersed objects. There is an opportunity, in other words, to integrate more fully insights from work on the historical geographies of knowledge into the writing of global histories.[47] If such integration tends to sit awkwardly with the grander ambitions of global histories by calling attention to the sometimes disruptive and resistant particularities of place and patterns of movement and through multiplying sites and forms of mobility in ways that defy easy summary, it can also enrich the spatial vocabulary available to the historian and encourage more careful thought about the meaning and use of geographical categories.

The pursuit of a more explicit geographical sensibility corresponds well to the fact that the chapters in this collection do not, in the end, operate with a fixed definition of global knowledge or a unitary understanding of what a global history of knowledge endeavours might entail. Indeed, for some contributors, a global orientation remains implicit and in others is ultimately relegated to the margins in terms of analysis and empirical focus. Putting the global 'in its place' or tracking the mundane and material movement of 'universal' claims can make macro-scale accounts seem fragile, fragmentary and finite – that is to say, not global in the sense in which the term is often used. For all that, the chapters, taken together, might be said to replicate the kind of spatial reach that can make the turn to global histories suggestive, transformative and inclusive. Indeed, the collection offers a provisional and partial outline map of the global knowledge economies that emerged and developed during the long nineteenth century. From this, it can be suggested that those dynamic economies relied upon, and fed into, global economic, political, cultural and technological regimes. At the same time, it also seems apparent that they were unevenly governed and pervasively if not exhaustively shaped by the sharp imbalances of power associated with intercultural contact in an imperial

---

47   For a full account, see Livingstone, *Putting Science in its Place.* For an overview, see Diarmid A. Finnegan, 'The Spatial Turn: Geographical Approaches in the History of Science', *Journal of the History of Biology* 41(2) (2008): 369–88; and for a recent set of essays informed by a concern with the spatialities of past scientific knowledge, see David N. Livingstone and Charles W.J. Withers, eds, *Geographies of Nineteenth-Century Science* (Chicago: University of Chicago Press, 2011).

age. These general points are worth bearing in mind in turning to synopses of the book's sectional divisions and individual chapters.

* * *

The chapters that follow this introduction avoid a narrow definition of what global knowledge is or was, and question as much as utilise the concept. This, combined with the contributors' varied disciplinary backgrounds and thematic concerns, has produced a collection in which global knowledge is understood not just as knowledge *of* the globe, but as knowledge facilitated and disseminated *by* the processes of imperialism, exploration and exploitation which connected the global world of the nineteenth century. A range of additional factors unites the various contributions. Methodologically, for instance, all take the form of empirically grounded case studies that marry detailed archival research with broader thematic and conceptual reflection. Relatedly, all proceed on the assumption that nuanced case studies attending to local specificity are necessary in order to fully understand the broader processes by which global knowledge was constructed and contested. In this sense, the volume is characterised by a 'bottom up' perspective, which uses local specificity to shed light on global structures and processes, revealing the latter to be lived and experienced phenomena rather than abstract historiographical concepts.

Further commonalities between chapters are identified through the use of three sectional headings that signal some of the crosscutting themes animating the volume as a whole. The first section, 'encounters and identity', draws attention to the intimate connections between the processes of knowledge making and self-fashioning. Each of this section's chapters are centrally concerned with the fleeting or more sustained encounters experienced by British and American observers with natural and cultural environments profoundly different and distant from their home territories. Such encounters were mediated by a range of pre-formed identities but could also unsettle a coherent sense of self.

The interplay between distant influences and intimate encounter is readily apparent in Nuala Johnson's reconstruction of a year in the life of the colonial wife, botanist and artist, Charlotte Wheeler Cuffe. Cuffe's views of Burma, materially expressed in watercolour paintings, were the product not simply of her inherited sense of class and gender identity that travelled with her from Ireland to Burma but were also enlivened by the physicality of being in the landscape and by her interactions with Burmese people. Cuffe's own quite private knowledge-making activities, while certainly shaped by conventional colonial assumptions, cannot be fully accounted for without a painstaking excavation of the non-discursive and 'more-than-imperial' aspects of her observational practices.

The dynamic between colonial conventions and global movements and ostensibly more local exigencies and encounters also played out in the case of the two antiquarian fieldworkers that take centre stage in Polina Nikolaou's chapter. To understand the archaeological field investigations of Robert Lang and Luigi di Cesnola, Nikolaou argues that attention needs to be paid to three overlapping spheres of influence: the institutions that set the intellectual framework for an emerging archaeological science, the Ottoman colonial authorities and Cypriot informants in the field. It was precisely encounters and negotiations with the latter two groups that proved crucial to the construction of Cyprus as an open field for archaeological investigation. The 'global' norms communicated by the British Museum were also essential in allowing knowledge to gain the necessary credibility to permit the transformation from field notes to scientific text. Here, of course, a certain scalar complexity is evident as we observe interactions between two different colonial regimes, British and Ottoman, both of which operated from a distance but in unequal ways. The 'global' knowledge of Lang in particular was not only grounded in close working relations with field observers but also mediated by colonial actors operating with regional as well as imperial interests in mind.

As well as attempting to overcome the frequently conflicting priorities of different groups at a range of scales, the pursuit of knowledge might be thought of as an impossible quest for a settled 'global' identity. As Diarmid A. Finnegan demonstrates in his chapter on the Belfast linen merchant and spider specialist, Thomas Workman, the well-worn route-ways marked out by commercial travel in the late-Victorian period could provide profoundly disturbing encounters that undermined a sense of mastery of global affairs. In order to cope with the disorientating effects of long-distance travel, Workman employed a classic technology of the self, diary writing. This was coupled with a growing interest in arachnology, a scientific pursuit that provided Workman with a set of descriptive routines that helped him not only to make sense of spiders wherever encountered but also to have confidence that the world was patterned in stable and knowable ways. For all that, Workman's global knowledge, whether of commercial opportunities or tropical spiders, did not always fare well even in the ostensibly more secure and confirmed spaces he occupied in Belfast.

As Angela Byrne shows in the final chapter in this section, the sheer fragility of global knowledge was yet more marked in the case of fur traders exchanging information with metropolitan centres a century or more earlier. Operating at the outer edges of a Euro-American intellectual community, knowledge produced through the practices of cultivation in Rupert's Land reflected and refracted the hugely vulnerable character of remote existence in the far north. Yet knowledge was pursued to shore up a scientific identity that might be recognised by cultivators of science operating according to ostensibly global standards at the metropolitan centre. If this, exceptions aside, tended to efface any transformative encounters with indigenous groups, it also provided a way to retain an identity otherwise imperilled by the drudgery and dangers of coping with extreme isolation. Attempts to produce a form of knowledge stable enough to keep open communication with London largely worked against the making of a more comprehensive and inclusive account of the sub-arctic.

If a concern with encounter and identity helps to capture something of the inner workings of knowledge made or marketed on a global stage, the focus of section two on collection and display sheds further light on the mechanics behind the construction of global knowledge during the long nineteenth century. Sarah Millar's narrative of three celebrated voyages to explore the Pacific in the early nineteenth century provides striking examples of how scientific collecting at sea expressed as well as frustrated the global ambitions of ocean exploration. By taking the re-fitted naval ship as a space of operation in the production of a global ocean science, Millar shows how collecting specimens was at once a means of securing warrant for knowledge claims made at sea and a practice that could disrupt a crew's social cohesiveness. If a ship naturalist or crew member could in principle collect specimens from anywhere on the world's oceans and transport them to distant centres of knowledge, they remained thoroughly enmeshed in, and constrained by, the social and material orderings of shipboard existence.

A moving ship, then, both enabled and checked the making of global knowledge. Similarly, as Caroline Cornish demonstrates, the long-distance movements of a museum object could engender a sense of having the whole world in immediate view and disrupt that sense through the changing and sometimes conflicted meanings of an object severed from the cultural setting in which it was manufactured. Even when set within a stable museological framework, the potential for incommensurable meanings remained. The object in question – a model of an Indigo Factory sourced from West Bengal – continued to 'move' even while remaining firmly 'in place'. If anything the model, rather than communicating the confident global vision of colonial endeavour, confirmed a sense of uncertainty and impotence in the face of a ceaselessly changing world economy.

The Museum of Economic Botany that housed and displayed the model Indigo Factory was certainly informed by a global vision in terms of its collecting policies and curatorial ambitions. This was true, too, of museums established and managed by local scientific societies in an earlier period. Enrolling and equipping local citizens whose careers took them to distant places to collect and donate natural and ethnographic objects for display was a common goal. As Jonathan Jeffrey Wright details, this objective was given particularly clear expression by founding members of the Belfast Museum. In practice, this meant exploiting the connections forged through the imperial careering of Belfast-born collectors, and it made manifest a version of cosmopolitanism informed by a Paleyite natural theology that not only provided collecting with a religious rationale but also underwrote a vision of humanity as a single and unified global community. Yet this global vision relied on more prosaic concerns and proved to have a limited shelf life. The 'exotic' nature of objects from distant places was thought important for attracting a large local audience, particularly one composed of members of the working classes. Success in this area was one way to deflect criticisms of elitism and to demonstrate the local value of a museum composed of what amounted to a rather eclectic set of collections that was, in real terms, far from global in any representative or comprehensive sense. That eclecticism was among the reasons why the global collecting policy of the Museum's founders was increasingly subject to criticism in the latter half of the nineteenth century.

Local objects began to be regarded as of greater pedagogic and scientific value than 'foreign' ones.

Whatever its fate, the desire to 'collect the globe' and put it to order was arguably one of the nineteenth century's most characteristic intellectual products. As Robert Mayhew argues, the construction in the nineteenth century of global 'imaginaries' made through practices of collection and display was strongly rooted in shifting standards of 'evidencing' that marked the decades around 1800. The rise of what Mayhew describes as 'empirical globalism' was fully registered in the changes made to the second edition of Thomas Malthus's *An Essay on the Principle Population.* The 'binary geography' of the globe sketched in the first edition divided the world into two relatively distinct areas. In one, positive checks to population growth (hunger, war, disease) dominated while in the other, under the aegis of more advanced societies, preventive checks (late marriage, self-restraint, birth control) were to the fore. This global view was ostensibly based on information gathered from authoritative sources. In the first edition, however, Malthus does little to substantiate his claim that his account relied on the careful collection of evidence from recognised authorities. In the expanded second edition, however, Malthus goes to some length to rectify this by incorporating and acknowledging information from across the world gathered by trusted observers. One of the interesting ironies that Mayhew uncovers is that despite the wealth of information about the globe that Malthus collected and displayed in the second edition, his binary division between primitive and progressive parts of the world not only remained but was, if anything, sharpened. Undoubtedly, the collection and textual arrangement of 'global information' by Malthus were decidedly geo-political.

Like Mayhew, Karen Salt's chapter keeps the politics of collection and display distinctly in view. Salt, in examining how artefacts, texts and images were deployed to counter mid-nineteenth-century efforts to undermine Haiti's status as a *bona fide* nation-state also adds further weight and nuance to the argument that the meaning of displayed objects or images are never static but have an irreducible recalcitrance. Read in this way, Salt's chapter offers a helpful segue into the third section of the book that gathers together the final three chapters under the rubric of circulation and translation. While certainly concerned with how particular objects and images displayed the legitimacy of Faustin I's imperial rule and of Haiti's place within the polity of nations, Salt also interrogates the politics and power manifest in the trans-Atlantic circulation of representations, positive and pejorative, of the Emperor and the territory he ruled. In tracking the movement of objects and images designed to defend Haitian sovereignty across the Atlantic and between different media and contexts of display, Salt challenges scholars to attend to the 'competing logics' inserted into knowledge circuits otherwise assumed to serve the interests of the more politically powerful.

The ways in which the dynamics of circulation could subtly destabilize assumptions about the provenance of authoritative knowledge are also apparent in Sarah Hunter's chapter on the missionary medicine of doctors and nurses associated with the Dublin University Mission based in Hazaribagh, western

Bengal during the late nineteenth and early twentieth century. Hunter is quite clear that commonplace assumptions about the superiority of 'western' knowledge were operating with full force in the rhetoric and practices of the University Mission. On this basis, the founding of hospitals and training centres in Hazaribagh and the surrounding district might be regarded as, *prima facie*, a standard case of the diffusion of 'western' medical knowledge to a colonial setting. What Hunter discerns, however, is that as medical knowledge achieved wider circulation, particularly among Indian medical practitioners, certain transformations occurred that cannot be understood in diffusionist terms. Hunter cautiously suggests that a more 'organically global' medical knowledge may have emerged that was not simply the result of the transfer of expertise from Dublin to Chota Nagpur.

In the cases of Faustin Soulouque and the Dublin University Mission the circulation of knowledge might also be productively understood in terms of translation both in a literal and more expansive sense. Faustin I couched his claims to sovereign rule in the 'display language' of world fairs or other exhibitionary complexes even while contesting how that language operated within largely European or North American cultural contexts. The translation of 'missionary medicine' into programmes and practices deemed more suitable for the culture and conditions that obtained in North East India involved a far-from smooth or unidirectional process of transformation. In the final chapter, Louise Henderson investigates such complexities further by taking textual translation as a central focus. By examining the production of German and French editions of David Livingstone's massively popular *Missionary Travels*, Henderson underlines the importance of emerging regulatory frameworks based on international copyright laws for understanding the intense contestations over textual authenticity occasioned by widening the circulation of texts through linguistic translation. The figure of Livingstone had a potential market that extended well beyond the Anglophone world but Henderson remains sceptical about talk of a 'global Livingstone' given the multiplication of meanings attached to his life through the contingent processes of translation, abridgement and piracy.

Taken together, the book's chapters open up some of the analytical possibilities latent in the category of global knowledge and, in more basic terms, operate across and between very different sorts of epistemic and cultural spaces at a global scale. At the same time, the arguments made present some significant challenges to how or whether knowledge can be thought about in global terms, not least because of the general historiographical lean towards the specificities and politics of the creation, circulation and contestation of knowledge claims within a set historical period. Global knowledge turns out to be a category that can be good to think with and against, but which should also provoke a certain sceptical distance from any easy evocations of globality.

PART 1
Encounters and Identity

Chapter 1

Global Knowledge in a Local World: Charlotte Wheeler Cuffe's Encounters with Burma 1901–1902

Nuala C. Johnson

Women's Natural (or Unnatural) Histories

In April 1922 Lady Charlotte Wheeler Cuffe, known affectionately through her lifetime as 'Shadow', was elected a Fellow of the Royal Geographical Society. Women were officially admitted as members from 1913 onwards, and by 1922 they represented 7.7 per cent of the Society's total membership.[1] About half these members were explorers/travellers, a quarter scientists/teachers and the remaining quarter professional women. This honour bestowed on Wheeler Cuffe was in recognition of her contribution to plant hunting and exploration, botanical illustration, and anthropological knowledge accumulated about Burma during the quarter of a century (1897–1922) she spent there with her husband as part of the colonial service. As Avril Maddrell has noted, 'Those who travelled with husbands or family on imperial duty … represent a continuation of the strong link between the RGS and state institutions and interests'.[2] Moreover Wheeler Cuffe also represented a burgeoning cohort of female naturalists who officially, and more often unofficially, were participating in and producing accounts of their work in natural history from the early-Victorian period onward.[3] From travelogues and specimen collections to botanical drawings and compendia of flora women were the producers of significant bodies of knowledge about the natural world both at the local domestic scale of their home lives as well as at a

---

1 Personal communication with Sarah Evans regarding female membership of the RGS in 1922.

2 Avril Maddrell, *Complex Locations: Women's Geographical Work in the UK 1850–1970* (Oxford: Wiley-Blackwell, 2009) 35.

3 Ann B. Shteir, *Cultivating Women, Cultivating Science: Flora's Daughters and Botany in England, 1760–1860* (Baltimore: Johns Hopkins University Press, 1996); Pnina G. Abir-Am and Dorinda Outram, eds, *Uneasy Careers and Intimate Lives: Women in Science 1789–1979* (New Brunswick, NJ: Rutgers University Press, 1987); Patricia Phillips, *The Scientific Lady: a Social History of Women's Scientific Interests 1520–1918* (London: Weidenfeld and Nicolson, 1990).

more global scale through travel or overseas residency, often as part of Britain's wider colonial project.[4]

While the distinction between the amateur naturalist and the professional scientific botanist gathered pace in Britain after 1870, remnants of the earlier undifferentiated category of natural historian persisted well into the twentieth century.[5] For women in particular, this tradition of knowledge production, through polite botany, remained of import as their access to the professionalising centres of botanical and zoological studies continued to be limited.[6] Among all the areas of scientific endeavour that women could engage with botany was regularly considered the 'feminine science par excellence', cultivating an interest in the outdoors, disciplining the body/mind and facilitating a culture of self-improvement.[7] However, detailed studies of Victorian women naturalists also indicate that they stepped outside these conventional interpretations of naturalised feminine pursuits in ways that confronted questions about gendered scientific authority and the practices of observation and documentation of the natural world.[8] For instance, Marianne North, Constance Gordon Cumming and Theodora Guest, who all travelled extensively to the American West, did engage in efforts to have their work taken seriously: 'Among their strategies, they submitted their paintings to public display; they provided Latin names and Linnaean terminology for different species ... they collected biological specimens, sometimes for public collections'.[9] And thus they attempted to be part of the wider production of natural history knowledge and in particular to interact with one of its major centres of accumulation and calculation: Kew Gardens.[10] In addition Isla Forsyth has argued,

---

4   Barbara T. Gates, *Kindred Nature: Victorian and Edwardian Women Embrace the Living World* (Chicago: University of Chicago Press, 1998); Dorothy Middleton, *Victorian Lady Travellers* (London: Routledge and Kegan Paul, 1965); Alison Blunt, *Travel, Gender and Imperialism: Mary Kingsley and West Africa* (New York: Guilford, 1994).

5   David E. Allen, *The Naturalist in Britain: a Social History* (London: Allen Lane, 1976); Nicolaas Rupke, *Richard Owen: Victorian Naturalist* (New Haven: Yale University Press, 1994).

6   Vera Norwood, *Made From This Earth: American Women and Nature* (Chapel Hill: University of North Carolina Press, 1993).

7   Cheryl McEwan, 'Gender, Science and Physical Geography in Nineteenth Century Britain', *Area* 30 (1998): 219.

8   Antonia Losano, 'A Preference for Vegetables: the Travel Writings and Botanical Art of Marianne North', *Women's Studies* 26 (1997): 423–48; Karen M. Morin, 'Peak Practices: Englishwomen's 'Heroic' Adventures in the Nineteenth Century American West', *Annals of the Association of American Geographers* 89 (1999): 489–514.

9   Jeanne K. Guelke and Karen M. Morin, 'Gender, Nature, Empire: Women Naturalists in Nineteenth Century British Travel Literature', *Transactions of the Institute of British Geographers* 26 (2001): 313.

10   Daniel P. Miller, 'Joseph Banks, Empire and 'Centres of Calculation' in Late Hanoverian London', in *Visions of Empire*, ed. Daniel P. Miller and Peter H. Reill (Cambridge: Cambridge University Press, 2010): 21–7.

'The *where* of scientific practice at times was liberating, the colonies in particular affording women space for practicing science'.[11] For Charlotte Wheeler Cuffe, her extended residency in Burma provided opportunities for exploration and travel that she would not have enjoyed at home.

The complex ways in which science has been produced historically has also been the focus of some recent debates on global histories of science where the challenge 'to reconsider the globe presents an opportunity to think in fresh ways about issues of commensurability, translation and circulation' and where 'connected histories will uncover the web of linkages and intermediaries that made science travel'.[12] Eschewing approaches which focus exclusively on the local, national or imperial as the unit of analysis, the turn to a global approach is more centrally situated around the zones of contact, the routes of mobility and the mechanisms of inclusion or erasure that went into the making of scientific knowledge across the world.[13] In many ways, paradoxically, this maps on to a parallel move by those working on the historical geographies of science who have foregrounded the 'spatial turn' in grappling with both the material and intellectual production, circulation and consumption of scientific knowledges.[14] Here site, location, situatedness have purchase in framing our understanding of how particular layers of knowledge emerge, gain currency, and circulate in the public sphere while other forms of knowledge remain peripheralised, unrecognised or confined to the private arena.[15] Charlotte Wheeler Cuffe represents an interesting instantiation of these processes as her work in natural history was to some degree a private pursuit practiced in Burma, yet she also connected into wider networks of professional botany. Among the things she did over her lifetime in Burma was to produce several hundred paintings, sketches and illustrations of plant life as well as of the Burmese landscape and people; she collected plants and sent seeds/specimens back to her family and to the botanic gardens in Glasnevin Dublin; on commission by the Forestry Department she planned and designed a botanic garden at Maymyo; she corresponded with curators at Kew and Glasnevin as well as maintaining a lively correspondence with family and friends at home; and she acted as hostess to many travellers and government officials visiting Burma including Frank Kingdon-Ward,

---

11   Isla Forsyth, 'The More-Than-Human Geographies of Field Science', *Geography Compass* 7/8 (2013): 529.

12   Sujit Sivasundaram, 'Introduction', *Isis* 101 (2010): 96.

13   See special issue of *Isis*, 'Focus: Global Histories of Science', *Isis* 101 (2010): 95–158, which contains a series of essays which address this issue.

14   For an overview of this literature see Diarmid A. Finnegan, 'The Spatial Turn: Geographical Approaches in the History of Science', *Journal of the History of Biology* 41 (2008): 369–88.

15   David N. Livingstone, *Putting Science in its Place* (Chicago: University of Chicago Press, 2003); Charles W.J. Withers, *Placing the Enlightenment: Thinking Geographically about the Age of Reason*, (Chicago: University of Chicago Press, 2007); Diarmid A. Finnegan, *Natural History Societies and Civic Culture in Victorian Scotland* (London: Pickering and Chatto, 2009).

who would later write numerous travel books about the region.[16] In this sense her performing of knowledge created the possibility of knowledge advancement, and she acted as a conduit for its exchange and transfer.

At the height of Britain's globalised empire, Charlotte Wheeler Cuffe became intimately acquainted with the land, life and culture of Burma, and in her travels around the country, she echoes Baigent's claim that 'The act of travelling, entering the public sphere in so manly a fashion, was still transgressive for a woman in the 1890s'.[17] During her lifetime in Burma she explored areas of the land relatively unknown to European botanists and she painted both the plants and the landscape of these regions as well as actively botanising. Unlike other female naturalists, Wheeler Cuffe's legacy is contained in her floral and landscape paintings of Burma and the botanical garden she established there, rather than in a corpus of published work. But it is through the lively correspondence she maintained during her time in the colonies that we can unravel the practices of producing such knowledge and her relationship with the cultural and natural world she encountered in the tropics. As geographers have noted 'Official scientific publications are often void of corporeal experiences', but the private letters and day-diaries she produced offers us an opportunity to get behind the underlying motives, experiences and practices that stimulated her engagement with the natural history of Burma and its people.[18] In this chapter therefore the aims are threefold. First, I wish to situate her in the context of a colonial wife operating within a socially and intellectually gendered milieu at the height of Britain's empire in the early twentieth century. Second, I explore her relationship with the landscape and people of Burma as she produced works of natural history and anthropology and I investigate her connections with the native population in this process. Third, I seek to interpolate this account with an analysis of her practices in the performance of field excursions, where she presented, through her letters, detailed descriptions, drawings and judgments about the material and existential circumstances surrounding these travels. Although she spent almost twenty-five years in Burma this chapter will focus primarily on one period, her first year based in Toungoo (1901–1902), in order to provide a detailed, microanalysis of her experience in the tropics.

Charlotte Isabel Wheeler-Cuffe 'Shadow' (1867–1967): Moving to the Tropics

Born in Wimbledon on 24 May 1867, Charlotte Isabel Williams was a granddaughter of the Reverend Sir Hercules Langrishe, third baronet of Knocktopher in

---

16   Charles Lyte, *Frank Kingdon Ward: the Last of the Great Plant Hunters* (London: John Murray, 1989).

17   Elizabeth Baigent, 'Travelling Bodies, Texts and Reputations: the Gendered Life and Afterlife of Kate Marsden and Her Mission to Siberian Lepers in the 1890s', *Studies in Travel Writing* 18 (2014): 39.

18   Forsyth, 'The More-Than-Human', 530.

Co. Kilkenny. While she came from London, and her mother continued to live there through her daughter's life, Charlotte's connections to Ireland were solidified in 1897 when she married Otway Wheeler Cuffe (1866–1934). He was the son of Major Otway Cuffe, the third son of the first baronet and he succeeded his uncle as third baronet in 1915. The family seat was a Leyrath, a few miles from Kilkenny city. Otway Wheeler Cuffe trained as a civil engineer at the Royal Indian Engineering College in Cooper's Hill and in 1889, he joined the Indian Public Works Department [PWD hereafter]. He and Charlotte spent the following 24 years of their lives stationed in Burma, but made regular trips home when leave was granted. He was appointed Executive Engineer in 1906 and later in 1913 became Superintending Engineer in Burma before retiring back to Leyrath in Ireland in 1921. He was also an honorary *aide-de-camp* to the Viceroy of India between 1911 and 1918.

It was during this quarter of a decade in Burma that Charlotte expanded her botanical hunting and drawing skills and produced several hundred watercolour illustrations of the local flora.[19] While she had not undertaken any formal training in art, like many women in her social position, gardening, sketching and painting formed part of the panoply of 'hobbies' to which she engaged from a young age. Her principal paintings were of Burmese orchids, completed between 1902 and 1921. She also illustrated native rhododendrons and other species that she encountered in her travels. She rarely used pencil to outline her paintings and unlike professional botanical illustrators, her orchid studies are exemplars of painting directly from life in the field. She situated the plants in their natural habitats, for example epiphytes, growing from the branches of trees, rather than drawing species crafted to demonstrate flower stamens and pistils, a style so beloved of plant taxonomists.[20] She continued a Victorian tradition in which 'discrete categories of what is scientific, proto-scientific, or popular, commercial, or fine art' had not been fully demarcated.[21]

In a letter to the National Botanic Gardens in 1937 she informed the Director that she would like the paintings she had donated to be permanently held there after her death. It is through her correspondence, alongside the paintings, that we can gain some insight into her life as a botanical hunter and collector in Burma in the early twentieth century and her role as a European in the exploration of some of the remoter parts of this region of Southeast Asia.[22] The Wheeler Cuffes spent much of their time resident in central Burma but they travelled extensively around the country, Otway in his capacity as civil engineer and Charlotte to accompany

---

19   *The Art of Flowers: National Botanic Gardens, Glasnevin Bicentenary Exhibition 1995 Catalogue* (Dublin: National Botanic Gardens, 1995).

20   *Plant Treasures: Two Hundred Years of Botanical Illustration from the National Botanic Gardens, Glasnevin* (Dublin: National Botanic Gardens, 2002).

21   Guelke and Morin 'Gender, Nature, Empire', 313.

22   Patricia Butler, *Irish Botanical Illustrators and Flower Painters* (Suffolk: Antique Collectors Club, 2000).

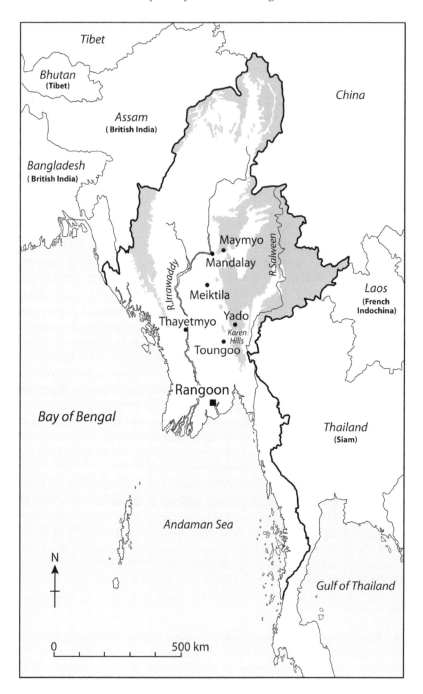

Figure 1.1   Map of Charlotte Wheeler Cuffe's Burma
Credit: Libby Mulqueeny, Queen's University, Belfast.

Figure 1.2   From the flood gauge, Thayetmyo 1900
Source: Charlotte Wheeler Cuffe.
Credit: Courtesy of the National Botanic Gardens, Glasnevin.

and assist him. She also, however, undertook expeditions on her own or in the company of other colonial 'wives' where she independently engaged in plant hunting and the drawing of species new to her.[23] During the 24 years in Burma, the Wheeler Cuffes were stationed in a variety of different places. They initially settled in Thayetmyo in central Burma but they also spent periods in Toungoo, Rangoon, Meiktila, and Maymyo – as directed by the PWD and Colonial Office.

Charlotte immediately took to living in the tropics and was enchanted by the Burmese landscape and the possibilities for botanising. Her experience challenged a tradition which regarded the tropics as enervating and full of danger particularly for women. In an early letter after her arrival in Burma in 1897 she wrote to her mother about their journey to Thayetmyo, a settlement 185 miles north of Rangoon, where they were initially posted. She described the trip taken by steamer from Rangoon where the landscape immediately impressed her 'Such a river! Over a mile wide and a current like a millrace. Pretty steep wooded hills on the W. and lower ground with palm trees and rice fields on the E'. On arriving at their new house she commented 'the place is charmingly pretty and this house though not actually on the river bank is not far behind. The ground all around the station is just like a very pretty park'. Their predecessor, Mr Gobbett, had left his two dogs and a pony and trap for their use until their own arrived. Although of aristocratic stock Charlotte was somewhat surprised by the number of servants provided as she opined 'I have not clearly made out yet how many are servants and how many are "hangers on", but there is about a dozen I think'. Others at the colonial settlement included the Deputy

---

23   E. Charles Nelson, 'The Lady of the Rhododendrons – Charlotte Wheeler Cuffe 1867–1967', *Rhododendrons* (1981–82): 33–41.

Commissioner, Mr Carter, and his wife, the Surgeon and other acquaintances of Otway's from home. Even though they had journeyed over 5,000 miles from the other side of the earth, they were instantly part of a wider social network with a global reach. She immediately related that Mr Carter was going to assist her in making a garden of her 'compound', to domesticate it in a settlement which had 'a nice church ... also a club and a library – we are so civilized'. Thus the institutions and emblems of the social milieu she inhabited at home simultaneously marked her dislocation to the tropics. The colonial context, did not of course, go unnoticed and on the question of adapting to the new physical environment she remarked: 'The mosquitoes here are terribly big and hungry and have tasted me rather badly but they say one gets inoculated in time and the bites don't inflame as they do at first'.[24]

Charlotte seems to have quickly adjusted to her new environs and readily took up developing a garden and exploring the surrounding countryside. In October 1897, she recounted one of the first trips she undertook to Mindon, 46 miles from Thayetmyo. She commented that 'gloriosa' grew everywhere and she 'often brings home sprays of [it] ... when we are riding through the jungle'.[25] Observing and recording the flora of Burma was an instant preoccupation. Moreover, she actively played a role in her husband's work as engineer for the District and assisted him in preparing estimates for a road-building project to Mindon. Her drafting skills were employed from the start, where, over the following months, she made 'a map of the whole road from here to Mindon, scale 1 inch to 1 mile ... Of course it was signed off by Otway [to go to the Chief Engineer], but as I have made the map, he said that my initials must go somewhere "unbeknownst", so we put a lovely monogram as the end of the north point instead of the usual arrows point'.[26] She was therefore, from the outset, an actor in the production of local geographical knowledge about the area undertaken to assist in the colonial administration of Burma.

In her early years in Burma Charlotte developed an appetite for travelling and assisting her husband in his work as well as starting her 'career' of painting the flora of Burma. She did not restrict herself to the conventional role of the colonial wife, confined to the compound. She challenged the orthodoxy of 'women's restricted mobility' by engaging directly with her husband's work as engineer responsible for developing the transport network, and undertaking individual knowledge generating projects especially in relation to the botanical character of the region and its peoples.[27] If we now turn to 1901–1902, five years into their residency in Burma, we begin to appreciate and identify how the Wheeler Cuffes, and Charlotte in particular, began to build up a body of knowledge about the place that was both projected onto the colony but also reliant upon the expertise of a

---

24　Letter from Charlotte Wheeler Cuffe [CWC hereafter] from Thayetmyo to Mrs Williams, July 28 1897, National Botanic Garden Archive [hereafter BG Archive].

25　Letter from CWC from Thayetmyo to Mrs Williams, 8 October 1897. BG Archive.

26　Letter from CWC from Thayetmyo to Mrs Williams, 21 August 1898. BG Archive.

27　Guelke and Morin 'Gender, Nature, Empire', 312.

panoply of native agents, other European settlers, local domesticated animals and wildlife as well as the diverse range of Burmese cultural traditions.[28] Despite the asymmetric power relations between native and newcomer during the height of Britain's overseas empire, a dialectical relationship nonetheless existed that made possible the accumulation of a knowledge bank that would ultimately travel back and forth between Burma and home.

In the Shadows of the Hills of Central Burma

By mid-1901 the Wheeler Cuffes had moved from Thayetmyo about 85 miles southeast to Toungoo. Transfers were a characteristic of life for colonial administrators, sometimes welcomed, but often disruptive of established networks. Nonetheless on moving to Toungoo Charlotte quickly announced that she had a tennis court installed adjacent to their house which 'is quite a success and has been played on three afternoons this week and is pronounced decidedly good'. Such leisure facilities could be important in nurturing new social networks when colonial families moved. She also employed a gardener to assist in the development of the new plot. In December 1901 she and Otway planned a trip over the mountains to Yado, northeast of Toungoo, using 'all ponies and one elephant for tents and heavy luggage'. They would not return until close to Christmas. In preparation for this longer excursion they undertook a short two-night trip 'taking only ponies to test our new pack equipment and make sure everything is all right – a break down at the "back of beyond" would not be pleasant!'.[29] Preparation for field visits was vital to ensure their success, as carrying out fieldwork necessitated the engagement of native labour and animals as well as being in readiness to confront the challenges of the natural environment. Independent travel in this context required the mobilisation of other human and non-human agents who had the necessary local expertise.[30] On 6 December Charlotte wrote from Leiktho describing the first leg of their trip to Yado. Having had to delay departure due to the death of one of their elephants, they travelled by train from Toungoo to Yedashà and were met by Mr Petley's Karen servant. The Petleys were coffee planters who had estates on the hills but their plantations had suffered from the same disease that had afflicted the crop in Ceylon and were thus struggling financially. From there they took a bridle path to Karen Choing crossing the River Sittang and arriving at a little house owned by the Petleys. They employed 10 coolies to assist in their onward journey 'up the most precipitous places you ever saw' where they camped in the jungle. The following day they travelled through a mountain pass over 4,000 feet,

---

28   Felix Driver, 'Hidden Histories Made Visible? Reflections on a Geographical Exhibition', *Transactions of the Institute of British Geographers* 38 (2013): 420–35.

29   Letter from CWC from Thayetmyo to Mrs Williams, 24 November, 1901. BG Archive.

30   Sarah Whatmore, *Hybrid Geographies* (London: Sage, 2002).

with their ponies, coolies and about 600 pounds of equipment, of which 40 pounds belonged to Charlotte including her 'painting materials'.[31]

On 13 December she wrote from Yado, indicating that their party had reached their destination safely. The group eventually included: two elephants, four pack ponies, their personal ponies Spar and Monkey, as well as three of their own servants including their cook, a PWD timekeeper, a head constable, and two Burmese policemen who were armed 'to act sentry at night when we camped – not for fear of 'humans' as much as tigers and leopards'. Of the landscape they passed through she made the following observations:

> Our march of 30 miles to this place was over a series of steep ridges of hills, all thickly covered with jungle – tree ferns, palms, wild plantains and every sort of tropical tree and plant you can think of – in fact one of the Kew hot houses spread up and down precipices with torrents in the gorges – would give you the only illustration I can think of.

The path they took was the old Karen and Shan hill track. As it was a well-travelled pathway by locals and thick with dew, Charlotte reported that it was very slippery in places, making the trek all the more difficult for animals and people. Their journey was made even more challenging as they had to cross the only biggish river on their trek, the Thonk yè gat or 'drinking water river, so called from a legend that a Burmese king drank of it on his travels and praised it for its purity'. The path descended at an angle of 60 degrees, she reported, making it particularly difficult for the pack mules and elephants, some of who had to have their load carried by the party. The landscape between Toungoo and Yado, was composed of a network of mountains running mostly north and south, and after crossing the river, they began a very steep ascent to 'suddenly come to the rim of a lovely level valley – obviously an old lake basin – walled in by steep limestone pinnacles of hills, and lying so to speak, in a cleft near the top of the range'. While the topography proved a challenge to traverse she does not mention climate as a barrier to travel. The valley, about a mile across at its widest was, as she described, a huge paddy land stretching for several miles. Yado was a mission station and they stayed in the mission house adjacent to the school run by Baptists. At the time of their visit she reported that 'there is no white missionary here just now', but the house had been put at their disposal by the senior American missionary for the entire region. Painting formed part of the trip as Figure 1.3 indicates.

Found near Yado in the Karen Hills, this species, *Acriopsis javanica*, an epiphytic orchid, is in her depiction attached to a dead bamboo. A plant with dense, onion-shaped pseudobulbs, and small cream or pinkish flowers, born on wiry, branched pinnacles, it is painted by Charlotte in a naturalistic manner, yet evocative of the appearance of the species in the wild. While the painting was completed in March 1902 it must have been fashioned from sketches taken in the field during this trip.

---

31   Letter from CWC from Leiktho to Mrs Williams, 6 December, 1901. BG Archive.

Figure 1.3 *Acriopsis javanica* 1902
Source: Charlotte Wheeler Cuffe.
Credit: Courtesy of the National Botanic Gardens, Glasnevin.

Charlotte also reported that there was a Catholic mission in the area but it had a little village to itself and their priest was Italian, as were the two Catholic priests they visited in Leiktho. Ironically though, meeting Europeans presented its own challenges. The priests could not speak English, French, or Burmese, all languages that Charlotte could converse in, while she couldn't speak much Italian, so they had to enlist the services of a Karen to conduct a polyglot conversation: 'Karen, Burmese, English and Italian'. Without the native expert communication would have been very limited between the Wheeler Cuffes and the Italian priests. She was impressed by interior of their church which 'was a surprise – a really well painted altar piece (the work of a former priest) of a life-sized figure of Our Saviour in an attitude of benediction'.[32] While colonial engineers like Otway were creating a physical infrastructure to facilitate the management of an empire, Christian missionaries were also significant actors in efforts to culturally transform Burma.

She wrote again from Leiktho on December 19 as they were on their homeward journey or as she put it 'on their way back to civilization once more'. Clearly Toungoo represented a far different experience than the remoter mountain villages of the surrounding countryside. To reach Leiktho they rode over a pass of around 5,600 feet where they 'did a prismatic compass survey of the peaks from the top and over the other side into the Shan country'. Surveying the landscape as they traversed it became a common practice for the Wheeler Cuffes, with Charlotte's drafting skills utilised to good effect. During this part of the expedition they also came upon a village inhabited by Shans, Padonongs and Karen. She observed: 'The Shans are a fine healthy, pleasant looking people, Padonongs extraordinary beleaguered savages, and the Karens rather furtive looking puny creatures, who always seem to be hiding something; though there are some fine specimens'. These anthropological observations conveyed her position as colonial actor, equipped to judge the character of the indigenous ethnic groups by sight alone. Although she learned Burmese and relied on local knowledge to assist her and her husband in a myriad of ways, she periodically asserted her cultural superiority in the production of knowledge about the peoples of Burma by declarations like this. Her description of their 'caravan' reinforced her position as leader of the 'pack', while, at the same time, indicating the Wheeler Cuffes' reliance on native expertise. She headed the group on her pony Spar, with Otway behind her on Monkey. The Head Constable and the PWD timekeeper followed them, also both on ponies, and constituting what she called the 'cavalry'. The 'infantry' behind comprised two policemen, the chaprassi [official messenger or servant in the British raj] dressed in a 'brown coat and putties and a scarlet turban', carrying Otway's rifle and acting as baggage escort. These were followed by a few more ponies carrying boxes and overseen by the cook. At the rear were the two elephants, walking more slowly than the rest of the party. Their daily routine was as follows: the caravan travelled until about 10:30am before they halted in a shady spot, had breakfast and rested for a couple of hours. The ponies then went ahead of the main party and the trek continued until a suitable camping place was found. She and Otway spent overnights in two real tents, 7 ft square, one to sleep in and the other to wash and dress in while 'the servants have a big tarpaulin and waterproof sheet, and the ponies do well under this with their thick rugs. We have a huge camp fire, and a folding board,

---

32   Letter from CWC from Yado to Mrs Williams, 13 December, 1901. BG Archive.

on the top of the three pack saddles, makes a capital table'.[33] Clearly the Wheeler Cuffes enrolled the help of many indigenous people and this echoes the claim that the routine of life in the field is 'distributed, relational and encompassing many intersecting geographies and histories'.[34] While this colonial family had the surveying and cartographic knowledge acquired in Europe, in rural Burma without the assistance of local aides and their knowledge base; animals to transport their equipment and provisions; the social networks of missionaries and other white settlers, travel around the country would have been greatly restricted. And while the social hierarchies were maintained at the discursive level, their dependence on the skills and know-how of the native people and their animals to ensure their safety and mobility was ever present.

Christmas of 1901 was spent in Toungoo where the architecture of European life in the tropics was re-established after their trip to Yado. Her mother sent her linen-bound sketchbooks from London and she and Otway began undertaking 'a series of panoramic sketches, drawn to scale and marked with their compass bearings from the tops of hills and passes lately, and plotting the results on a larger scale for a map. It is rather interesting and gives a capital idea of the country'.[35] This exercise in cartography was enabling the Wheeler Cuffes to present a bird's eye view of Burma and to make the country more accessible to western ways of knowing. Mapping, surveying, cross-sectional drawing are all well documented parts of colonial projects, and producing this type of visual knowledge for one of the remoter regions of Britain's empire was significant.[36] To translate the Burmese landscape for European consumption meant utilising the vocabulary and syntax familiar and understood by westerners, while at the same time acquiring that information through the assistance of local labour.

After Christmas they undertook another excursion, with their pack ponies, to Thandaung, 23 miles from Toungoo. While Otway spent the first two days with the surveyor and coolies, inspecting and correcting the plans for the proposed Thandaung cart track, and laying out the foundations for a small water supply weir, on the third day he and Charlotte undertook an excursion up the local peak Thandaung gyi, located behind the Circuit house where they were staying. The peak rose to 4,882 feet and presented a wonderful panoramic view which she likened to a 'regular S African "kopje" of granite boulders forming the top most peak'.[37] They ascended by pony as far as they could and then walked the remainder of the way, accompanied by Mr Dale, the local PWD representative who 'in spite

33   Letter from CWC from Leiktho to Mrs Williams, 19 December, 1901. BG Archive.

34   Hayden Lorimer and Nick Spedding, 'Locating Field Science: a Geographical Family Expedition to Glen Roy, Scotland', *British Journal for the History of Science* 38 (2005): 13–33.

35   Letter from CWC from Toungoo to Mrs Williams, 5 January 1902. BG Archive.

36   Matthew Edney, *Mapping an Empire: the Geographical Construction of British India 1765–1843* (Chicago: University of Chicago Press, 1997).

37   Letter from CWC from Toungoo to Mrs Williams, 12 January 1902. BG Archive.

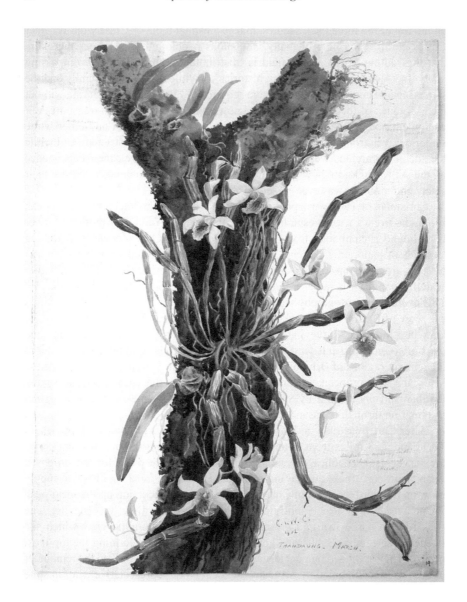

Figure 1.4   *Dendrobium aureum* 1902
Source: Charlotte Wheeler Cuffe.
Credit: Courtesy of the National Botanic Gardens, Glasnevin.

of the fact that he weighs about 15 stone and has only one foot, having lost the other in a railway accident when he was an engineer of the great Indian lines' succeeded in reaching the summit.[38] Charlotte sketched the whole panorama to scale, with Otway giving her the bearing at each point by prismatic compass.[39] She also painted an epiphytic orchid, the yellow-flowering *Dendrobium aureum* based on observations made on the trip (Figure 1.4).

This is a species found in lowland evergreen or primary montane forests. She depicted the plant attached to the trunk of a tree in a naturalistic composition, where the detail of the tree's foliage and bark compliment her intricate representation of the orchid. This family of plants had been gaining increasing popularity among European gardeners as they saw first-hand various tropical species in the hothouses of botanical gardens.[40] Orchids were also fascinating to botanists as they represented one of the two largest families of flowering plants and were to be found in a huge range of different habitats and climates. Clearly for Charlotte orchids were also extremely alluring and worthy of her attention amid the myriad of other plant families in Burma.

Although Thangaung was supposed to act as 'a hill station and Sanatorium for Lower Burma' she observed that it consisted of the Circuit House, a mission school run by a 'not very scrupulous Yankee' and a couple of other 'tumble down shanties built by people who thought they were going to make their fortunes over coffee'.[41] In the past troops were taken there in the hot season but the place had not proved especially healthy. This survey for a proposed new road, however, was being undertaken with a view to rejuvenating the station once again as a site of refuge from the summer heat. Hill stations had proved to be important for colonial residents across Britain's Indian empire and a significant ingredient for survival. They became not only climatic refuges for colonial settlers but they also became magnets for the social and cultural conduct of empire. As Kenny remarks 'the imperial hill station reflected and reinforced a framework of meaning that influenced European views of the non-Western world'.[42]

By 1902 Charlotte had completed a number of orchid paintings which she informed her mother 'I only lend ... for examination and identification, and get them back again. I must do some more soon. They are just beginning to come into flower. I have a huge Vanda pteris out now – great painting flowers three inches across'.[43] She clearly recognised the scientific value of her botanical drawings for

38   Letter from CWC from Toungoo to Mrs Williams, 12 January 1902. BG Archive.

39   Letter from CWC from Toungoo to Mrs Williams, 12 January 1902. BG Archive.

40   Nuala C. Johnson, *Nature Displaced, Nature Displayed: Order and Beauty in Botanical Gardens* (London: I.B. Tauris, 2011).

41   Letter from CWC from Toungoo to Mrs Williams, 12 January 1902. BG Archive.

42   Judith Kenny, 'Climate, Race and Imperial Authority: the Symbolic Landscape of the British Hill Station in India', *Annals of the Association of American Geographers* 85 (1995): 694.

43   Letter from CWC from Toungoo to Mrs Williams, 12 January 1902. BG Archive.

plant taxonomy, and, although she was not officially employed as plant artist she was happy to have her images deployed to that end. She would later go on to sell some of her Burmese paintings. At home she continued to cultivate her garden with the installation of a 'grand Burmese apparatus' for drawing water from the river beside their house into the vegetable garden. Using pulleys, counterweights and hollowed out bamboos, the system enabled irrigation channels to be inserted in the vegetable plot, hence enhancing growth.[44]

In late January she and Otway undertook another week's excursion to Bonmadi, south of Toungoo along the old Burmese line of road, with a view to investigating the possibility of building a road to connect the town with the river and the railway station. But the local people wanted a road following the old 'trade route … along which are innumerable villages, whereas the other way there are none'. Official government planning was confronted by the desires of the local population for connectivity between existing settlements. While Otway spent some of his time deer stalking accompanied by the local head man, Charlotte enlisted a local villager with whom to go exploring 'and discovered from him that this old water course is nothing more nor less than the former bed of the Sittang, which changed its course to the west ages ago'. The legend and belief held by the local people, she reported, was that

> about 1,500 years ago, the sea covered all the great alluvial plain of Lower Burma and that a great earthquake raised the level and formed the present enormous area of paddy land which stretches between the boundary of the mountains. Then they say the Sittang flowed by Bonmadi and afterwards for some unknown cause, changed its course 10 miles to the west. I believe this is all perfectly true and borne out both by geology and history.[45]

Local knowledge formed the basis of her interpretation of the topography she observed and she did not question its authenticity, as it conformed to the understanding of earth sciences she brought with her from Europe and thus represented corroboration of her views rather than 'new' insights provided by native informants.

From Burma to Home and Back Again

In June of 1902 the Wheeler Cuffes returned home on three months leave. They travelled on the SS Derbyshire, a ship they had used on previous visits home and their journey took a familiar route via Colombo, Suez, Port Said and Marseilles, from where they travelled overland to London.[46] They spent the summer visiting

---

44 Letter from CWC from Toungoo to Mrs Williams, 19 January 1902. BG Archive.
45 Letter from CWC from Toungoo to Mrs Williams, 26 January 1902. BG Archive.
46 Letter from CWC from Toungoo to Mrs Williams, 4 May 1902. BG Archive.

family in England and Ireland before departing back to Burma in August. Among their return luggage were three cases of wine, a cask of whiskey and shoes from Harrods.[47] While Britons were regularly travelling long-distance across the empire, luxury goods were also circulating and accompanying them. The Wheeler Cuffes were part of this globalised network in the transport of commodities not readily available in the colonies. Mary Curzon, Vicereine of India between 1899 and 1905, for instance, was involved in 'complex transnational networks of production and consumption that tied together' the imperial space she occupied in India and the fashion houses of Europe which provided her with suitable clothing for her role.[48]

During the voyage Charlotte became very ill and while Otway returned to Toungoo she remained in Rangoon until she was fit to travel.[49] The pace of her recovery was slow and in October, although back in Toungoo, she apologised to her mother for not being well enough to write to Professor Dunstan about the orchids as, she declared, 'I can't collect my brains sufficiently to think business'.[50] Professor Wyndham Dunstan was a chemist and Director of the Imperial Institute which had been founded to 'investigate and publicise new or little known mineral and vegetable resources of the countries of the empire'.[51] His expertise in pharmaceutical chemistry and the medicinal use of plants may account for his correspondence with Wheeler Cuffe and is indicative of her increasing connections with professional scientists. She also noted that she had received a letter from a Mr Scott O'Connor, who wanted one of her orchid paintings to illustrate a book on Burma, but due to his re-assignment to Assam the publication had been delayed.[52] Clearly her botanical drawings were beginning to attract some attention from the wider community.

As Otway was regularly away on work-related business, she moved to Maymyo, outside Mandalay, to fully recuperate at a friend's house. She requested from her mother

> to let me have my last little sketch book back. It isn't full and has only one or two decent sketches … It is on rough paper and has survey notes in it on the backs of the sketches and a rough pencil outline of a Padong woman with rows and rows of brass rings round her neck – I want the survey notes and that sketch especially.

---

47 Letter from CWC from SS Lancashire, Port Said to Mrs Williams, 20 August 1902. BG Archive.

48 Nicola J. Thomas, 'Embodying Imperial Spectacle: Dressing Lady Curzon, Vicereine of India 1899–1905', *Cultural Geographies* 14 (2007): 394.

49 Letter from Otway Wheeler Cuffe from Kokine, Rangoon to Mrs Williams, 7 September 1902. BG Archive.

50 Letter from CWC from Toungoo to Mrs Williams, 5 October 1902. BG Archive.

51 T.A. Henry (2004) 'Wyndham Dunstan', *Oxford Dictionary of National Biography*, accessed online at http://www.oxforddnb.com/view/article/32938 on May 7, 2014.

52 Letter from CWC from Toungoo to Mrs Williams, 5 October 1902. BG Archive.

She had been asked by another colonial official, Mr Lowis employed by the Commission, to 'draw outlandish people for the ethnological work he is doing for the gazetteer, and I may not get another shot at a Pandong, for they are very shy, and I have notes as to the colouring of her garments'. Clearly her artistic skills and deepening acquaintance with Burma's different ethnic groups were gaining attention. Moreover this request mirrors a much longer history of Europeans representing not only a natural history of the tropics, but also engaging in the visual depiction of ethnographic subjects. The drawing, painting and later photographic image of indigenous peoples emerged as central tools in the anthropologist's kit for, what Stepan has described as the, 'picturing of tropical nature'.[53] In addition to her sketching material Charlotte also asked her mother to send her 'some seeds of the large cape gooseberry. The small one grows like weed here but the large one is much nicer'. Not only then was she providing information about Burma's flora to people at home but also she was simultaneously introducing plants sourced in Britain to grow in the Burmese environment.

As an active woman, her long recuperation, which kept her housebound, was frustrating. She complained, 'one gets tired of reading and knitting all day'. Her weight had reduced to 6st and 8lbs and as she explained 'It is very cold at night here now and I find it hard in the day to keep warm, getting no exercise. Warm clothes don't take the place of exercise a bit in getting my circulation going'.[54] By November Charlotte was fit enough to return to Toungoo and she stopped en route to stay with the Adamson's [he was the Mandalay Commissioner] in Mandalay where 'the Burmese buildings and the Palace and all are wonderfully picturesque, particularly when you get the background of the blue hills behind them looking eastward'. The weather proved more attractive as she claimed it 'a far pleasanter climate here than at Maymyo at present. It is too cold for Eastern manners and customs up there now and damp inside and one doesn't have European appliances. Here it is dry and cool and delightful'.[55] Unlike those who feared that the climate of the tropics induced all manner of disease, for Wheeler Cuffe recovering from her illness would be facilitated, in her view, by transferring from the cooler temperatures of Maymyo to the warmer ones of Toungoo.[56]

When she arrived back in Toungoo she was happy to report that the plan for their new house and the estimate associated with it had been approved. She was

---

    53   Nancy Leys Stepan, *Picturing Tropical Nature* (London: Reaktion Books, 2001). See also Jill H. Casid, *Sowing Empire: Landscape and Colonization* (London: University of Minnnesota Press, 2005).

    54   Letter from CWC from Maymyo to Mrs Williams, 31 October 1902. BG Archive.

    55   Letter from CWC from Mandalay to Mrs Williams, 23 November 1902. BG Archive.

    56   David Kennedy, 'The Perils of the Midday Sun: Climatic Anxieties in the Colonial Tropics', in *Imperialism and the Natural World*, ed. John M McKenzie (Manchester: Manchester University Press, 1990), 118–40; David N. Livingstone, 'Tropical Climate and Moral Hygiene: the Anatomy of a Victorian Debate', *British Journal for the History of Science* 32 (1999): 93–110.

particularly delighted as the design was of their own and 'quite unlike the standard ordinary type'. She had also been

> busy getting the garden into order and potted up 20 big pots of Eucharis Lilies – that is, I sat and superintended – The vegetable seeds are doing well but the idiotic man mixed up the labels so I can only guess what plants are. Cardoons and artichokes galore, but I don't know which is which. Is there any way you can tell by leaf? And how big should cardoons be before you earth them up?[57]

Domesticating the garden with flowers and vegetables that suited a European gastronomy and aesthetic contrasted with her love and appreciation of the Burmese landscape and flora. The social order of the colony was maintained through this juxtaposition of the ordered space of the colonial garden and the 'native' spaces outside the compound of British settlement where indigenous plants and local agricultural practices prevailed. By Christmas her health had improved so much that she was ready to order many things from Harrods and Harveys for her new house. She acknowledged receipt of her sketchbook containing the image of the Padong woman which Mr Lewis had requested for inclusion in his gazetteer.[58] In her final letter of 1902 Charlotte thanked her mother for arranging for her orchid paintings to be framed. Although they had been produced in and of Burma, their final preparation for display had been carried out in London. And while she received a book from a friend on the 'Culture of Greenhouse Orchids' from which she was 'getting some hints as to potting and propagation generally ... unfortunately there are very few Burmese ones in it, which rather defeats me'.[59] If the circulation of knowledge was taking place at a globalised scale, the content of that knowledge itself was, in areas like botany and natural history more widely, necessarily regionalised. Species of orchids, native to different parts of the globe, required different growing conditions and planting regimens, and consequently that made necessary understanding the varied geography of their distribution. A general text on orchids grown in greenhouses in temperate climates was of limited utility to a woman seeking to propagate native species in a tropical environment. In this context, indigenous knowledge, obtained from the Burmese people, would ultimately be more helpful than a book produced in Britain and transported around the globe.

Conclusion

Charlotte Wheeler Cuffe achieved much during her quarter century in Burma. She painted the landscape and plants she observed extensively during her time there. She explored relatively uncharted territories, often in the company of her

---

57 Letter from CWC from Toungoo to Mrs Williams, 14 December 1902. BG Archive.
58 Letter from CWC from Toungoo to Mrs Williams, 20 December 1902. BG Archive.
59 Letter from CWC from Toungoo to Mrs Williams, 28 December 1902. BG Archive.

husband but also with others. With no formal training in either botany or art, she acquired her knowledge in the field and used other's expertise when available and when necessary. Her trips through the Burmese countryside pre-date those of Frank Kingdon-Ward and her position as the wife of a senior official allowed her access to a range of places otherwise fairly inaccessible to a white, European woman living in one of the remoter regions of south-east Asia. From an analysis of one year of her time in Burma a few suggestive conclusions can be drawn. First, the role of women in the construction of natural history knowledge, while receiving some attention in historical geography, has been relatively neglected. Charlotte Wheeler Cuffe clearly is indicative of a woman who did actively produce geographical and anthropological knowledge about the world she encountered in Burma. From her field excursions though it is also clear that native agents and their animals were central to both the pragmatics of working in field as well as being a source of knowledge about the flora, fauna and cultures she came across. Second, as a colonial wife she broke away from the customary occupations associated with such women in Britain's empire. She accompanied her husband regularly on trips, as well as travelling alone. Her skills in drafting, surveying and field observation were all important for the practice of his work as engineer. But the trips also provided her with ample opportunity to botanise and prepare sketches. Third, as she was part of a globalised empire the knowledge that she did produce was heavily mediated through the prism of colonialism and calibrated through her class and gendered positions in particular. But also significantly her knowledge was consistently a collaborative exercise conducted in dialogue with and along the contact zones between European and Burmese populations.

Chapter 2

Archaeology, Empire and the Field: Exploring the Ancient Sites of Cyprus, 1865–1876

Polina Nikolaou

In 1862 two Frenchmen, Henry William Waddington and Count Melchior de Vogue, visited Cyprus while travelling to the Levant as part of Napoleon III's policy of sending nationally organised scientific missions to the Eastern Mediterranean.[1] Waddington and de Vogue conducted the first systematic archaeological explorations on the island. Until then, Cyprus was rarely visited by European travellers. The island did not form a part of the Grand Tour, and was perceived as belonging neither to the ancient Greek civilisation, nor the Near Eastern civilisation. Thus, it was viewed as being of minimal importance in antiquarian pursuits. The discoveries made by the French explorers were, however, to change this, encouraging further attention to be paid to the island's ancient remains and initiating a period of intense excavation and exportation of Cypriot antiquities.[2] The excavations were mostly organised by travellers and settlers who came to the island in connection with the expansion of European Empires.[3] Indeed, excavating became a popular occupation among the foreign consuls stationed on the island. Among the most vigorous and prominent of these foreign consuls were Robert Hamilton Lang and Luigi Palma di Cesnola, whose collections were acquired by the British Museum and the Metropolitan Museum of Art in New York respectively.

Lang and Cesnola exemplify the process whereby westerners visited a part of the (supposedly) underdeveloped world, as the Ottoman province of Cyprus was considered to be, and became the main drivers of collecting and interpreting

---

1   Rita Severis, 'Edmond Duthoit: An Artist and Ethnographer in Cyprus, 1862, 1865', in *Cyprus in the 19th Century AD: Fact, Fancy and Fiction*, ed. Veronica Tatton-Brown (Oxford: Oxbow Books, 2001), 93–106.

2   The bulk of the collections that were exported in the latter half of the nineteenth century were displayed in metropolitan museums in Europe.

3   Robert H. Lang, 'Reminiscences – Archaeological Researchers in Cyprus', *Blackwood's Magazine* 177 (1905): 622–39; Elizabeth Goring, *A Mischievous Pastime. Digging in Cyprus in the Nineteenth Century* (Edinburgh: National Museums of Scotland, 1988).

archaeological specimens.[4] Lang (1836–1913) was a Scottish financier with an upper class education. He went to Cyprus in 1861 and until his departure from the island in 1872 he acted as an occasional Consul and Vice-Consul of the British government. In addition, during his residence in Cyprus, Lang engaged in farming and collecting antiquities. Cesnola (1832–1904) was Lang's contemporary in archaeological explorations. He was an Italian-born naturalised American citizen who acted as the American consul on the island. Cesnola's sole activity – besides his consular post – was conducting large-scale excavations in different sites in Cyprus. Both individuals lacked prior antiquarian knowledge and neither were acquainted with Cypriot artefacts before their arrival on the island. Nevertheless, their explorations initiated the transformation of Cypriot archaeology. In the words of the eminent British archaeologist John Linton Myres (1869–1954), Cypriot archaeology was transformed 'from a mischievous pastime into a weapon of historical science'.[5] This process of transformation was to peak with the British Museum exhibitions of the 1890s, but mapping Lang and Cesnola's activities in the field highlights a crucial period in the development of archaeological practice in Cyprus.

Lang and Cesnola's stories present some obvious similarities, however, as Charles W.J. Withers and Diarmid A. Finnegan note, there was 'no such thing as a single amateur identity'.[6] Archaeological explorations were, at one and the same time, social, economic, and cultural phenomena, and the particular identities of the different practitioners involved helped shape scientific conduct in the field.[7] The idea that the production of knowledge is a complex, social and situated practice has, of course, long been established.[8] The 'spatial turn' of the last twenty years has led historical research to focus on how particular localities affect the production and circulation of science.[9] Following the 'spatial turn', the current trend in global histories of science has argued for a 'history of the shifts and reinventions of a variety of ways of doing science across the world' that accounts for the connections

---

4   Henrika Kuklick and Robert E. Kohler, 'Introduction', *Osiris* 11 (1996): 1–14.

5   John Linton Myres, *Handbook of the Cesnola Collection of Antiquities from Cyprus* (New York: Metropolitan Museum of New York, 1914), xv.

6   Charles W.J. Withers and Diarmid A. Finnegan, 'Natural History Societies, Fieldwork and Local Knowledge in Nineteenth-Century Scotland: Towards a Historical Geography of Civic Science', *Cultural Geographies* 10 (2003): 335.

7   David C. Harvey, 'Broad Down Devon: Archaeological and Other Stories', *Journal of Material Culture* 15 (2010): 345–67.

8   Jan Golinski, 'Is It Time to Forget Science? Reflections on Singular Science and Its History', *Osiris* 27 (2012): 19–36.

9   For a review of the literature see Diarmid A. Finnegan, 'The Spatial Turn: Geographical Approaches in the History of Science', *Journal of the History of Biology* 41 (2008): 369–88 and Simon Naylor, 'Introduction: Historical Geographies of Science – Places, Contexts, Cartographies', *British Journal for the History of Science* 38 (2005): 1–12. See also David N. Livingstone and Charles W.J. Withers, eds, *Geographies of Nineteenth Century-Science* (London: University of Chicago Press, 2011).

and disconnections of science.[10] The focus of analysis here shifts from specific spatial units to networks of knowledge. Following Sujit Sivasundaram, networks of knowledge are understood here as webs 'of linkages and intermediaries that made science travel'. Such webs connected disparate contact zones and were 'forged through multiple and sometimes contradictory affiliations with regions, nations, and empires'.[11] Building on recent work, this examination of Lang and Cesnola's archaeological activities seeks to foreground the particular imperial and antiquarian networks underpinning the development of Cypriot archaeology in the mid to late nineteenth century. In so doing, it reconstructs the informal corresponding networks of Lang and Cesnola. As Hayden Lorimer states, small stories make better sense of 'big words or strange and distant deeds'.[12] Lang communicated his archaeological activities regularly to Charles T. Newton (1816–1894), then Keeper of the Greek and Roman Department of the British Museum. Likewise, Cesnola corresponded frequently with his friend Hiram Hitchcock (c.1833–1900) in Hanover (New Hampshire, USA), informing him of every detail of his stay in Cyprus and of his occupation with the collection of antiquities. Using the letters that passed between these men, the consuls' activities and the operation of archaeology in colonial Cyprus will be traced in four thematic sections: in the first, the motives for engaging in the collection of antiquities are considered; in the second and third, encounters with the Ottoman colonial authorities and with the local population are discussed; and, in the fourth, the transformation of the discovered antiquities into archaeological knowledge is explored.[13]

Robert Hamilton Lang, Luigi Palma di Cesnola and the Motives for Cypriot Field Archaeology

Sir Robert Hamilton Lang, K.C.M.G. was born in Scotland in 1836 and before taking the post of a clerk in a marine insurance office in Glasgow at the age of 14 he studied Greek for a year at the University of Glasgow.[14] He arrived in Cyprus in 1861 as a clerk working for a merchant firm with Levantine connections and in 1863 took the position of the Director of the Imperial Ottoman Bank branch in Larnaca,

---

10   Sujit Sivasundaram, 'Sciences and the Global: On Methods, Questions, and Theory', *Isis* 101 (2010): 155.

11   Sujit Sivasundaram, 'Introduction: Global Histories of Science', *Isis* 101 (2010): 96–7.

12   Hayden Lorimer and Nick Spedding, 'Excavating Geography's Hidden Spaces', *Area* 34 (2002): 300.

13   Felix Driver, 'Editorial: Field-Work in Geography', *Transactions of the Institute of British Geographers* 25 (2000): 267–8.

14   Elizabeth McFadden, *The Glitter and the Gold: A Spirited Account of the Metropolitan Museum of Art's First Director, the Audacious and High-handed Luigi Palma di Cesnola*, (New York: Dial Press, 1971).

a post he retained until his departure from the island.[15] Lang resided in Cyprus for twelve years and served, during this time, as a Vice-Consul of the British Empire on four different occasions and as a Consul in 1871–1872.[16] Demetrius Pierides, a leading local antiquarian and Lang's colleague at the Imperial Ottoman Bank, first introduced Lang to Cypriot antiquities. According to Lang, his antiquarian interests were also quickened by the visit of Waddington and Count de Vogue who discussed the discoveries they had made on the island with him.[17] Lang eventually became the most prolific collector of Cypriot antiquities on behalf of the British Museum.[18] In Lang's own telling, his involvement in archaeological digs was motivated by financial concerns:

> a peasant, from the site of the ancient Salamis, brought me for sale a gold coin which the sock [sic] of his plough had uncovered. It was in perfect preservation, and its beauty at once fascinated me ... I bargained with the peasant, who had asked £10, and finally purchased the coin for £5. General Fox had asked me to send him any interesting coins which I might acquire, and so I sent him my new and first acquisition. To my surprise, by return of post I received a letter of the most grateful thanks, and a cheque for £70. The coin was rare ... I had thus in my hands a profit of £65.[19]

Even if we view with caution the specific details of this transaction, it nevertheless suggests that Lang's interest in collecting ancient Cypriot relics was linked to an awareness of the commercial value of 'exotic' objects.[20] In this, Lang was far from unique: many European consuls resident on the island were preoccupied with commercial concerns and the trade in antiquities was one of their most profitable activities.[21] Indeed, by the mid-1860s Cyprus was acknowledged as 'a rich mine' for collecting antiquities since it provided large quantities of material culture

---

15   Lang, 'Reminiscences', 1905.

16   Robert H. Lang *Cyprus: Its History, its Present Resources and Future Prospects.* (London: MacMillan and Co., 1878), v. Lang was acting as a Vice-Consul from December 1861 until June 1862, from May until November 1864, from October 1865 until April 1866, and from April 1868 until January 1869.

17   Lang, 'Reminiscences', 1905. In his writings Lang mentioned his meetings with Waddington and de Vogue only in passing, and gave no particulars.

18   Thomas Kiely, 'Charles Newton and the Archaeology of Cyprus', *Cahiers du Centre d'Etudes Chypriotes* 40 (2010): 231–51.

19   Lang, 'Reminiscences', 623.

20   For other colonial contexts see Londa Schiebinger and Claudia Swan, eds, *Colonial Botany: Science, Commerce and Politics in the Early Modern World* (Philadelphia, Pa: University of Pennsylvania Press, 2005).

21   Michael Given, 'The Fight for the Past: Watkins vs Warren (1885–6) and the Control of Excavation', in *Cyprus in the 19th Century AD*, ed. Tatton-Brown, 255–60; correspondence between A. Billioti and C.T. Newton in British Museum Greek and Roman Department Archives Original Letters (hereafter BM GR OL), Vol. 1869–1872.

with considerable monetary value, and financial considerations also provided the primary motivation for Cesnola to engage with excavations.[22]

Like Lang, Cesnola did not have any prior archaeological training, though his social background differed from Lang's. He was an Italian-born naturalised American citizen who acted as the American, Greek and Russian Consul on the island, and is perhaps the most notable antiquarian and collector of this period.[23] A cavalry officer, he immigrated to the United States in the early 1850s, where he took part in the American Civil War as a Union Officer and was imprisoned for a period of time. By the end of the war he was in a very poor state financially and he begged for an official position. It was as a result of this that he was appointed as the Consul in Cyprus in 1865, but the appointment was not considered as a success. Cyprus was an insignificant consulate at the time and was not favoured amongst politicians and diplomats.[24] In 1871 the American Consulate in Cyprus was abolished and in early 1872 Cesnola left the island; however, he returned to Cyprus in 1873 and between then and 1876 conducted extensive excavations. As a member of the British Consulate, Lang was acquainted with Cesnola and he appears to have been on good terms with him and to have provided him with information about the island. Cesnola's low income (his annual salary consisted of US$1000, from which $425 were deducted for the payment of the Consulate's employees) forced him to find other means of gaining profit – the export of Cypriot antiquities provided one such means.[25]

But there was also another factor driving Cesnola to engage in the collection of Cypriot antiquities. As he frankly informed his friend, the New England antiquarian Hiram Hitchcock, he returned to the island in 1873 'for more work, glory and money!'[26] The large-scale excavations Cesnola undertook during the period 1873–1876 were initially funded by a financial contribution from the Metropolitan Museum of New York and they provided Cesnola with an opportunity to build a

---

22 Lang, *Cyprus*, 327; Max Ohnefalsch-Richter *Ancient Places of Worship in Kypros. Catalogued and Described … A Dissertation, etc. [Translated from 'Die Antiken Cultusstäten auf Kypros', etc. With plates.]* (Berlin, 1891). Although Cesnola was an American-naturalised citizen of Italian descent, his collecting practices in Cyprus are contextualised within the European and British archaeological and collecting values since he operated within their framework as well.

23 L.P. di Cesnola to H. Hitchcock (7 February 1869), Dartmouth College Archives (hereafter DCA), MS-68, box 2, f. 2.

24 McFadden, *The Glitter and the Gold*, 1971.

25 On arrival in Cyprus Cesnola got involved with the commerce of local wine, but this did not generate the required profit, and so he turned his attention to finding antiquities. L.P. di Cesnola to H. Hitchcock (7 October 1866), DCA, MS-68, box 2, f.2. See also Anja Ulbrich, 'An Archaeology of Cult? Cypriot Sanctuaries in 19th century Archaeology', in *Cyprus in the 19th Century AD*, ed. Tatton-Brown, 93–106.

26 L.P. di Cesnola to H. Hitchcock (5 November 1873), DCA, MS-68, box 2, f. 3, page 1 (original emphasis).

metropolitan reputation as an active and industrious man of 'science'.[27] From the early nineteenth century the emerging middle class transcended the respectability deriving solely from social rank by placing emphasis on morality, sobriety, duty and work.[28] The 'field' became a space where individuals belonging to lower-ranked classes could gain longed-for social respectability.[29] In particular, nineteenth-century archaeology had a quality of heroic individualism – as shown from the examples of Arthur Evans, Heinrich Schliemann, and Augustus Henry Lane-Fox Pitt-Rivers – and of being a scientific enterprise.[30] In a manner that is similar to other field sciences, such as geography, archaeology's glamour derived, in part, from its association with adventure, danger, physical challenges and exotic places.[31] That a similar glamour attached to Cesnola's activities is, perhaps, suggested by the observations of Stuart Poole, keeper of the British Museum's Department of Coins and Medals, that the scientific value of Lang's collections lay in the fact that they were gathered 'with the utmost care under the eye of the discoverer who was not deterred by the extreme heat of the summer'.[32] Thus, in addition to financial reward, Cesnola's archaeological activities enabled him to develop a reputation as a man of learning and upon returning to the United States he was appointed as the first Director of the Metropolitan Museum of Art in New York.

Cesnola and Lang can be seen, broadly, as adventurers who collected in the spirit of the 'era of commodities, the era of equivalence, exchange and capitalism'.[33] Their concern with financial profit reflects one of the main traits of the early large-scale archaeological endeavours in Cyprus: such activity was part of a more general process of commercial exploitation or resource-harvesting conducted by foreign consuls residing in Cyprus.[34] This is revealed particularly clearly by their

---

27   The collection he gathered during the period 1865–1871 was sold to the said museum. *New York Times*, 11 December 1882, Metropolitan Museum of New York Central Archives.

28   Anne Secord, 'Corresponding Interests: Artisans and Gentlemen in Nineteenth-Century Natural History', *The British Journal for the History of Science* 27 (1994): 383–408.

29   On field sciences in this period, see Felix Driver, *Geography Militant* (Oxford: Blackwell, 2001); Jane R. Camerini, 'Wallace in the Field', *Osiris* 11 (1996): 44–65.

30   Christopher Evans, '"Delineating Objects": Nineteenth-Century Antiquarian Culture and the Project of Archaeology', in *Visions of Antiquity: The Society of Antiquaries of London, 1707–2007*, ed. Susan Pearce (London: Society of Antiquaries of London, 2007), 267–305.

31   Stephanie Moser, 'On Disciplinary Culture: Archaeology as Fieldwork and Its Gendered Associations', *Journal of Archaeological Method and Theory* 14 (2007): 235–63.

32   Robert H. Lang and Stuart Poole, 'Narrative of Excavations in a Temple at Dali (Idalium) in Cyprus', *Transactions of the Royal Society of Literature XI*, 2nd ser. (1878): 54.

33   Yannis Hamilakis, 'From Ethics to Politics', in *Archaeology and Capitalism, From Ethics to Politics*, ed. Yannis Hamilakis and Philip Duke (Walnut Creek: Left Coast Press, 2007): 15–40, 16. Cesnola was, by the end of the nineteenth century, a highly controversial figure whose work was disputed by contemporary archaeologists.

34   L.P. di Cesnola to H. Hitchcock (7 February 1869), DCA, MS-68, box 2, f. 2.

selection of sites to excavate. Having located sites with ancient ruins, collectors had to decide where, in particular, they should undertake intensive excavations. Cesnola and Lang tended to choose tomb sites, as those sites typically produced greater quantities of antiquities than did the sites of ancient temples.[35] Such choices were fundamentally connected with the exportation of Cypriot relics to metropolitan museums.[36]

Nevertheless, it would be too simplistic, and also somewhat unjust, to characterise Lang and Cesnola as mere looters of ancient relics. As has been shown, collecting antiquities was a social, financial, cultural and leisure activity for the foreign residents of the island. Cesnola once noted that he was digging because he did not have anything else to do on the island outside his official business and, after the novelty of collection waned, Lang abandoned the search for antiquities, leaving Cesnola with a monopoly.[37] In simple terms, digging for antiquities in Cyrpus was, for a certain class of resident, a 'fashionable amusement of the day'.[38] But in addition to being conditioned by the particular identities of the collectors, such as Lang and Cesnola, this 'fashionable amusement' was conditioned by the legal framework set by the Ottoman Empire, the then colonial sovereign of the island. Thus, if the archaeological explorations of Lang and Cesnola are to be fully contextualised it is necessary to reconstruct the particular 'biographies of place' of nineteenth-century Cyprus and foreground the ways in which Ottoman authority circumscribed, or sought to circumscribe, archaeological practice.[39]

Colonial Encounters: Negotiating Archaeology with the Ottoman Authorities

Cyprus was under the direct rule of the Ottoman Empire during most of the nineteenth century and was administered by the Caimakam, the provincial Ottoman Governor.[40] Until 1869, the Ottomans did not have any official regulations regarding the excavation or exportation of antiquities. According to G.R.H. Wright, there is no indication of the existence in Cyprus, or in the Ottoman Empire in general, of any secular state law (Qanun) on the prohibition of the excavation or

---

35   Luigi Palma di Cesnola *A Descriptive Atlas of the Cesnola Collection of Cypriote Antiquities in the Metropolitan Museum of Art, New York.* (Boston: James R. Osgood and Company, 1885).

36   Georges Perrot and Charles Chipiez, *A History of Art in Phoenicia and its Dependencies* (translated and edited by Walter Armstrong) (London: Chapman and Hall, 1885).

37   L.P. di Cesnola to H. Hitchcock (10 June 1868), DCA, MS-68, box 2, f. 2; McFadden, *The Glitter and the Gold,* 1971.

38   Myres, *Handbook,* xiv.

39   Naylor, 'Introduction', 11.

40   The island was ceded to the British in 1878 as part of the Convention of Defensive Alliance.

exportation of antiquities; indeed, no reference is made to antiquities at all.[41] If any regulation regarding antiquities existed the only possible mention of it would be in firmans.[42] The firman (or farman) was a decree issued in the name of the Sultan, which carried his official cipher (tughra). It typically referred to a specific affair and was an official ordering about a certain activity, which overruled the existing law on the matter.[43] In their correspondence Lang and Cesnola make no reference to interference from the Ottoman authorities, and they appear to have carried out their excavations without any complications.

The first regulation referring explicitly to antiquities was enacted in 1869.[44] This was the first case of a generalised official attempt to formalise policies and attitudes towards antiquities and can be contextualised within the broader reformation measures the Ottoman Empire enacted in this period. According to the 1869 regulation all individuals that wished to excavate had to address their request to the Ministry of Public Instruction: if a foreign citizen wanted to dig, he had to obtain a special imperial decree (Irade) or a firman. The 1869 restrictions did not, however, lead to a reduction of the collection of antiquities in Cyprus. By this point a thriving market in antiquities had been established and the regulations of 1869 simply served to make the collecting of such material more complicated.

In 1871 Cesnola informed Hitchcock that the Ottomans had forbidden all diggings.[45] In theory *all* excavations were prohibited, but in practice it was only British consuls who were officially prevented from digging. Thus, Lang observed that the consuls resident on the island in this period who were unable to obtain a firman from Constantinople permitting the excavation of antiquities were the British consuls.[46] When Lang himself applied for a firman at this time his request was refused on the grounds that the High Porte intended to establish an Ottoman Imperial Museum.[47] The British Embassy did not interfere in the matter on Lang's behalf. By contrast, Cesnola's firman was renewed yearly, permitting him to dig across the whole island.[48] According to the Vice-Consul of Rhodes, the British officials got 'much less than the

---

41   A farman was an organic law of the land and was viewed as complimentary to the Islamic law. G.R.H. Wright, 'Archaeology and Islamic Law', in *Cyprus in the 19th Century AD*, ed. Tatton-Brown, 261–6. The Ottoman Empire was an Islamic state (essentially a theocratic one) ruled by the Islamic (religious) Law (Shari'a).

42   Wendy M.K. Shaw, *Possessors and Possessed: Museums, Archaeology, and the Visualization of History in the Late Ottoman Empire* (Berkeley: University of California Press, 2003).

43   Wright, 'Archaeology and Islamic Law', 2001.

44   Nicholas Stanley-Price, 'The Ottoman Law on Antiquities (1874) and the Founding of the Cyprus Museum', in *Cyprus in the 19th Century AD*, ed. Tatton-Brown, 267–75.

45   L.P. di Cesnola to H. Hitchcock (25 April 1871), DCA, MS-68, box 2, f.3.

46   Lang, *Cyprus*, 1878; A. Billioti to C.T. Newton (29 July 1869), BM, GR OL, Vol. 1869–1872, fol.45.

47   Lang, *Cyprus*, 1878.

48   Lang, *Cyprus*, 1878.

Americans'.[49] It is important to remember that this was a period of great European influence in the High Porte, and that the Ottoman Imperial Museum was established under the direction of the French.[50] Once established, the Ottoman Imperial Museum made efforts to stop the granting of firmans permitting the excavation and exportation of Cypriot antiquities, and donating archaeological artefacts to the museum became an important means of currying favour in order to obtain favourable firmans.[51] On paper, the Ottoman regulations regarding the excavation for antiquities were strict, but, in reality, private individuals continued to dig in Cyprus, both with and without firmans.[52] Alfred Billioti, the British Vice-Consul at Rhodes, noted that Cesnola was able to excavate freely on the island without the interference of the Turkish authorities.[53] Lang also continued his digging operations and C.T. Newton attributed this to his influence in Cyprus, which did 'better than a firman'.[54] As the Director of the Ottoman Imperial Bank branch in Cyprus, Lang had close relations with Turkish officials and he carried on his excavations quietly and the local authorities did not interfere with his work.[55]

From the fragmentary information we get from the consuls' correspondence, it appears that it was through a process of negotiation with the Ottoman authorities that archaeology took place on the island of Cyprus.[56] This process of negotiation is illustrated neatly in one particular episode in which Cesnola was involved.[57] In 1870, following his usual practice, Cesnola let his diggers excavate at Athienou (a village in the district of Larnaca) whilst he was in a nearby town, where the consuls resided. On the day of a consular meeting Cesnola received a message from a local informant apprising him of the discovery of a gigantic stone head and sculptures by his diggers and requesting that a cart be sent immediately for their removal. News of these findings soon spread and various individuals arrived at the site of the excavation, including the landowner, police officers who guarded the objects in the name of the Sultan, locals and Cesnola himself. Cesnola's account

---

49   A. Billioti to C.T. Newton (29 July 1869), BM, GR OL, Vol. 1869–1872, fol.45.

50   Shaw, *Possessors and Possessed*, 2003. The reformations (Tanzimat) was attributed to the major French influence in the High Porte during the period 1856 to 1871. The French influence on the internal affairs of the Ottoman Empire had broader geopolitical implications. The Ottoman Government sought French assistance for counterbalancing the emerging Russian and Austrian influence on the Turkish territories, which included getting financial aids in the form of loans from France (and Britain).

51   L.P. di Cesnola to H. Hitchcock (17 January 1872 and 27 April 1975), DCA, MS-68, box 2, f. 3.

52   A. Billioti to C.T. Newton (29 July 1869), BM, GR OL, Vol. 1869–1872, fol.45.

53   A. Billioti to C.T. Newton (29 July 1869) BM, GR OL, Vol. 1869–1872, fol. 45.

54   C.T. Newton to R.H. Lang (2 June 1869), BM, GR LB, Vol. 1861–1879, fol. 175.

55   R.H. Lang to C.T. Newton (15 September 1869), BM, GR OL, Vol. 1869–1872, fol. 361.

56   See Debbie Challis, *From the Harpy Tomb to the Wonders of Ephesus: British Archaeologists in the Ottoman Empire 1840–1880* (London: G. Duckworth and Co, 2008).

57   Luigi Palma di Cesnola, *Cyprus: Its Ancient Cities, Tombs and Temples* (London: John Murray, 1877).

of this episode depicts vividly the sense of competition that existed between the consuls: on hearing that two other consuls were going to the site Cesnola's 'mule sped on, *ventre à terre*'.[58] The Caimakam also sought to take possession of the discoveries and Cesnola thus purchased the land, fearful that further excavations would be prohibited or that the discoveries would be seized. He then conveyed the objects discovered to the American consulate in Larnaka, employing two local police officers to protect them. Cesnola's plan was successful; the site was deserted by the time the Caimakam arrived and since the antiquities were already stored inside the American consulate they could not be seized.

This multifaceted incident might be seen to bear out Gosden and Knowles' argument that colonial relations involved new forms of physical and social action in collecting practices.[59] The whole spectrum of colonial Cyprus's society was connected through the search for antiquities but not officially organised as such. Cesnola established a network of individuals around him that included Ottoman officials and local informants. Stable relationships were formed around archaeological explorations between foreign consuls and Ottoman authorities, who gained different things from each other. The Ottoman officials gained financial profits and Cesnola gained space to conduct his excavations without interference. This diverse network of individuals facilitated the movement of the objects from their discovery sites to places that were perceived to be safe for their storage, and it becomes clear that the fate of these antiquities was dependent on the network of colonial authorities and local people that Cesnola employed.

Despite the regulation of 1869, foreign consuls continued to exploit Cyprus's potential as a field for antiquarian discovery. Consequently, a more robust antiquities law was enacted in 1874.[60] As with the previous regulations, the Antiquities Law included Cyprus but was not enforced adequately. After its enactment, excavations for antiquities were carried out all over the island both legally and illegally, and in 1875 Cesnola acquired a new firman allowing him to dig.[61] This relative lack of regulation created a context in which thousands of objects could be removed from Cyprus. Cesnola, alone, was able to export a collection that amassed around 35,000 antiquities of various types, such as life-sized statues and coins, to Europe and North America. Likewise, Lang sold his extensive collection of statuary discovered at the ancient site of Idalion to the Greek and Roman Department of the British Museum in the early 1870s.

---

58   Cesnola, *Cyprus*, 121 (original emphasis).

59   Chris Gosden and Chantal Knowles, *Collecting Colonialism: Material Culture and Colonial Change* (Oxford: Berg, 2001), 22.

60   This law may be viewed as part of a broader political reformation movement of the Tanzimat. Stanley-Price, 'The Ottoman Law', 2001.

61   According to Cesnola he acquired the new firman after sending antiquities to the Ottoman Imperial Museum as a gift. L.P. di Cesnola to H. Hitchcock (27 April 1875), DCA, MS-68, box 2, f. 3.

Local Encounters: Digging with the Cypriots

Cesnola's description of the discovery of the antiquities at Athienou, and the discussion of the particular legal framework in which such excavations occurred, brings to the fore the collective character of archaeology. The field was a space inhabited by a socially diverse population. Until recently, this diverse population has been obscured by dominant narratives of the mythical age of exploration, which foregrounded the heroic individual.[62] 'Minor figures' such as laboratory assistants or non-European assistants in colonial settings were ignored in traditional histories of science that privileged the genius and individuality of the scientist. More recent work has, however, presented field exploration as a collective enterprise that involved a range of different relationships and interactions.[63] Cypriot archaeology was no different. Archaeological explorations involved a range of participants, including labourers who would dig, field assistants who would negotiate with the diggers and archaeologists who would organise proceedings. Lang and Cesnola established two main types of relationships during their work in the field: personal networks with Ottoman colonial authorities; and networks of relationship with the local population. As interactions with the colonial authorities have already been considered, attention will now turn to the local population and to the contribution it made to archaeological discovery in the 1860s and 1870s.

Reflecting on his archaeological experiences in his *Cyprus: Its Ancient Cities, Tombs and Temples* (1877) Cesnola remarked that the most difficult part of excavating on the island was finding ancient sites to dig.[64] Due to the lack of reference to ancient towns in the work of classical writers and the sparsity of material remains above ground, the knowledge of local residents was of critical importance and excavators utilised such knowledge to locate tombs and sanctuaries where excavations could be undertaken.[65] This practice of exploiting local knowledge is illustrated neatly by Cesnola's claims that his success was based partly on his 'inside track' with local people, while the excavations undertaken by others were unsuccessful, 'on the account of their ignorance of the island'.[66] But the contribution of locals was not limited simply to providing information on where to dig. Lang and Cesnola also required the labour of locals in order to excavate the sites and extract antiquities. Cesnola informed Hitchcock that 20 diggers were working for him at the site of Amathus.[67] While such references appeared casually

---

62 Hayden Lorimer, 'Telling Small Stories: Spaces of Knowledge and the Practice of Geography', *Transactions of the Institute of British Geographers* 28 (2003): 200.

63 See Felix Driver and Lowri Jones, *Hidden Histories of Exploration: Researching the RGS-IBG Collections* (London: Royal Holloway, University of London, 2009); Harvey, 'Broad Down, Devon', 345–67.

64 Cesnola, *Cyprus*, 1877.

65 Cesnola, *Cyprus*, 1877.

66 L.P. di Cesnola to H. Hitchcock (21 January 1874), DCA, MS-68, box 2, f. 3, page 4.

67 L.P. di Cesnola to H. Hitchcock (6 August 1875), DCA, MS-68, box 2, f. 3.

in the consuls' correspondence, the viability of the excavations was dependent on the availability of a local labour force and excavations were conducted when the local workers were not engaged with other agricultural activities. Indeed, on one occasion, Lang informed C.T. Newton that he had had to 'suspend temporarily the operations on account of the harvests', since the locals would be occupied there and could not dig for him.[68]

As is revealed by the account, discussed earlier, of Cesnola's excavation at Athienou, clear hierarchical structures were an essential part of the social organisation of excavation sites. Collectors employed locals to act as intermediaries between them and the diggers. These local intermediaries facilitated communication between supervisors and diggers, directed the diggers on minor issues and informed their employers of findings and discoveries. In Cesnola's case, the employment of intermediaries proved highly effective: employing men 'on the spot' enabled him to move between digging sites and collect a large amount of antiquities from his local supervisors.[69]

An examination of the archaeological activities of Lang and Cesnola, thus foregrounds certain similarities of collecting practice. Both men had similar motives in searching for antiquities, and both employed a range of 'third parties', establishing colonial and local networks to facilitate the regular operation of the excavations. But there was, however, one crucial difference in the collecting practices of Lang and Cesnola: the methodologies they used to record their findings.[70]

Performing Archaeology in the Field

The fundamental importance of recording and methodologies is revealed clearly in a law-suit filed against Cesnola by Gaston Feuardent in 1883 in New York City.[71] Feuardent, a well-known French art dealer, accused Cesnola of tampering with antiquities and attaching miscellaneous fragments to certain statues displayed as part of the Cesnola Collection in the Metropolitan Museum of Art in New York.[72] This law-suit sparked debates among academic archaeologists on both sides of the Atlantic. W.J. Stillman, a notable photographer of archaeological remains,

---

68   R.H. Lang to C.T. Newton (27 April 1869), BM, GR OL, Vol. 1869–1872, fol. 359, page 3.

69   Cesnola, *Cyprus*, 1877.

70   For their respective feelings towards personal competition see R.H. Lang to C.T. Newton (27 October 1869), BM, GR OL, Vol. 1869–1872, fol. 362 and L.P. di Cesnola to H. Hitchcock (17 January 1872), DCA, MS-68, box 2, f. 3.

71   Feuardent was a member and regular speaker of the American Numismatic and Archaeological Society in New York and provider of antiquities for the Louvre and the British Museum. McFadden, *The Glitter and the Gold*, 1971.

72   Gaston L. Feuardent, 'Tampering with Antiquities', *The Art Amateur* 3 (1880): 48–50.

submitted a report to the American Numismatic and Archaeological Society in 1885 stating that the Cesnola Collection's utility to students of archaeology was greatly diminished.[73] While conceding that some of the objects it contained were of interest, Stillman argued that the Cesnola collection was not scientifically valuable because the objects were not accurately linked with their place of origin and this made the critical task of determining the place of Cyprus in archaeology difficult.

Contemporary newspapers reported extensively on the episode. The *New York Times*, for example, commented on Cesnola's misconduct by referring to the work of the French antiquarian Georges Colonna-Ceccaldi, whose *Monuments Antiques de Chypre, de Syrie et d'Égypte* (1881), supported the accusation that Cesnola attached false provenance to his findings.[74] In the second chapter of his *Monuments* Ceccaldi discussed his visit to the ancient site of Golgoi. Cesnola claimed to have discovered a temple at Golgoi, but Ceccaldi concluded that no traces of such a temple existed there.

The accusations levelled against Cesnola show that proximity to the antiquities was not, by itself, sufficient to bestow authority on claims of archaeological expertise. Ancient sites were not merely the locus of antiquities' discovery. Crucially, they also provided credibility and authority for scientific claims.[75] In simple terms, it was place which made objects scientifically valuable, for if it could be linked to a specific location an object could then be classified and compared with objects discovered at other locations.[76] Cesnola was, however, poor at recording the locations where his objects were discovered and he was regularly criticised for his excavation records.[77] When discussing the famous Amathus bowl, for instance, J.L. Myres, highlighted the shortcomings of those records which Cesnola did produce.[78] Although said to have been discovered by Cesnola in 1875 in a partially

---

73   William J. Stillman, *Report of W. J. Stillman on the Cesnola Collection*, (New York: Thompson and Moreau, 1885). Stillman was a notable journalist and photographer who studied antiquities. He attained distinction in the archaeological field, was invited to become one of the founding members of the Hellenic Society and was a member of the American Numismatic and Historical Society.

74   *New York Times*, 11 December 1882, Metropolitan Museum of Art (NY), Central Archives.

75   Lawrence Dritsas, 'From Lake Nyassa to Philadelphia: A Geography of the Zambesi Expedition, 1858–1864', *British Journal for the History of Science* 38 (2005): 35–52; Thomas F. Gieryn, 'City as Truth-Spot: Laboratories and Field-Sites in Urban Studies', *Social Studies of Science* 36 (2006): 5–38.

76   Christopher Evans, '"Delineating Objects"', 267–305; David G. Hogarth, *A Wandering Scholar in the Levant* (London: MacMillan, 1896).

77   For a similar occasion in the Victorian scientific community see Stuart McCook, '"It May Be Truth, But It Is not Evidence": Paul du Chaillu and the Legitimisation of Evidence in the Field Sciences', *Osiris* 11 (1996): 177–97.

78   The Amathus bowl is exhibited at the British Museum. John Linton Myres, 'The Amathus Bowl: A Long-Lost Masterpiece of Oriental Engraving', *Journal of Hellenic Studies* 53 (1933): 25–39.

despoiled chamber-tomb in Amathus, Myres ascertained that Cesnola's account of the discovered spot 'exaggerated the depths at which the chamber-tombs were found in Cyprus'.[79]

By contrast, Lang published his findings at the ancient site of Idalion in the *Journal of the Anthropological Institute of Great Britain and Ireland* and in the *Transactions of the Royal Society of Literature XI* (2nd series), and his work has been accepted as being the first scientific work in Cypriot archaeology according to modern criteria.[80] The different reactions which the work of Lang and Cesnola has elicited brings to the fore the crucial issue of the performance of archaeology in the field and the reception of the produced knowledge by contemporaries. Two intertwined practices are involved in fieldwork: being physically in the field and the epistemological framework brought into the field.[81] As Cesnola's case shows, the first practice alone was not a prerequisite of scientific authority. The methodologies employed to record discoveries – that is, the epistemological frameworks imposed on discoveries – were equally important and it was this that set Lang apart from Cesnola as a 'man of science'.

During his excavations at Dali (the modern site of ancient Idalion), Lang engaged in a constant correspondence with Charles T. Newton, the British Museum Keeper. This correspondence had a functional role in the conduct of Cypriot archaeology. Correspondence was the medium by which Lang was provided with the epistemic framework for collecting, namely the 'cosmopolitan' knowledge of taxonomy, and was facilitated by the expansion of the British Empire.[82] Newton's acquaintance with British Consuls was not confined to those residing in Cyprus, but was extended across the Eastern Mediterranean Sea. While acting as a Vice-Consul in Mytilene and occasional Consul in Rhodes from 1852 to 1859, Newton established what Thomas Kiely has termed a 'prosopographic web'.[83] For instance, he often corresponded with Alfred Billioti, the acting Vice-Consul at Rhodes, who informed him about the developments in the area regarding excavations.[84] Thus, while Newton was never personally or physically involved in excavations undertaken on

---

79   Myres, 'The Amathus Bowl', 25; Cesnola, *Cyprus*, 255.

80   Robert H. Lang, 'On Archaic Survivors in Cyprus', *Journal of the Anthropological Institute of Great Britain and Ireland* 16 (1887): 186–8; Lang and Poole, 'Narrative of Excavations', 1878; Kiely, 'Charles Newton', 2010; Goring, *A Mischievous Pastime*, 1988.

81   Driver, *Geography Militant*, 2001.

82   Robert E. Kohler, 'History of Science: Trends and Prospects', in *Knowing Global Environments: New Historical Perspectives on the Field Sciences*, ed. Jeremy Vetter (New Brunswick, NJ: Rutgers University Press, 2011): 230.

83   Mytilene was a town on the island of Lesbos in the North Aegean. Kiely, 'Charles Newton', 238; Lucia P. Gunning, *The British Consular Service in the Aegean and the Collection of Antiquities for the British Museum* (Farnham: Ashgate, 2009).

84   Sir Alfred Billioti (1833–1915) was of Italian origin and joined the British Foreign Service and served as a Consul in various places in the Eastern Mediterranean. He was an active antiquarian conducting many excavations in the Levant. For example, see A. Billioti to C.T. Newton (5 October 1869) BM GR OL, Vol. 1869–1872, fol. 48.

the island, he nevertheless played an important role in the development of Cypriot archaeology by corresponding with the consuls resident in Cyprus and proffering archaeological advice.[85] Through his letters, Newton offered a form of what Christopher Evans has described as 'site- mentoring', whereby excavators without formal training in excavation techniques were trained by the experts.[86] Newton was, indeed, a leading archaeological expert. He not only had extensive field work experience, but was well published. His publications on his fieldwork were widely accepted by the academic community and he held key positions in highly respected institutions, including the British Museum, where he served as curator and, later, Keeper of the Greek and Roman Department, and University College, London, where he was appointed, in 1880, as the first professor of archaeology.[87]

Through Newton's epistolary 'site-mentoring' Lang was able to collect and record his findings in compliance with modern archaeology. Lang informed Newton of the discovery of a considerable amount of statuary in Dali, sharing his belief that he might have stumbled upon the celebrated temple of Venus of Idalion.[88] He asked Newton's advice on the archaeological value of the antiquities discovered and of the excavation site before continuing the excavation. Newton encouraged Lang's efforts, advising him to continue digging and to keep his collection together as archaeological findings had to remain in groups in order to be scientifically comparable with objects found in other places.[89] Newton urged Lang to record his findings *in situ*, in other words to record the provenance of the objects and the exact find spots, and to pursue the lines of the foundation as far as they would carry him and urged him to photograph and make a plan of the ruins.[90]

To paraphrase Innes M. Keighren and Charles W.J. Withers, the correspondence between Lang and Newton was an embodied practice whose performance was predisposed by the broader scientific imperative of archaeology.[91] Newton's

---

85 Kiely, 'Charles Newton', 231–51.

86 Evans, '"Delineating Objects"', 270. A famous example was Augustus Henry Lane-Fox Pitt-Rivers who was site-mentored by Canon Greenwell, the perceived expert in barrow-digging of the day.

87 Newton discovered the celebrated Mausoleum of Halicarnassus whose remains were later displayed in the British Museum Ground Floor. R.C. Jebb, 'Sir C. T. Newton', *The Classical Review* 9 (1895): 981–5. See also Kiely, 'Charles Newton', 231–51.

88 R.H. Lang to C.T. Newton (27 April 1869), BM GR OL, Vol. 1869–1872, fol. 359, page 3.

89 C.T. Newton to M. Feuardent (18 August 1869), British Museum, Greek and Roman Departmental Archives Letterbooks (hereafter BM GR LB), Vol. 1861–1879, fol. 189. See also Charles Thomas Newton, *Essays on Art and Archaeology* (New York: MacMillan, 1880).

90 C.T. Newton to M. Feuardent (18 August 1869), BM GR LB, Vol. 1861–1879, fol. 189; C.T. Newton to R.H. Lang (2 June 1869), BM GR LB, Vol. 1861–1879, fol.175.

91 Innes M. Keighren and Charles W.J. Withers, 'Questions of Inscription and Epistemology in British Travelers' Accounts of Early Nineteenth-Century South America', *Annals of the Association of American Geographers* 101 (2011): 1333.

advice on the precise recording of archaeological findings points to the disciplinary transformations occurring in archaeology during this period. The institutionalisation of the discipline provided the field workers with a particular way of seeing.[92] As modern archaeology gradually specialised, it emphasised a scientific curriculum, while the knowledge it produced became more esoteric and the discipline became dependent on scientific data management.[93] These changes affected the conduct of archaeology; however, the methods of excavating or recording antiquities in the field, although crucial to the discipline, were not standardised.[94] The archaeological methodologies of stratigraphy (the recording and understanding of soil sequences), established in Denmark during the first half of nineteenth century, were not the norm, and typological artifact classification was in its early stages.[95] Apart from the recognised need to adopt an empirical methodology reliant on physical data, there was no single methodology of excavating and reporting ancient relics.[96]

Archaeology in the nineteenth century was a personal project with fieldwork, illustration and classification being fundamental to its claims for authority.[97] The different ways in which these practices were employed by Lang set the knowledge provided from his collections apart from that provided by Cesnola as 'credible'. Through the communication of local and cosmopolitan knowledge, Lang was able to record his findings in accordance with the contemporary archaeological standards. Cesnola's records, on the other hand, were widely disputed. The places of origin of the objects he gathered could not be verified and his mere proximity to the excavation sites did not suffice to provide him with epistemic authority. Viewed alongside each other, the cases of Lang and Cesnola serve to foreground the contested nature of the field in the production of authoritative knowledge.

Conclusions

Lang and Cesnola commenced a period of extensive excavations which initiated the interpretation of Cyprus's ancient past. Mapping their every-day activities in the field brings to the fore the historical and spatial processes that shaped the

---

92   Philippa Levine, *The Amateur and the Professional: Antiquarians, Historians and Archaeologists in Victorian England 1838–1886* (Cambridge: Cambridge University Press, 1996); Lorraine Daston, 'On Scientific Observation', *Isis* 99 (2008), 97–110.

93   Louise Steel, 'The British Museum and the Invention of the Cypriot Late Bronze Age', in *Cyprus in the 19th Century AD*, ed. Tatton-Brown, 160–67.

94   Evans, '"Delineating Objects"', 267–305.

95   Hogarth, *A Wandering Scholar*.

96   Evans, '"Delineating Objects"', 267–305.

97   Noah Heringman, *Sciences of Antiquity: Romantic Antiquarianism, Natural History, and Knowledge Work* (Oxford: Oxford University Press, 2013); Sam Smiles, 'Record and Reverie: Representing British Antiquity in the Eighteenth Century', in *Enlightening the British: Knowledge, Discovery and the Museum in the Eighteenth Century*, ed. Anderson et al. (London: British Museum Press, 2003), 176–84.

'start' of Cypriot archaeology and illustrates the extent to which knowledge in this period was produced via the cooperation of various individuals brought together by the local and global colonial scene and the place-dependence of the communication networks these individuals established. In order to cope with the practical involvement of the local colonial regime in the conduct of archaeology the collectors had to negotiate with the Ottoman authorities, and they had also to make use of the local population in pursuing their explorations. Locals informed the collectors on where to dig and they performed the physical practice of excavating. More broadly, corresponding with the Keeper of the British Museum allowed Lang, in contrast to Cesnola, to transform the field site into a bounded space and to produce scientific knowledge in line with modern archaeology's rhetoric. These examples of the particular 'social' character of Cypriot archaeology show that the experience of the excavation sites was dependent on the employment of global and local networks. It is through the examination of these networks that a multi-vocal account of colonial science, distant from the simplistic one-way European models, becomes possible. Finally, in relating the biographies of place with the daily conduct of archaeology on the island the contingent character of these networks is demonstrated, bearing out Miles Ogborn's assertion that networks are local at every point.[98] It becomes evident that the analysis of both the historical and spatial contexts of each 'connection' and 'disconnection' involved in the making of archaeology is necessary.

---

98 Miles Ogborn, *Indian Ink: Script and Ink in the Making of the English East India Company* (Chicago: University of Chicago Press, 2007).

Chapter 3

Webs of Science, Webs of Commerce: The Life-Worlds of a Merchant Naturalist

Diarmid A. Finnegan[1]

In 1899, the Belfast merchant and naturalist Thomas Workman (1843–1900) took as the subject of his Presidential address to the Belfast Natural History and Philosophical Society 'incentives for the study of natural history'. In his opening remarks he noted that despite the 'tremendous development' represented by Darwin's *On the Origin of Species* science, like a traveller entering a spiral tunnel, appeared to have lost its way. Workman was nevertheless confident that looking back at the end of the century the onward and upward course of scientific knowledge was now evident giving grounds for optimism at the cusp of a new era. What was more, the rewards of 'struggling on towards the light' of scientific truth would be truly epochal.[2]

Drawing primarily on private journals and correspondence, this chapter examines Workman's quest for global knowledge through a reconstruction of his own travels, physical and intellectual. Particular attention will be given to Workman's accounts of his long-distance trips to the United States, Brazil and South East Asia and the spaces in which they were created and communicated. These trips were undertaken to promote linen and muslin goods in overseas markets and to pursue his interest in global natural history. They followed a well-established commercial circuit that was an integral part of an expanding British maritime world. The specific texture of Workman's globetrotting emerged from his own combination of economic and scientific ambitions and helped to produce knowledge of the world marked by the exigencies of commerce and the practicalities of late-nineteenth-century long-distance travel. It was also a form of knowledge that was put to local use to secure Workman's reputation as a respected man of commerce and of science. More specifically, Workman's interest in the distribution of tropical spiders and his strenuous efforts to tailor his textiles to suit perceived tropical needs brought into commercial and cultural contact very

1   I am grateful to the Deputy Keeper of Records, Public Records Office of Northern Ireland and the Belfast Natural History and Philosophical Society for granting permission to use the archival material on which this chapter is based.

2   Thomas Workman, 'Incentives for the Study of Natural History', *Proceedings of the Belfast Natural History and Philosophical Society* [*BNHPS* hereafter] (1899): 18.

different worlds and occasioned a series of negotiations between contrasting scientific, political and religious realities.

The account of Workman's life offered here, profoundly shaped as it was by the inter-weaving of commerce and science through travel, connects with recent scholarship that follows, in diverse ways, a global turn. Historical geographers and cultural historians have examined the importance of mobility and migration in understanding projects of empire in the nineteenth century.[3] The importance of maritime travel and trade in the nineteenth century has also been given increased attention, not least within the context of attempts to explore the causes and consequences of an expanding 'British world'.[4] At the same time, historians of science have been calling for studies that extend beyond local and national contexts and examine the movement of science across borders and over global space.[5] This has manifested itself not just in the investigation of the movement and mutation of scientific ideas across different cultural and epistemic domains but also in work on the mobilisation of scientific texts, objects and instruments.[6]

The primary purpose of this chapter is not, however, to construct a 'global history' of science and commerce. Rather, it examines how a particular and piecemeal global vision was formed, enacted and reconstituted in and through particular sorts of spaces and movements. Workman's world-making activities, and his own 'life-worlds', are localised and historicised rendering them, to borrow from John Law, 'something ... broken, poorly formed, in patches [and] very small and elusive'.[7] At the same time, adopting this miniaturising approach does not mean

---

3   David Lambert and Alan Lester, eds, *Colonial Lives Across the British Empire: Imperial Careering in the Long Nineteenth Century* (Cambridge: Cambridge University Press, 2006); Tony Ballantyne and Antoinette Burton, eds, *Moving Subjects: Gender, Mobility and Intimacy in an Age of Global Empire* (Urbana: University of Illinois Press, 2009).

4   For a survey, see Glen O'Hara, '"The Sea is Swinging into View": Modern British Maritime History in a Globalised World', *English Historical Review* 124 (2009): 1109–34. See also, David Killingray, Margarette Lincoln and Nigel Rigby, eds, *Maritime Empires: British Imperial Maritime Trade in the Nineteenth Century* (Woodbridge: Boydell Press, 2004) and David Lambert, Luciana Martins and Miles Ogborn, 'Currents, Visions and Voyages: Historical Geographies of the Sea', *Journal of Historical Geography* 32(3) (2006): 479–93.

5   James E. Secord, 'Knowledge in Transit', *Isis* 95 (2004): 654–72. Sujit Sivasundaram, 'Sciences and the Global: On Methods, Questions and Theory', *Isis* 101 (2010): 146–58. For another view, see Carla Nappi, 'The Global and Beyond: Adventures in Local Historiographies of Science', *Isis* 104 (2013): 102–10.

6   See, for example, Marwa Elshakry, *Reading Darwin in Arabic, 1860–1950* (Chicago: University of Chicago Press, 2013) and Marie Noëlle Bourguet, Christian Licoppe and H. Otto Sibum, eds, *Instruments, Travel and Science: Itineraries of Precision from the Seventeenth to the Twentieth Century* (London: Routledge, 2002).

7   John Law, 'And if the Global were Small and Noncoherent? Method, Complexity and the Baroque', *Environment and Planning D: Society and Space* 22 (2004): 18.

that fixed locations become stable reference points for understanding Workman's unfolding biography. Following Latour, the local has to be 'redistributed' as much as the global 'localised'.[8] Workman's travels, his science and the venues in which he spoke were constructed from a mass of often-fragile connections to other spaces and actions. In recounting the ways in which Workman's confident global vision was set within, and was threatened by, the contingencies of various connected spaces of knowledge this essay sketches the geography of a life.[9] While following a chronological line, the chapter pauses at certain formative spaces to explore how Workman attempted to hold in focus a world on the move. The starting point is a newly-built warehouse in Bedford Street, Belfast.

Bedford Street, Belfast, September 1852

Thomas Workman was nine when the British Association for the Advancement of Science made its first visit to Belfast in early September 1852. His father, Robert, made his new warehouses on Bedford Street available to the Association for the evening soiree. The 'Great Lower Room' provided an excellent venue for the hundreds of members of the Association to promenade and converse. The large hall, designed to store muslin and linen textiles for export, was tastefully decorated and filled with scientific objects and instruments to instruct and entertain the guests. On the walls 'colossal diagrams, maps and charts' depicted the geological structure of the globe.[10] A distribution centre for textiles produced from both local and internationally sourced fabrics and sent to trading houses in different parts of the world had been converted into a temporary museum, art gallery and space of scientific conversation about a wider world.

The day before, the Association's President, Colonel Edward Sabine had addressed members from in front of the pulpit of May Street Presbyterian Church. The vision of science that Sabine presented was self-consciously global. The onward march of the 'magnetic crusade'; progress towards calculating the figure of the earth through trigonometrical survey; the worldwide study of tides and temperatures – all spoke clearly of science's growing command of global phenomena.[11] Sabine touched too on the increasing facility with which scientific ideas moved across the globe. He singled out for praise the government of the United States for encouraging the free exchange of scientific texts across national

---

8   Bruno Latour, *Reassembling the Social: An Introduction to Actor-Network-Theory* (Oxford: Oxford University Press, 2005).

9   On life geographies, see Stephen Daniels and Catherine Nash, 'Lifepaths: Geography and Biography', *Journal of Historical Geography* 30 (2004): 449–58 and Miles Ogborn, *Global Lives: Britain and the World, 1550–1800* (Cambridge University Press, Cambridge, 2008).

10   'Soiree and Promenade', *The Belfast News-Letter*, 3 September, 1852, 2.

11   'The President's Address', *The Belfast News-Letter*, 3 September, 1852, 1.

borders by exempting them from Customs Duty. In his vote of thanks, the astronomer Thomas Romney Robinson echoed Sabine's support for the global spread of scientific enlightenment and called on the inhabitants of Belfast to take to 'remote regions the power of intellect and the energy of strength'. Belfast, 'the heart of Ireland' and 'the centre of Ireland's life', was well placed to participate in a global project of scientific education.[12]

This rhetoric of global knowledge, influence and transformation found echoes in the instruments and charts on display the following day in Robert Workman's warehouses. In a more intimate way, it also resonated with Robert Workman's own combination of scientific interests and commercial ambitions. Thomas Workman was being raised in a household surrounded by scientific conversation and supported by global trade in textiles. The family home on Windsor Avenue, South Belfast, had been given the name Ceara after Robert's favourite Brazilian port.[13]

Workman's schooling further encouraged an interest in science. He was educated at the Royal Belfast Academical Institution, a school with a celebrated reputation for cultivating a taste for natural philosophy and natural history.[14] In 1858 he entered Queen's College, Belfast as a non-matriculated student, allowing him to attend lectures and use the library.[15] By 1861 he was participating in local scientific societies.[16] In 1863 he became a founding member of the Belfast Naturalists' Field Club.[17] At the same time he took on increased responsibility in his father's textile business. When Robert Workman died in 1870, T. and G.A. Workman Ltd. was passed to Thomas and his brother George.[18] The Bedford Street warehouses remained the focal point and Thomas took over his father's role in securing orders from merchant houses across the world. Yet even before he had full responsibility, Thomas had the opportunity to travel and learn the science and art of observing alien natural, cultural and economic environments.

Camp Supply, Indian Territory, January 1870

Just a few months before his father's death, Thomas departed from Londonderry on the SS *Nestorian* to embark on a tour of Canada and the United States. Although Thomas met with some business contacts, the trip's primary goals were

---

12   'The President's Address', *The Belfast News-Letter*, 3 September, 1852, 2.

13   Margaret A.K. Garner, *Robert Workman of Newtownbreda, 1835–1921* (Belfast: William Mullan & Son Ltd, 1969), 19.

14   For his schooling, see 'The Death of Mr Thomas Workman, J.P. of Belfast', *Northern Whig*, 14 May, 1900.

15   *The Queen's University Calendar* (Dublin: Alexander Thom and Sons, 1859), 101.

16   'Belfast Polytechnic Association', *The Belfast News-Letter*, 26 April, 1861, 3.

17   A. Albert Campbell, *Belfast Naturalists' Field Club: Its Origin and Progress* (Belfast: Hugh Greer, 1938), 15.

18   'Deaths', *The Belfast News-Letter*, 21 September 1870, 1.

self-improvement and a search for personal adventure. It also provided an opportunity to learn how to observe a world strikingly different from the one Thomas knew in Belfast. The observational practices Workman honed in the United States, especially in the American West, were those he would later employ to fuller effect to acquire knowledge of natural history and create economic opportunities in places far distant from Belfast. In Workman's life, as we shall see, the American West marked both the beginning and the end of his travels. His first encounters with that 'mythic' frontier land were formative and followed a common and well-established practice of using the American West as a space of self-culture and as a preparation for global travel.[19]

Workman's largely touristic descriptions of North America were recorded in a journal written up as a series of letters to members of his family and illustrated with his own hand-drawn images and photographs either taken by him or acquired while on tour. For the first part of his journey, much of the descriptions and sketches conveyed the dramatic spectacles of waterfalls and charismatic fauna. Whatever the appeal of these vistas and natural historical novelties, Workman frequently punctuated his narrative with a promise of more exciting scenes to come. It was encounters with Native Americans that represented for Workman the climax of his travels in the United States.

Workman's first sighting of indigenous groups came on 7 January 1870 a mile outside Camp Supply, a US army post located in 'Indian Territory', Oklahoma. His initial impressions, written up in his letter-book journal, were marked by disappointment and unmet expectations. With further meetings and more intimate encounters, Workman added more positive portraits of Native Americans. Throughout, his descriptions oscillated between bestialising Native Americans and romanticising their character and lifestyle. On the one hand they were 'like gipsies' living in dirt and squalor. The 'Indian children resemble[d] barrel monkeys in every respect except the tail'.[20] On the other hand, their 'vices' were apparently derivative, an effect of contact with Europeans. Their worst fault, Workman concluded was, 'that they are inveterate beggars'. But it was an acquired rather innate habit that only emerged after prolonged residence among European settlers. A romanticising impulse was reflected in Workman's note in his journal that he 'did not consider Fenimore Cooper's Indians overdrawn … I have seen many quite as good as those he depicts'.[21] Any sense of ambivalence or uncertainty in

---

19 On the American West as a staging post on the global itineraries of European travellers, see David M. Wrobel, *Global West, American Frontier* (Albuquerque: University of New Mexico Press, 2013); and Monica Rico, *Nature's Noblemen: Transatlantic Masculinities and the Nineteenth-Century American West* (New Haven: Yale University Press, 2013).

20 Thomas Workman, Illustrated Notebook, 'Letters from the Far West', 88, D2778/1/1A Acc 9353, Public Record Office of Northern Ireland (hereafter cited as PRONI).

21 Workman, 'Letters from the Far West', 110.

Workman's ethnographic descriptions was largely masked by the confident prose and pictorial practices of the assured roaming observer. His fleeting encounters with Native Americans may have disturbed the exhibitionary logic of Workman's illustrated narrative but they did not dislodge it.

The largely non-reflexive mode of Workman's narrative, in many respects typical of contemporary British attitudes and accounts of indigenous Americans, reasserted itself even in reports of circumstances where his own life was either seriously endangered by, or depended upon, the actions and decisions of non-Europeans.[22] When he found himself involved in a US cavalry expedition to pacify 'marauding' Kiowa, Workman made the most of the opportunity to witness the rituals and costumes of war. Once the two parties finally met on a 'large plain', some twenty miles from Camp Supply, Workman, after initially being overcome with fear at the possibility of slaughter on both sides, joined the officers in meeting Kiowa chiefs in council. Then, after an agreement not to fight was eventually reached, Workman circled up to the top of hill and 'look[ed] down at the extensive plain traversed by the white coated Indians'. It was, he noted, 'a sight that I think will be vividly photographed in my memory for the remainder of my life'.[23] He was, with relative ease, safely behind the camera – on this occasion metaphorically – confirming his position as a detached and elevated onlooker apparently removed from the intricate and imbalanced play of power that seemed to unfold independently of his presence. The apolitical tenor of Workman's discourse was reinforced when he sketched a subsequent meeting between the interpreter John Smith and Spotted Wolf, Chief of the Cheyenne. Observing an apparent bond between the two men, Workman remarked, 'so you see, there can be true friendship between whites and Indians' (see Figure 3.1).[24]

As well as providing his family with a full account of his adventures, Workman's tour of the United States, which subsequently took in New Orleans, Washington DC and New York, supplied material for talks to Belfast's learned societies. On his return home Workman delivered a lecture to the Belfast Naturalists' Field Club entitled 'a month on the Prairies'. The centrepiece was an account of the standoff between the US Cavalry and the Kiowa, 'in all their savage magnificence of war plume and paint'.[25] In lecture form, Workman's observations were presented as a form of 'global' knowledge – that is, knowledge readily detached from the particular circumstances in which it was first acquired and reproduced for a Belfast audience. As such, Workman's lecture did not foreground the stubborn incongruities of intercultural contact or the labour involved in producing representations in text and image of the mid-West.

---

22   Kate Flint, *The Transatlantic Indian, 1776–1930* (Princeton: Princeton University Press, 2009).

23   Workman, 'Letters from the Far West', 107.

24   Workman, 'Letters from the Far West', 111.

25   *Annual Report of the Belfast Naturalists' Field Club*, 7 (1870): 50–51.

**Figure 3.1   Thomas Workman, Watercolour Sketch showing meeting
between John Smith, Interpreter and 'Spotted Wolf, Cheyenne
Chief', 17 January 1870**

Source: Illustrated Notebook, between f. 110/111, PRONI, D2778/1/1A Acc 9353. I would
like to acknowledge the Deputy Keeper of the Records, PRONI, for granting permission
to publish.

At the same time, of course, Workman's confident posture, which seems
to confirm a Victorian stereotype, was a fragile achievement. The use of
photographs and sketches, along with textual description, may have in James
Ryan's words, 'produced a sense of an expanding and all-encompassing global
vision', but Workman was also implicated in the overwhelming multiplication
of views that, to continue Ryan's argument, 'not only collected portions of the
geography of the world, [but also] ensured its endless proliferation'.[26] Further,
if Workman regarded his letter-book as a composite memory device which, as
he put it, 'vividly photographed' scenes that he had fleetingly encountered, he
was, in Elizabeth Edwards' terms, borrowing a precarious 'mode of reassurance

26   James R. Ryan, 'Photography, Visual Revolutions, and Victorian Geography', in
Geography and Revolution, ed. David N. Livingstone and Charles W.J. Withers (Chicago:
University of Chicago Press, 2005), 230.

against the instabilities, messiness and unknowability of the colonial endeavour and experience'.[27]

The pattern of commercial and scientific travel, recorded in journals illustrated with sketches and photographs was one that Workman adopted for the rest of his life. His travels to Brazil in 1880–1881 and to South East Asia in 1883, 1888, 1890 and 1892 provided abundant opportunities for commercial expansion, scientific fieldwork and journal writing. After his first three trips to the tropics, Workman sifted through his material, selecting what he thought would best serve a lecture for local scientific societies, most usually the Belfast Natural History and Philosophical Society.[28] His growing interest in photography meant that the camera became relatively more important as a recording device and his changing interests in natural history also altered his observant practices.[29] What provided stability and continuity was a confidence that the world could be described and understood using the same basic techniques of looking and interpreting. Running in parallel was a conviction that universal commerce was possible despite the highly variant conditions, cultures and social contexts that his global travels had to negotiate.

Between Sea and Shore: Brazil, January to March 1881

When Workman addressed the Belfast Natural History and Philosophical Society in 1899, he contrasted the spiralling upward progress of observers using the locomotive power of science to the 'creatures of circumstance' who, like Tennyson's mariners in his poem the Lotus Eaters, can find no peace 'in ever climbing up the climbing wave'.[30] It was an apt analogy for someone familiar with the sense of disorientation produced at sea. As a yachting enthusiast, Workman knew well the thrill and terror of battling against heavy weather, of being tossed about by a churning sea. His experience of the steamship lines that crisscrossed the globe also underlined epistemic as well as existential instabilities. His own life described an attempt to overcome such disorientation and move in the direction

27 Elizabeth Edwards, 'Photographic Uncertainties: Between Evidence and Reassurance', *History and Anthropology* 25(2) (2014): 184.

28 Thomas Workman, 'A Recent Visit to Brazil', *Proceedings of the BNHPS* (1883): 15–18; Thomas Workman, 'Eastern reminiscences', *Proceedings BNHPS* (1885): 21; Thomas Workman, 'Eastern Reminiscences: Aden, India and Burmah', *Proceedings BNHPS* (1886): 12–15; Thomas Workman 'Eastern Reminiscences: China and Manila', *Proceedings BNHPS* (1887): 28–33; Thomas Workman, 'A Visit to Singapore', *Proceedings BNHPS* (1889): 30–34.

29 A 'clever amateur photographer', Workman was elected to the committee of the Ulster Amateur Photographic Society on Monday 14 December 1885. 'Ulster Amateur Photographic Society', *The Belfast News-Letter*, 15 December 1885, 7.

30 Workman, 'Study of Natural History', 18–19.

of truth. Nearly two decades earlier, his first trip to the tropics had provided an opportunity to see beyond the climbing wave.

On 15 January 1881 Workman embarked from Liverpool on the Red Cross Line Steamer *Paraense* bound for Pará, Brazil. This was Workman's first trip to the tropics and it presented exciting commercial and scientific opportunities. He was also consciously following not only his father's footsteps but also those of the naturalist, Alfred Russel Wallace. A month earlier, Workman – who had turned to a specialist study of spiders three years earlier on the advice of the arachnologist Octavius Pickard-Cambridge – had written to the celebrated naturalist asking for advice on the best means of collecting spiders in Pará.[31] In his reply, Wallace advised Workman to concentrate on gardens and waste grounds rather than virgin forest and noted that local white rum, Cachaça, was cheap and abundant enough to be used to preserve spiders in bottles. Giving 'negro and Indian boys' 1/2d would also pay dividends, as they 'will bring you any quantity of strange spiders'.[32] Wallace's advice was based on the understanding that Workman would spend only ten days in Pará before travelling on to other Brazilian ports. Even then, it proved impracticable as Workman moved rapidly from one stop off to another.

Being at sea proved to be as significant to Workman as making contact with merchant houses and exploring local flora and fauna. During the several stages of the voyage, Workman negotiated the unsettling effects of occupying a mobile and heterotopic space.[33] This was evident, for example, in his encounters with Roman Catholic passengers. In the confined space of the ship's cabin or smoking room, these conversations took on a particular significance. As a devout Presbyterian, Workman was particularly intrigued by the 'exotic' beliefs subscribed to by dedicated Roman Catholics. On one occasion, as the SS *Paraense* sailed away from Tenerife into the mid-Atlantic, Workman found himself discussing the 'confessional, absolution and papal infallibility' with a 'heterodox' Roman Catholic.[34] A month later, Workman shared a cabin with a 'frocked friar'. In his description Workman commented that this 'strange bedfellow ... was not dressed in the dirty slovenly way in which you see European friars but in a long brown and purple coat made of the best cloth'. In broken Portuguese, Workman explained that he was 'a Zwinglian Calvinist' but observed that the friar seemed to 'confound all the 3 sects Calvinist, Zwinglian and Lutherans'. Defeated, Workman

---

31   On Thomas Workman's turn to spiders see, 'A Contribution Towards a List of Irish Spiders', *Entomologist* 8 (1880): 125.

32   Alfred Russel Wallace to Thomas Workman, 17 December 1880, D2778/5/G/1/6, PRONI.

33   On the effects of steamship travel on self-identity and global knowledge, see Tamson Pietsch, 'A British Sea: Making Sense of Global Space in the Late Nineteenth Century', *Journal of Global History* 5 (2010): 423–46. On the ship as a space of alternative orderings, see William Hasty and Kimberly Peters, 'The Ship in Geography and the Geographies of Ships', *Geography Compass* 6 (2012): 660–76.

34   Thomas Workman, Copy Letter Book vol. 1, 36–7, D2778/1/4, PRONI.

wished for 'some language in which we could communicate as I would like to know what he thinks of us'.[35] His strong sense of religious identity generated a curiosity about other forms of Christian belief only partly sated by tantalising shipboard conversations and encounters. Brief and broken though they were, these conversations playfully worked through tensions and differences that, in the context of Belfast, carried a much stronger political charge.[36]

Arguably, however, it was the movement between ship and shore that exercised the most profound influence on Workman's impressions of the tropical world. His first sighting of the tropics came when the SS *Paraense* steamed up the mouth of the Amazon River. In his journal, Workman recorded that 'today my mental barometer has gone up two or three inches. For some time back I have been thinking that travelling was a nuisance and the Tropics a delusion but the excitement of ... coming up the river has greatly cheered me up'.[37] Employing a standard trope borrowed from a long-established 'cult of the tropical picturesque', he recorded how the scene also overwhelmed him.[38] He was 'quite unable to realize the fact that I am in the mighty Amazonas River that I have so often read of and wished to see'.[39] On awaking the next morning he wrote that 'I can hardly tell what I have seen it has been such a maze'.[40] Once back at sea after his stay in Pará, Workman lapsed into a more melancholic mood and was infected by homesickness. Writing to his wife 'Meg', Workman wondered whether 'you are as anxious to see me as I am to get home to you'. The fatigue produced by 'too much novelty and hard work' was taking its toll and Workman determined that he would not embark on a long journey again without his spouse. [41]

The remaining accounts of his travels to Brazil are thick with descriptions of the shoreline made while on board one of the coastal steamers that served the country's port cities. His journal entries oscillate between sketches of familiar and strange maritime scenes, mediated by comparisons with similar seascapes nearer to home. In this, Workman was operating within what Derek Gregory and James Duncan have described as the 'complex dialectic between the recognition

---

35   Thomas Workman, Copy Letter Book, vol. 3, 32, D2778/1/4, PRONI.

36   For a comparable case, see Jonathan Jeffrey Wright, '"The Perverted Graduates of Oxford": Priestcraft, "Political Popery" and the Transnational Anti-Catholicism of Sir James Emerson Tennent', in *Transnational Perspectives on Modern Irish History*, ed. Niall Whelehan (London: Routledge, 2014), 127–48.

37   Thomas Workman, Letter Book, vol. 1, 45.

38   On the 'tropical picturesque', see David Arnold, '"Illusory Riches": Representations of the Tropical World, 1850–1940', *Singapore Journal of Tropical Geography* 21 (2000): 6–18.

39   Workman, Letter Book, vol. 1, 46.

40   Workman, Letter Book, vol. 1, 48.

41   Thomas Workman to Margaret [Meg] Workman, 16 February 1881, D2778/1/4, PRONI. Workman married Margaret Elliot Hill, daughter of James Hill, Collector of Inland Revenue, on 19 June 1872 at Fisherwick Place Presbyterian Church. *The Belfast News-Letter*, 20 June 1872, 1.

and recuperation of difference' characteristic of travel writing as a genre.[42] When exploring rock pools near to the port of Recife, Workman pointed out the similarity between the coral and seaweeds and British varieties. At the same time, his journal records the 'brilliancy' and 'richness of colouring' of tropical fish making such seaside naturalising a must 'if anyone should come to a tropical country'.[43] This movement between homely comparisons and a tropical 'excess' was also apparent in a description of the coastline south of Rio de Janeiro. It was, Workman suggested, 'wonderfully like the west coast of Scotland'. In Workman's estimation, the explanation lay in the former action of glaciers. The coastline around Santos, despite now being clothed in tropical vegetation had 'in some long forgotten time [been] strangely different' with glaciers streaming down the hills.[44] It was a worldwide process of glaciation, as then understood, that bridged the temperate/tropical divide. A combination of what James Duncan has called the 'shock of the familiar' and an account of a planetary-wide natural process helped Workman manage the unsettling effects of tropical difference.[45]

However effective such stabilising strategies proved to be, the liminal space between ocean and land also provided some of Workman's most unnerving personal experiences. When Workman arrived off Fortaleza, the main port city of the province of Ceará, he discovered that, due to the heavy surf, it was only possible to get to land by using 'native rafts' or Jangada. When Workman later narrated his experience, he commented that he had 'tried to follow [Captain Kaas] from the platform and would have done so all right but for the wretched niggers who made a wish to carry me ashore'.[46] Falling into the surf, he blamed the helpers and exposed a strongly racialist attitude more often articulated in his journals using the conventional vocabulary of comparative racial description. When he returned to the coastal steamer to travel further down the coast he made sure to avoid being carried and managed to 'get on the Jangada without help and without accident'.[47] The sense of threat to personal integrity occasioned by the material and cultural liminality of the surf zone had passed and Workman was able to regain the self-control associated with the objective observer. It was a return to a sense of a stable self with its felt capacity to keep in view a unified world. That this episode occurred on first arrival off the coast of Ceará, the name of Workman's family home on Windsor Avenue, underlines its power and poignancy.

When Workman did spend time on land, his primary task was commercial. As a result, he was limited not only by the set timetables of steamship lines but also by

---

42   Derek Gregory and James S. Duncan, eds, *Writes of Passage: Reading Travel Writing* (London: Routledge, 1999), 5.

43   Workman, Letter Book, vol. 1, 46.

44   Workman, Letter Book, vol. 3, 59.

45   James S. Duncan, 'Dis-Orientation: On the Shock of the Familiar in a Far-Away Place', in *Writes of Passage*, ed. Duncan and Gregory, 151; 157.

46   Workman, Letter Book, vol. 1, 98.

47   Workman, Letter Book, vol. 3, 3.

the rhythms of business transactions. These dealings were recorded in a separate journal, written as a series of letters to his brother. Workman was careful to note the merchant houses that expressed most interest in the samples that he brought from Bedford Street. The ultimate goal was to secure significant orders and open up new markets in Brazil for his firm. This required sensitivity to local economic conditions and business practices in a way that echoed the attention to detail that marked Workman's descriptions of his fleeting encounters with tropical nature. It was vital to quickly work out what type and pattern of textile would sell in the Brazilian market. Heavy linen was of little obvious use. Lighter fabrics spun from finer yarn would cost less to import due to reduced custom charges and would be more suitable for local conditions.

The importance of timely actions and precise specifications necessary to secure orders was evident in Workman's instructions to his brother written during his stay in Brazil. In one letter to his brother, after reflecting on the fragile economic climate in Ceará following a severe drought and famine, Workman requested the making of 'a web with our own weavers 52 inches wider with 1/8 inch split in the middle for cutting up and have it bleached when I return'. In issuing this instruction Workman was rapidly shuttling between the demands of an uncertain Brazilian market and the technicalities of textile production in Belfast.[48] Science and commerce were here brought intimate contact through a technical interest in two types of web – the natural snare of a spider and the fabric specifications for his 'own weavers'. In his scientific reporting and business speculations Workman operated with a conviction that both types of web could achieve a global distribution. Among the species of spiders that Workman managed to collect in Brazil were several that had a trans-continental distribution. Perhaps his most treasured find in Brazil was a species first collected in Java. Workman already had in his possession one that had been sent from Madagascar. The webs it created were clearly adequate for survival in Brazil, a fact that resonated with Workman's conviction that, whatever the challenges of oceanic travel and crossing material and cultural thresholds, his own woven fabric could be successfully exported to a tropical economy.[49] What remains to be explored, however, are the ways in which Workman's strenuous efforts to collapse distance and overcome the friction of natural and cultural difference played out in his home-worlds.

Belfast Museum, College Square North, February 1889

On 5 February 1889, Thomas Workman entertained members of the Belfast Natural History and Philosophical Society with a talk on 'A visit to Singapore'. Illustrated

---

48 Thomas Workman, Copy Letter Book, vol. 2, 22, D2778/1/5, PRONI.

49 Placing spider's webs and the webs of human weavers within the same frame echoes a literary tradition that more explicitly did the same thing in mythological mode. See Katarzyna and Sergiusz Michalski, *Spider* (London: Reaktion Books Ltd, 2010), 60–65.

with Workman's own 'limelight photographic views', the lecture presented a vivid picture of the 'Liverpool of the East', allowing the listeners to participate in a typically late-Victorian mode of vicarious travel.[50] During his lecture, Workman offered a panoply of views that ranged from topographic scenes to descriptions of 'Hindoo funeral processions' and Chinese graveyards. Rhapsodizing about the tropical scenery (it was, Workman noted, 'impossible to paint'), Workman decried the spoiling effects of the numerous defensive structures around Singapore – a consequence of a recent geopolitical standoff between Britain, Russia and France.[51] These fleeting views transported the audience to the far side of the globe and reinforced Workman's well-established reputation as a credible scientific witness of tropical flora, fauna and human culture. Workman's efforts to transport his listeners to far-flung exotic places were echoed in the collection of 'foreign birds' in cases positioned around the lecture hall. Birds of Paradise from Malaysia, a 'parson bird' from New Zealand, penguins from the Antarctic and a Bengal Vulture were among the stuffed specimens on display.[52]

Workman was intimately acquainted with the small lecture room at College Square North and with his audience (restricted as it was to members of the Society). He had been a Society shareholder since 1877 (by transfer from his mother) and a member of the Council since 1878.[53] His first paper to the Society was on spiders and was delivered in January 1878. Over the next decade and more, Workman gave several accounts of his trips to Brazil and South East Asia and instructed Society members on the respiratory organs of animals, Irish spiders and island biogeography. He was also one of the Society's most significant donors. His collection of Irish spiders, deposited in 1880, was among his more important gifts. He also regularly presented objects gathered during his travels. The annual report for 1889 records Workman's contributions of, 'a number of land and freshwater shells from Singapore, a flint knife from New Guinea, two quartzite knives from North Queensland, several nests of trapdoor spiders, specimens of land and freshwater shells from Madagascar and Brazil, insects and fish from North Australia, a snake from Burma and several snakes from North Australia'.[54]

In submitting this eclectic set of objects, Workman followed a tradition of collecting that the Society had, in its earlier years, done much to foster.[55] In terms of global reach if not quantity, Workman's donations ranked him alongside earlier Belfast travellers such as Gordon Augustus Thomson (1799–1886) and James Emerson Tennent (1804–1869). This was one reason why Workman was elected

---

50   'Belfast Natural History and Philosophical Society', *The Belfast News-Letter*, 6 February, 1889, 7.

51   Thomas Workman Lecture Notes, D2778/1/5, PRONI.

52   *Visitor's Guide to the Belfast Museum* (Belfast: BNHPS, 1880), 6.

53   Belfast Natural History and Philosophical Society List of Shareholders and Members, p. 65, D3263/C/6, PRONI.

54   *Proceedings of the BNHPS* (1889): 9.

55   See Jonathan Jeffrey Wright, Chapter 7, this volume.

President of the Society in 1898 and it demonstrates the importance of the Society to Workman's self-fashioning as a civic worthy and reputable naturalist. Yet Workman's efforts to confirm and consolidate his standing as a virtuous scientific citizen were not entirely in keeping with how associational science in civic as well as intellectual terms was assessed. By the 1880s, 'foreign' objects did not carry the same significance as they had in earlier years. Priorities had begun to change and this altered the value of Workman's efforts to sustain a tradition of donation of objects gathered from the 'four quarters of the globe'.

By the 1880s the Society's museum was not considered adequate for a rapidly growing industrial town. Although it had been extended in 1880 to accommodate the collections of the Belfast antiquarian Edward Benn, this did not meet the requirements of those calling for a museum that was free to the public.[56] Linked to this was the fact that technical education was being promoted as an urgent concern and it proved difficult for the Society to align its aims with moves to improve Belfast's provision of more practical learning. A museum of industrial arts was considered by some as a much more urgent desideratum.[57] The opening of the Free Public Library on Royal Avenue in October 1888, paid for by the adoption of the Free Libraries Act by the Town Council, further altered the standing of the Society's Museum. While prominent members of the Society had strongly backed the adoption of the Act and had hoped, given the right conditions, to donate its collections to the new museum, this did not occur.[58] In part, this was due to lack of funds. The new publicly funded library included an art gallery but there was insufficient room for the Society's collections. Accompanying these practical challenges was resistance to rate-supported public museums among other members of the Society. For John Brown, the Society treasurer, 'handing over' the collections was not only, in legal terms, questionable but would also have a 'weakening effect' on voluntary efforts to create a successful local museum.[59]

Alongside such local wrangling over the relative merits of municipal control and voluntarism, changing policies of collection and display associated with provincial museums across Britain and Ireland were having an effect. The 'foreign collections' of a number of museums increasingly occupied a kind of

---

56   On the extension, see 'Annual Meeting of the Belfast Natural History and Philosophical Society', *The Belfast News-Letter*, 10 June, 1880, 6.

57   On the need for a public museum displaying industrial arts, see, for example, 'Progress of Science and Art in Belfast', *The Belfast News-Letter*, 1 January, 1880, 5; and William Nicholl, 'Belfast Free Museum', *Northern Whig*, 16 April, 1890.

58   See 'Natural History and Philosophical Society', *The Belfast News-Letter*, 20 June, 1883: 6; Robert Patterson and Robert Lloyd Praeger, 'Belfast Museum', *Northern Whig*, 12 April, 1890; William Gray, 'Belfast Museum', *Northern Whig*, 13 April, 1890. Gray was an untiring champion of the idea of a rate-supported public museum. See William Gray, *Science and Art in Belfast* (Belfast: Northern Whig, 1904).

59   John Brown, 'Belfast Museum', *Northern Whig*, 8 April, 1890. Brown was a significant figure in Belfast's scientific culture. He was elected to the Royal Society in 1902.

third rank below local and 'typical' collections of natural history and antiquities.[60] This certainly appears to be true of the Belfast museum. As the annual report of the Belfast Society in 1881 put it, 'the council still keep steadily in view the leading idea that a local museum should, in the first place, be illustrative of the antiquities, geology, flora and fauna of the country in which it exists'.[61] The value of foreign objects remained in terms of attracting visitors on Easter Monday when the Museum was opened to the public, or spicing up fund-raising conversaziones, but they were increasingly regarded as beyond the legitimate scope of regional museums, including those funded by private subscription. The only non-local items that retained a higher value were the proceedings of foreign learned societies sent in exchange for the Society's own publications. Fittingly, it was the Society's 'zealous librarian Mr [Thomas] Workman' who maintained those wider connections throughout the 1880s.[62] Yet for all Workman's zeal in establishing and maintaining transnational exchange, there was a growing sense that contributors to local natural history and technical education were regarded more highly than donors of foreign material in the reputational economy of civic society in late-Victorian Belfast.

What was true of Workman's donations was also true of his lectures on foreign travel. By the late 1880s, the Society was promoting heavily its courses of 'popular' lectures delivered by visiting speakers that aimed to attract non-members and educate a local public in science. This was one way to justify its existence and meet criticisms that its relevance was restricted to those who could afford the cost of membership. The small room on the first floor where Workman's lecture on Singapore was held could only accommodate an audience of two hundred and was cramped and poorly ventilated.[63] It was no longer considered adequate for accommodating a different style of lecture and a public rather than essentially private audience.[64] While still clearly appreciated by members, Workman's account of his 'visit to Singapore', and his donation of objects acquired during it, were not fully in tune with the shifting emphases shaping late-Victorian Belfast's scientific and civic culture.

None of this suggests that Workman himself was marginalised by the Society to which he dedicated so much of his time outside of business hours.

---

60   On these debates, see Diarmid A. Finnegan, *Natural History Societies and Civic Culture in Victorian Scotland* (London: Pickering and Chatto, 2009), 76–84; cf. Kate Hill, *Culture and Class in English Public Museums, 1850–1914* (Aldershot: Ashgate, 2005), 69–89. Note that Hill's study is of English municipal or rate-supported museums only.

61   'Natural History and Philosophical Society Annual Meeting', *The Belfast News-Letter*, 8 June, 1881, 3.

62   'Annual Meeting', *The Belfast News-Letter*, 8 June, 1881, 3.

63   The estimate of 200 is given in: William Nicholl, 'Belfast Museum', *Northern Whig*, 18 April, 1890. For commentary on the room's inadequacies, see 'Belfast Natural History and Philosophical Society', *The Belfast News-Letter*, 24 May, 1890, 7.

64   'Belfast Natural History and Philosophical Society', *The Belfast News-Letter*, 28 June, 1888.

It does highlight, however, that Workman had to negotiate a range of concerns in establishing and enhancing his standing locally. It is worth noting, too, that these negotiations were made more urgent by the fact that Workman, at the time of giving his lecture on Singapore, was fighting to preserve family honour in the face of sensational divorce proceedings involving his brother and business partner, George Augustus Workman. The case had been before the probate and matrimonial court in Dublin several times during 1888 and remained unresolved until April 1889.[65] With very public accusations of forged letters and cruel treatment levelled by George's estranged wife not only at her husband but also his close relations, the case presented a serious threat to the good character of the Workman family.

The prioritisation and valorisation of a particular version of 'the local' by the Philosophical Society was, perhaps, one of the reasons why Workman stored and displayed most of his collections of tropical spiders (including some significant type specimens) and other foreign exotica at his home in Helen's Bay on the southern shore of Belfast Lough.[66] This domestic museum was, however, much more than a temporary storehouse for under-valued collections. As an ostensibly more private place for dealing in global knowledge, it deserves fuller attention.

Craigdarragh House in the 1890s

Built in 1850 on the southern shore of Belfast Lough, Craigdarragh House was the residence of several wealthy merchants before the Workman family began to rent the property in 1881.[67] It was one of the several large houses built along the Lough shore and reflected, among other things, the growing wealth of Belfast industrialists. Designed by Charles Lanyon for the county Down landowner, Robert Francis Gordon, it was later purchased by the British diplomat and local worthy, Lord Dufferin, who rented it until Thomas Workman purchased the freehold in 1883.[68] By then Workman's family had grown in size (at that stage, there were three daughters and two sons) and his commercial interests were expanding. Set in an estate of some 26 acres, Craigdarragh House underlined Workman's growing status as a member of Belfast's business élite.

Like Victorian homes in general, Craigdarragh House was composed of 'a number of spaces intended for diverse purposes that [did] not fit neatly into the

---

65   For sample reports, see 'Dublin Law Reports', *The Belfast News-Letter*, 13 March, 1888, 8; and 'A Belfast Divorce Suit', *Freeman's Journal*, 29 April 1889, 6.

66   The vast bulk of Workman's collection of tropical spiders, which included a number of type specimens, were deposited with the natural history department of the Dublin Science and Art Museum after his death.

67   Workman is recorded as residing at Craigdarragh from that year. For example, 'Births, Marriages and Deaths', *The Belfast News-Letter*, 22 November, 1881, 1.

68   Charles Brett, *Buildings of North County Down* (Belfast: Ulster Architectural Heritage Society, 2002).

single category of the private'.[69] A key illustration of this was the fact that the house, as well as being a domestic space, functioned as a site of scientific interest. As early as 1883, Workman received members of the Belfast Naturalists' Field Club as part of a daylong excursion. As one of the founding members of the Club, Workman was well placed to play host and to talk the members through the 'many objects of interest' held at Craigdarragh not least the large collection of arachnids from Brazil, Java and Madagascar. As the report of the excursion noted, Workman also guided the excursionists down the glen that ran to the shore, a part of the grounds that had 'a tropical aspect' set off by the luxuriant ferns and 'festoons of ivy'.[70] The house and its grounds provided an opportunity to investigate local flora and fauna and glimpse the natural history of the tropics in a way that echoed other more public spaces explicitly designed to accomplish the same effect.[71]

In the time-honoured fashion of the Victorian naturalist, Workman's home also became the centre of an extensive network of scientific correspondence in ways that transgressed a clear demarcation between private and public.[72] Corresponding with experts was vital if Workman was to establish himself as an authority and make his mark on the developing discipline of arachnology. The hundreds of letters sent to Workman during that decade from arachnologists based in Europe, North America and New Zealand testify to Workman's emerging reputation as a specialist in tropical spiders.[73] In concentrating his efforts on Malaysia, Workman had also chosen a region that, in terms of its arachnids, had not been explored before except by the Hungarian entomologist, Carl Ludwig Doleschall.[74] This provided opportunities to find species new to science and to collect spiders that would fetch a high price in the marketplace of transnational intellectual exchange.

If Workman's scientific correspondence was an integral part of his efforts to participate in the wider public world of science, the domestic and familial setting where this was coordinated did significantly intrude. During the 1890s, Workman increasingly relied on the help of his daughters to manage his collection, including such mundane tasks as re-filling specimen bottles with 'spirits of wine'. One daughter, Margaret, provided more significant assistance using her artistic talents to provide detailed sketches of Malaysian spiders for Workman's book on that

---

69   Lynda Nead, 'Gender and the City', in *The Victorian World*, ed. Martin Hewitt (London: Routledge, 2012), 294–5.

70   *Annual Report and Proceedings of the Belfast Naturalists' Field Club* 2(4) (1883): 227.

71   See Nuala C. Johnson, *Nature Displaced, Nature Displayed: Order and Beauty in Botanical Gardens* (London: I.B. Tauris, 2011), 87–128.

72   On this general theme, see Janet Browne, 'Corresponding Naturalists', in *The Age of Scientific Naturalism*, ed. Bernard Lightman and Michael S. Reidy (London: Pickering and Chatto, 2014), 157–70.

73   Many of these are to be found in Thomas Workman Letter Book, E MSS WOR A1:1, Library and Archives, London Natural History Museum.

74   See Tamerlan Thorell to Workman, Montpellier, 3 March 1894, Workman Letter Book, 31.

subject. In a letter to Workman, Reginald Pocock, the British Museum's expert on arachnids, praised Margaret's figures and recipients of the first volume of Workman's *Malaysian Spiders* frequently singled out the 'extraordinarily good' illustrations.[75] Margaret's drawings also facilitated Workman's collaboration with the Swedish arachnologist Tamerlan Thorell who, on account of her precise depictions of tropical spiders, counted her a 'colleague' and sent her copies of his scientific papers.[76] In these ways, Margaret's sketches made in the drawing room of Craigdarragh House helped to consolidate Workman's scientific networks and reputation.

Beyond such scientific endeavour, Craigdarragh functioned in other ways as a space in which the global was domesticated and the domestic entrained in global flows. Workman's financial investments, for example, were strongly mediated by familial relations. In addition to his linen and muslin business, Workman helped to manage a large Belfast-based shipbuilding firm – Workman, Clark and Co. Ltd – established by his brother Frank in 1880. By the 1890s, Thomas was Director of the firm and the largest shareholder.[77] He also held shares in the Rangoon-based rice-trading firm, the Arracan Company Limited. William Hill, Workman's brother-in-law, was the Company's agent in Burma.[78]

Needless to say, Workman's efforts to maintain Craigdarragh as a hub for scientific exchange and as a place to manage the challenges of a 'global' economy occurred alongside activities that had a distinctly local horizon. The house, for example, was used to host a Sunday School run by Workman and his family before the erection of a new Presbyterian Church in the vicinity.[79] Workman used his influence and his respected position within the Presbyterian Church to lobby for a new church building in Helen's Bay and subsequently helped to finance the costs of erecting it.[80] He was involved in political concerns, not least with mass agitation over Home Rule. In the period leading up to his final trip to Singapore in 1892, a local committee nominated Workman as a delegate to the Ulster Unionist Convention held in Belfast in June of that year.[81] And he invested his capital in local property speculations. With his several brothers, Workman owned two

---

75   Reginald I. Pocock to Thomas Workman, British Museum, 6 May 1892, Workman Letter Book, p. 6; Lord Dufferin to Thomas Workman, Clandeboye, 16 November 1896, Workman Letter Book, 91.

76   Tamerlan Thorell to Workman, Montpellier, 29 March 1893, Workman Letter Book, 23.

77   'Death of Mr Thomas Workman, J.P., of Belfast', *Northern Whig*, 14 May, 1890.

78   See, for example, William Hill to Thomas Workman, Rangoon, 23 October 1885, Business Correspondence to Thomas Workman, D2778/5/A/3/8, PRONI.

79   'Death of Mr Thomas Workman', *Northern Whig*, 14 May, 1890.

80   Margaret Garner, *A History of Helen's Bay Presbyterian Church, 1896–1958* (Bangor: Abbey Press, 1958).

81   'Great meeting in Holywood', *The Belfast News-Letter*, 31 May, 1892, 7.

estates in the districts of Windsor (in Belfast) and Marlborough (then just outside the city boundary).[82]

At the same time, none of these activities were entirely cut off from the wider worlds associated with Workman's various commercial and scientific pursuits. Even the mundane matter of managing the grounds around Craigdarragh House intersected with Workman's expertise in tropical spiders. In 1896, on completion of his first volume on Malaysian arachnids, Workman sent a complimentary copy to his neighbour, Lord Dufferin. The reply was warm and admiring: 'I am really very grateful to you and proud that I have such a learned neighbour'.[83] This cordial exchange occurred not long after Dufferin and Workman faced each other in court over a 'right of way' dispute and this gift may have signalled, and enacted, an end to hostilities.[84]

Workman's life at Craigdarragh threaded together scientific collecting, commercial enterprise and local endeavour in what might be described as a 'buffered' but 'porous' domestic space.[85] Perhaps more than any other activity, creating a collection of tropical spiders underlined the unstable character Workman's attempts to be at home in a wider world. Immersed in 'spirits of wine' and placed in flattened phials by his daughters, the tropical spiders in Craigdarragh House could be safely studied and admired without risk or sense of danger. At the same time, the tropical spiders carried darker meanings that evoked some of the uncertainties and risks associated with trade and travel in the tropics.[86] The more threatening qualities of tropical spiders sat uneasily alongside domestic as well as intellectual ideals and pointed to the limits and fragility of efforts to domesticate 'global' realities.

The Death of a Merchant-Naturalist, May 1900

In 1900, a few days before Easter, Thomas Workman embarked on a long overdue return visit to the United States. As usual, his purpose was twofold. He had business matters to attend to but he also saw the trip as an opportunity to expand his knowledge, and his collection, of the world's spiders.[87] This trip, however, was

---

82   See Windsor Building Company Ltd and Marlborough Park Company, D2778/13, PRONI.

83   Frederick Hamilton-Temple-Blackwood [Lord Dufferin] to Thomas Workman, Clandeboye, 14 November 1896, Correspondence to Workman Letter Book, 91.

84   On the court case, see *Belfast News-Letter*, 2 August, 1895, 7.

85   I have borrowed these terms from Charles Taylor, *A Secular Age* (Cambridge, MA: Harvard University Press, 2007).

86   On the rise of the association between tropical spiders and Gothic horror in precisely this period, see Claire Charlotte McKechnie, 'Spiders, Horror and Animal Others in Late-Victorian Empire Fiction', *Journal of Victorian Culture* 17(4) (2012): 505–16.

87   'Death of Mr Thomas Workman', *Northern Whig*, 14 May, 1900.

tragically cut short. Part way through his tour, at St Paul Minnesota, Workman suddenly took ill and, shortly afterwards, died from 'affection of the heart'.[88] In lamenting the loss, John Brown, the incoming President of the Belfast Natural History and Philosophical Society and one of Workman's 'earliest Belfast friends' offered a 'tribute of esteem'. Workman, he declared, 'was one of the few business men in our city who found time for original scientific research'. His death was to be deplored, marking as it did the end of rare efforts to use 'far business connexions' to master a branch of natural history.[89]

Arguably, Workman's death marked more than the passing of a local civic worthy. It also signalled the steady decline of a mode of being. By 1900, the cultural logics that supported the combination of business, science and travel embodied by Workman were waning. The professionalisation and specialisation of science, operating as it did with new definitions of certified expertise, was working against assigning a significant role to amateur investigators. In business circles, science had increasingly become a form of 'shop talk' rather than a mark of respectability or a binding agent for strengthening commercial relationships.[90] Further reinforcing this disintegration was the steady retreat of the voluntarism that had been the lifeblood of the kind of associational culture so crucial to Workman's public reputation.[91]

The geography of Workman's life that has been sketched here, then, was a geography of its time. Even so, it prompts more general reflection. Workman's life can be said to demonstrate how global ambitions – or a desire to capture global patterns and a fully comprehensive view of things – form within, and are frustrated by, overlapping and often incommensurable spheres of action. In Workman's case, it is also clear that a search for global knowledge can operate in ways that smooth out the uneven and intricate textures of local realities even while they seek to capture them in a piece of text, a collection of spiders or a set of images. Certainly, his attempts to freeze-frame the fluid complexities of inter-cultural contact in the American mid-West tended to retreat to the safety of stereotyped views which were at best a simulacrum of 'global' or comprehensive knowledge. The maritime and littoral spaces he traversed in 1881 were represented *in situ* and back in Belfast in ways that eased the more turbulent realities of commercial and scientific travel in the tropics. Perhaps unexpectedly, holding the world steady in these ways did not guarantee Workman's success within Belfast's associational culture. In that context, he had to struggle to make knowledge of distant places resonate with local priorities. Finally, all of Workman's efforts to manage the spaces that were

---

88   'Death of Mr Thomas Workman, J.P.', *The Belfast News-Letter*, 14 May, 1900, 4.

89   John Brown, 'Address by the President', *Proceedings of the BNHPS* (1901): 17–18.

90   See James Secord, 'How Scientific Conversation Became Shop Talk', *Transactions of the Royal Historical Society* 17 (2007): 129–56.

91   For this trend see, for example, Anne B. Rodrick, *Self-Help and Civic Culture: Citizenship in Victorian Birmingham* (Aldershot: Ashgate, 2004).

central to his life and ambitions were strongly coloured by the composite domestic setting of Craigdarragh House. In that context, the familial and domestic basis of his commercial and scientific initiatives was more evidently on display.

When Workman delivered his final address to the Belfast Natural History and Philosophical Society in 1899, he offered the search for 'similarity in structure existing between animals from separate parts of the globe' as a primary incentive for studying natural history. It was also imperative to determine whether these similarities were the product of connections in the geological past. It was, Workman noted, spiders preserved in amber that provided the best evidence for achieving this goal. One spider, *Eriauchenius workmani* –with its elevated head or caput – provided a case in point. Until it was described and named after Workman by O.P. Cambridge in 1881, its unusual anatomy was only known in extinct species preserved in amber found on the shores of the Baltic Sea.[92] Such fossil spiders in amber or extant spiders in spirits were tiny fragments of de-contextualised knowledge similar to Workman's letter-books and lectures offering glimpses of encounters with distant places. If such knowledge might in one sense be termed global, it was also profoundly patchy and scattered. As such, it reflected the delicate and half-formed webs of science and commerce that were spun within and between the diffuse and intimate spaces that made up the geography of Workman's life.

---

92 Workman, 'Incentives', 21. O.P. Cambridge, 'On some new genera and species of Araneidea', *Proceedings of the Zoological Society of London* (1881), 768–70.

Chapter 4

Scientific Practice and the Scientific Self in Rupert's Land, c.1770–1830: Fur Trade Networks of Knowledge Exchange

Angela Byrne

Introduction

This chapter studies features of scientific practice, intellectual sociability and scientific identity formation at fur trading posts in Rupert's Land c.1770–1830.[1] Rupert's Land was the name given to the Hudson Bay watershed, the territory over which the Hudson's Bay Company (HBC) had exclusive trading rights in 1670–1870 – rights which were extended to the Pacific coast in 1821 with the amalgamation of the HBC and its former rival, the Northwest Company (NWC). In the period, the vast region now known as Canada (a term which then referred mainly to the present-day provinces of Ontario and Quebec) was the site of competition and cooperation between indigenous, British, French settler, American and Russian interests, for its lucrative fur trade and for its potential to contribute significantly to scientific and geographical knowledge. The HBC and NWC themselves played important roles in the gathering and dissemination of knowledge on North America, particularly after 1770.[2] Fur traders' identities as men of science were of course closely linked to the scientific practices they undertook in Rupert's Land, but were also dependent upon the local and global networks available to them as employees of a major trading company. This chapter examines fur traders' practices, exchanges and interactions to demonstrate the interrelated nature of global and local knowledge, while also considering the extent to which attempts to engage in the dominant form of global knowledge gathering represented as 'colonial science' hampered engagement with local, indigenous forms of knowledge.

Until the 1818 launch of British state expeditions into the Arctic, scientific activity in Rupert's Land found expression primarily in the day-to-day activities of fur traders and trading company surveyors. Rupert's Land differs in this respect

---

1   The research was supported by a Marie Curie/Irish Research Council CARA postdoctoral mobility fellowship, 2010–13.
2   See Ted Binnema, *Enlightened Zeal: The Hudson's Bay Company and Scientific Networks, 1670–1870* (Toronto: University of Toronto Press, 2014).

to other colonial spaces in the period;[3] Montreal and Quebec were the closest, albeit still far-removed, centres of intellectual activity, and EuroAmericans (in general, but with some notable exceptions) failed to acknowledge indigenous knowledge in the ways that had been achieved in parts of Asia. The region was, however, home to a network of fur traders who had been trained in surveying and cartography, and who gained personal and professional fulfilment in exploring the vast territory's natural history.

The everyday practices upon which scientific identities were fashioned and presented are detailed in fur trade correspondence and reports, providing insights into the operations of the sciences in Rupert's Land, as simultaneously site-specific and fundamentally rooted in the *habitus* of late-Enlightenment intellectual life. Scientific identities in this context were intimately connected with the circulation of knowledge – a process in which fur traders played no small part. Indeed, fur traders offer an important example of the development of identities which were scientific and therefore also 'global' or 'cosmopolitan'.[4] While scientific practices in Rupert's Land were of course contingent on local circumstances, such as climate and the resources to which traders had access (books, scientific instruments, paper, ink), the scientific identities springing from those practices functioned in a global context, deeply rooted in the *habitus* of British science. The dual contingency demonstrated here of local imperatives and global practices complicates the picture of colonial sciences as a whole, with trading post gardens offering an example of fur traders' participation in the sciences and demonstrating the reach of their activities in Rupert's Land, while also suggesting that scientific practices can exist and operate on dual planes of local and global importance. Locally, fur traders' activities were significant at individual and group levels, ensuring improved quality of life for the fur trading community and providing traders with sources of professional pride and personal meaning. Simultaneously, on an international scale, traders' scientific pursuits provided the HBC with practical information of crucial strategic importance to the development of their trading network.[5] The extent to which individual traders saw themselves as participating in a global enterprise is not explicitly stated, but is revealed through examination of the ways in which they presented their activities to fellow traders, the HBC committee in London, and scholarly societies. The knowledge networks studied here were fragile in many respects, their participants located as they were at a considerable remove from Anglophone centres of scholarship, but this chapter demonstrates how they played a vital if tenacious role in fur traders' intellectual life.

---

3   David Wade Chambers and Richard Gillespie, 'Locality in the History of Science: Colonial Science, Technoscience, and Indigenous Knowledge', *Osiris* 15 (2000): 221–40.

4   See Robert Mayhew, 'Mapping Science's Imagined Community: Geography as a Republic of Letters, 1600–1800', *British Journal for the History of Science* 38 (2005): 73–92, which also highlights instances of tension between cosmopolitanism and parochialism in the global scholarly community.

5   Ted Binnema demonstrates how some traders forged links between science and commerce in *Enlightened Zeal*, 52–3.

As demonstrated in more detail below, fur traders' understandings of Rupert's Land were communicated to HBC headquarters in London in the form of accounts of experimental gardening and meteorological and astronomical observations, couched in economic and imperial terms.[6] This chapter also demonstrates the extent to which fur traders operated a framework established by the HBC committee in London; certainly, committee requirements influenced the kinds of activities traders emphasised in their correspondence and official reports. Equally, while the levels of hybridity in knowledge achieved in parts of Asia, for example, were not replicated in Rupert's Land, the knowledge produced by fur traders was strongly conditioned by local contingencies.[7]

The scientific practices engaged in Rupert's Land, then, present a pathway into understanding the ways in which fur traders utilised and presented scientific knowledge and practices to both practical ends and to serve their own personal, professional and intellectual needs. The ensuing discussion identifies two aspects of fur traders' scientific practices – knowledge exchange in the form of correspondence and book circulation, and the development and creation of knowledge through experimentation with kitchen gardens at trading posts – and considers these practices in relation to the themes of global knowledge, encounter and exchange in terms of their everyday expression and self-fashioning purposes for traders.

The time period covered in this chapter was one of embedding and developing late Enlightenment approaches to exploration and global knowledge generation, closing during the early years of the long HBC governorship of George Simpson (b. Scotland 1786/7, d. Lower Canada 1860; governor 1821–1860), whose official correspondence emphasises the variety of important uses to which kitchen gardens could be put (discussed in more detail below). It was also a period in which knowledge of an expanding world was most often collected and synthesised not by states, but rather by scholarly societies, missionary organisations, companies like the HBC, and individuals of means. More specifically, the period saw an increase in British interest in northern regions (the Arctic and sub-Arctic), which, from the mid-eighteenth century, gained a reputation as a fruitful site for scientific discovery and research. Tempting cartographic blank spaces persisted above 70°N in North America and the scientific value of northern exploration was broadly acknowledged from 1818 with the launching of four decades of British state expeditions in search of a northwest passage.[8] The Admiralty instructed expeditionaries to concern themselves with 'the acquisition of knowledge'

---

6   Chambers and Gillespie, 'Locality in the History of Science', 221–40.

7   See examples cited in C.A. Bayly, *The Birth of the Modern World 1780–1914* (Oxford: Blackwell, 2004), 312–20.

8   Angela Byrne, *Geographies of the Romantic North: Science, Antiquarianism, and Travel, 1790–1830* (New York: Palgrave Macmillan, 2013), 20–32; T.H. Levere, *Science and the Canadian Arctic: A Century of Exploration, 1818–1918* (Cambridge: Cambridge University Press); Sverker Sörlin, 'Ordering the World for Europe: Science as Intelligence and Information as Seen from the Northern Periphery', *Osiris* 15 (2000): 51–69.

by making 'constant observations ... for the advancement of every branch of science'.[9] The aims of William Edward Parry's 1819–1821 expedition to the Northwest Passage, for example, included 'the improvement of geography and navigation, as well as the general interests of science' and equipped both of its ships with a range of scientific instruments. Official instructions to the expedition specified that geographical discovery, cartography, terrestrial magnetism and 'the advancement of science in general' were to be attended to.[10] Experiments were conducted on terrestrial magnetism and the aurora borealis; flora and fauna were catalogued; geographical information was gathered from First Nations and Inuit communities. The broader intellectual context of scientific interest in northern regions provides an important backdrop to the activities of the HBC and its employees in Rupert's Land.

Reading and Writing 'in the Heart of a Savage Land'

The long winters at trading posts presented ideal opportunities for the observance of routine and the maintenance of detailed records.[11] Fur traders recorded and shared scientific information through private correspondence, published narratives, official reports, personal diaries, and meteorological and astronomical logs. These activities formed the bases for the cultivation of scientific identities and the creation of an intellectual space within which those identities could be maintained.

Traders' journals served a wide range of functions, acting as daily meteorological and astronomical records, and repositories for geographical, ethnographical and natural history observations. The English-born HBC trader and surveyor Peter Fidler (1769–1822), for one, spent long winter days carefully copying and binding his detailed daily observations, which continue to form an invaluable resource.[12] Journals provided a means of keeping track of time – Fidler noted that NWC employees were often 'wrong in their account' as they 'seldom keep any Journal'.[13] Others, like the Scottish-born trader and naturalist Andrew Graham (?1730–1815), claimed to keep a journal 'for my own amusement, and to pass away an Idle hour in this Solitary part of the world' – but his journal contained refutations of

---

9   John Barrow, *Voyages of Discovery and Research within the Arctic Regions, from the Year 1818 to the Present Time* (London: John Murray, 1846), 11–12.

10   W.E. Parry, *Voyage for the Discovery of a North-West Passage from the Atlantic to the Pacific* (London: J. Murray, 1821), vi, vii, xxv, xvi.

11   *Eighteenth-Century Naturalists of Hudson Bay*, ed. Stuart Houston, Tim Ball, and Mary Houston (Montreal and Kingston: McGill-Queen's University Press, 2003), 95.

12   Peter Fidler, Journal at Nottingham House, Athabasca Lake 1803–1805. Typescript (Thomas Fisher Rare Books Library, University of Toronto: MS Coll 85, Fidler papers: item 1), ff. 9, 11.

13   Peter Fidler, 'A Journal from Isle a la Cross by way of Swan Lake a new Track to the Athapescow Lake in the Year 1791' in Peter Fidler, 'Journal of Exploration and Survey', 1790–1806 (Library and Archives Canada, Ottawa (hereafter LAC): HBC 4M3, E.3/1), f. 36.

Arthur Dobbs' criticisms of HBC failure to explore or to extend its trade, unlikely material for a purely personal record.[14]

Evidence for the wider circulation of fur traders' observations – and the mediation of their knowledge for wider audiences – exists, with Graham and Fidler providing good examples. Graham's observations on the natural history of the Hudson Bay area formed the basis for much of Thomas Pennant's *Arctic Zoology* (2 vols, 1784–1785) and the North American material in the third volume of John Latham's *A General Synopsis of Birds* (6 vols, 1781–1785). The importance of Graham's observations is further indicated in the extent to which they were plagiarised by his contemporaries, particularly in the celebrated manuscript notes of HBC surgeon Thomas Hutchins (d.1790) and Edward Umfreville's *The Present State of Hudson's Bay* (1790).[15] The Wernerian Society of Edinburgh heard extracts from Graham's meteorological register and Fidler's Clapham House meteorological journal of 1808–1809.[16] Fidler made painstaking astronomical calculations in his notebooks, creating his own tables upon which to base his own observations.[17] His calculations relating to the lunar eclipses of 14 and 28 April 1790 were published in an almanac that referred to him as the compiler's 'astronomical friend […] at York Fort, Hudson's Bay'.[18] His astronomical observations at Buckingham House in 1792–1793 were published alongside those of William Wales, David Thompson, Philip Turnor and James Cook in a volume that informed Aaron Arrowsmith's map of North America (1796).[19]

Broader recognition of fur traders' potential as a source of natural history and other information on remote regions was achieved by the late 1820s, with the Natural History Society of Montreal (NHSM) gaining permission from the HBC in 1828 to distribute a questionnaire amongst traders to guide their research and

---

14   Glyndwr Williams, 'The Hudson's Bay Company and its Critics in the Eighteenth Century', *Transactions of the Royal Historical Society* 5th ser. 20 (1970): 149–71.

15   *Andrew Graham's Observations on Hudson's Bay, 1767–91*, ed. Glyndwr Williams (London: Hudson's Bay Record Society, 1969), xxx–xxxvi, 388–96.

16   *Memoirs of the Wernerian Natural History Society: Vol. II For the Years 1814–16*, part 2 (Edinburgh: A. Constable and Co., 1818), 644–5, 649–50, 655.

17   P. Broughton, 'Astronomical Observations by Peter Fidler and Others in 'Canada' 1790–1820', *Journal of the Royal Astronomical Society of Canada*, 103 (2009): 141–51; Peter Fidler, Journal at Cumberland House 1806–7. Typescript (Thomas Fisher Rare Books Library, University of Toronto: MS Coll 85, Fidler papers: item 1), f. 19.

18   J. Partridge, *Merlinus Liberatus. Being an Almanack for the Year of our Redemption, MDCCXC* (Birmingham: Pearson and Rollason, [1790]), 39–40. Ted Binnema adds to this Francis Moore, *Vox Stellarum, or a Loyal Almanac for the Year of Human Redemption, 1790* (London: Company of Stationers, 1790), in 'Theory and Experience: Peter Fidler and the Transatlantic Indian', in *Native Americans and Anglo-American Culture, 1750–1850: The Indian Atlantic*, ed. Tim Fulford and Kevin Hutchings (Cambridge: Cambridge University Press, 2009), 155–70; 155, 167n.

19   *Result of Astronomical Observations, Made in the Interior Parts of North America* (London: A. Arrowsmith, 1794), 13–14.

knowledge gathering.[20] As demonstrated throughout this chapter, British explorers and scholarly societies in the period responded to traders' increasing visibility in Anglophone intellectual forums by recognising their geographical scope and their usefulness as a source of local knowledge.

Fur traders were also receptive to the broader currents of literature, current affairs and the sciences. Peter Fidler and, in a later period, Scottish-born James Hargrave (1789–1865, chief factor at York Factory on Hudson Bay), for example, read widely in newspapers, scholarly periodicals, histories and scientific texts conveyed to them by request from England, and which they then passed on to colleagues.[21] Requests for supplies included appeals for reading material. Fidler's entreaties for 'some Book or other to read' were as plaintive as his pleas for tobacco and ammunition, as he found 'the Days are now become […] I rather think long for want of something to read'.[22] He recorded his pleasure when Orkney-born fur trader Malchom Ross (c.1754–1799) and the English surveyor Philip Turnor (c.1751–1799) supplied him with bound magazines 'which will pass away several long hours – the remainder of the Spring or Rather winter'.[23] Fidler amassed an extensive library including titles in history, travel, economics, medicine and natural history.[24] Alongside essentials such as cloth, a fur cap and soap, Hargrave requested copies of the second and third volumes of Robert Southey's *History of the Peninsular War* (3 vols, 1823–1832), Thomas M'Crie's *Life of John Knox* (1812), John Jamieson's *An Etymological Dictionary of the Scottish Language* (1808), unspecified items from *Murray's Family Library* series (published 1829–1834), the novels of Walter Scott 'from Woodstock onwards' (1826), a file of the *New York Albion*, and John Clowes Tasso's 1581 poem, *Jerusalem Delivered* in the original Italian.[25] Fur trade libraries represent spaces of knowledge in their own

---

20   Suzanne Zeller, 'The Spirit of Bacon: Science and Self-Perception in the Hudson's Bay Company, 1830–1870', *Scientia Canadensis* 13 (1989): 79–101, here 88, and Byrne, *Geographies of the Romantic North*, 69, 97–8.

21   James Hargrave to C. Cummings, 8 July 1829 (Hargrave Collection, LAC: James Hargrave's Letterbooks, C-80: Rough Copies, Letters by Hargrave from 13 July 1829 – 24 June 1830); James Hargrave to J. Clarke, 21 Aug. 1829 (Letters by Hargrave from 13 July 1829 – 24 June 1830); and Fidler, Journal at Nottingham House, ff. 13, 41, 43, 59, 70.

22   Peter Fidler, 'A Journal of a Journey with the Chepawyans or Northern Indians, to the Slave Lake & to the East & West of the Slave River, in 1791 & 2' in Peter Fidler, 'Journal of Exploration and Survey', 1790–1806 (LAC: HBC 4M3, E.3/1), f. 84.

23   Fidler, 'A Journal of a Journey with the Chepawyans or Northern Indians', f. 86.

24   *Eighteenth-Century Naturalists of Hudson Bay*, 94; Judith Hudson Beattie, '"My Best Friend": Evidence of the Fur Trade Libraries Located in the Hudson's Bay Company Archives', *Épilogue* 8 (1993): 1–32.

25   James Hargrave to John Clowes, 27 Aug. 1828, and James Hargrave to C. Cummings, 8 July 1829 (Letters by Hargrave from 25th August 1828 – 11 July 1829); James Hargrave to J. Clarke, 21 Aug. 1829, and James Hargrave to John Clowes, 30 Aug. 1829 (Letters by Hargrave from 13 July 1829 – 24th June 1830).

right, and may be considered 'assemblages' representative of the paths of available knowledge and the networks of circulation of which fur traders were part.

Knowledge exchange also provided men like Hargrave with a vital outlet for self-expression. The society of labourers was not always to his liking, so he turned to his correspondents in other parts of North America for company and thoughtful occupation. On one occasion, he appealed to Roderick Mackenzie: 'Out of the middle of bawling Canadians, bands of Scotch that with their gabbling Gaelic rival a puddle of frogs in spring, and from among the still more distracting crowds of smoking half breed giglets I lift my voice unto thee'.[26] Epistolary sociability was a cornerstone of intellectual activity for men like Hargrave, whose dislike for the inland factories – which lacked the advantages of the regular communication with Britain enjoyed by coastal posts – was evident in his sympathy for colleagues located there. He assured another colleague:

> Situated as you are in the heart of a savage land I consider it the duty of a friend to write you every echo which reaches our ears from the great turmoil of civilisation. As part, I shall try and get you a Newspaper or two of the latest date, – and besides shall add as commentary every scrap of Nouvelles.[27]

Books and correspondence may have provided essential means of keeping 'the time from hanging too heavy on [traders'] hands', but they represented much more than entertainment.[28] The sharing of scientific and natural historical information and resources was an important aspect of the dispersed fur-trade sociability. Scientifically inclined fur traders sought to remain networked within a borderless republic of letters which was both ideally suited to, and challenged by, their remote locations. However, this ostensibly cosmopolitan desire had at its heart an insularity that paradoxically hindered any possibility for an alternative form of knowledge that could accommodate First Nations' or Inuit traditional knowledge.[29] The forms of intellectual sociability and knowledge exchange (and appropriation) in evidence in fur trade accounts represent a contradictory set of values and aspirations. Traders were hindered by European scientific conventions which closed any possibility of translating certain kinds of indigenous First Nations and Inuit information into a contribution to a global perspective. The limitations imposed by their location

---

26   James Hargrave to Roderick McKenzie [*sic*], 5 Dec. 1826 (Letters by Hargrave from Jany 3rd 1826 – 24th March 1827).

27   James Hargrave to [Alexander] Ross, 2 July 1828 (Letters by Hargrave from 2nd Dec. 1827 – 1st August 1828).

28   John Siveright to James Hargrave, 18 Apr. 1830, in *The Hargrave Correspondence, 1821–1843*, ed. G.P. de T. Glazebrook (Toronto: Champlain Society, 1938), 50–52.

29   There are notable exceptions to this. See, for example, Barbara Belyea, *Dark Storm Moving West* (Calgary: University of Calgary Press, 2007); J.R. Short, *Cartographic Encounters: Indigenous Peoples and the Exploration of the New World* (London: Reaktion, 2009); Byrne, *Geographies of the Romantic North*, 125–49.

and HBC secrecy inhibited the reach of traders' observations, resulting in a set of 'global' information which was fragile in nature and held together by the tenacity of scientifically minded traders. Their engagement with local forms of knowledge is addressed further in the following section.

Cultivating Scientific Identities at Trading Post Gardens

Traders' knowledge exchanges were not limited to the local and transatlantic circulation of books and manuscripts; transactions also occurred in the shape of exchanges of material specimens such as seeds across the same spaces. This section considers these specimens as representative of a form of knowledge less related to precise description or scientific classification than to the development of effective practices – in this instance, cultivation. Cultivating kitchen gardens provided traders with possibilities for scientific activity in the form of experimentation, observation, and innovation. In the Early Modern period, colonization and the sciences were mutually influential, as colonies provided spaces for horticultural experimentation.[30] The pattern of experimentation and 'discovery' taking place in early-nineteenth-century British Arctic expeditions occurred on a much smaller but continuous scale at fur trading posts. The transplantation of 'metropolitan' ways of knowing to northern imperial spaces and cultivation in unproven conditions demanded experimentation and offered the possibility of new discoveries. The dominant perception of the north as a space embodying scientific possibility provided the environmental backdrop and broader cultural context for the forging of scientific identities in northern imperial spaces, and experimentation presented traders with opportunities to contribute to the gathering and dissemination of British knowledge of Rupert's Land.

Early HBC correspondence and official instructions refer to the dispatch of 'severall sorts of seeds […] to enable you [traders] to make the Experiments', and the Company's hope that 'extraordinary care & ingenuity' would prevail among its servants to render the trading posts self-sufficient.[31] In 1696, the Company issued further instructions emphasising the usefulness of gardens for keeping settlers and traders 'healthful', while minimising their dependence on shipments of supplies from England.[32] In 1728, 24 garden spades and 12 shovels were sent to Albany Fort

---

30   Sarah Irving, *Natural Science and the Origins of the British Empire* (London: Pickering & Chatto, 2008), 48, 65; Binnema, *Enlightened Zeal*, 99–101.

31   Governor Nixon's Instructions, 29 May 1680, in *Copy-book of Letters Outward &c: Begins 29th May, 1680 ends 5 July, 1687*, ed. E.E. Rich (Toronto: Champlain Society, 1948), 9.

32   Beverley Soloway, 'The Fur Traders' Garden: Horticultural Imperialism in Rupert's Land, 1670–1770', in *Irish and Scottish Encounters with Indigenous Peoples: Canada, the United States, New Zealand, and Australia*, ed. Graeme Morton and D.A. Wilson (Montreal and Kingston: McGill-Queen's University Press, 2013), 287.

on Hudson Bay's southern shore.[33] HBC policy would increasingly emphasise self-sufficiency at trading posts, but some traders expressed indifference, with Richard Stanton of Moose Fort appealing to the Committee in 1738: 'As for garden seeds and gardening, I am almost a stranger to, as having spent my youth in your honours' service […] as for Mr Howy's being proficient in that branch, he will promote it to the utmost of his skill'.[34] The same traders became defensive, however, when their commitment to these initiatives was questioned, and suggested that the Company did not make adequate provision:

> I wonder who should inform your honours with such an untruth as we having less gardening than usual, for I am sure we have digged large spots of ground that never was digged before on purpose to sow our turnips, pease, beans and other seeds. […] Mr Howy has sent to a friend of his to send him a little new seed he can trust to, for one quarter of the seeds that your honours sends would do providing they were good and new. […] As for the soil there is no right natural earth here but what is occasioned from the many trees which […] has rotted upon the ground and turns into a sort of black mould not above nine inches deep […] but if the ground was naturally good, there is one thing your honours will be pleased to take note of, that is our season for digging and sowing, which is the middle and latter end of May when the frost is out of the ground and the river broke up; and that is the only time for the Indians coming to trade.[35]

Oral testimonies provided to the 1749 parliamentary inquiry into HBC activities asserted that attempts were made at cultivation, but it also emerged that basic good practice was not followed in that seeds were not harvested for sowing.[36] Some witnesses pleaded ignorance of such matters, and some traders were apathetic towards cultivation.[37] However, a notable change in traders' attitudes towards scientific participation can be discerned from c.1770, when references to kitchen gardens appear in fur trade records with increasing frequency.

While some early evidence of Company efforts to foster scientific pursuits among its employees exists, from c.1770, fur traders themselves responded to the broader intellectual trend towards scientific 'discovery' in exploratory

---

33 Letter of Joseph Myatt, 5 Aug. 1728 in *Letters from Hudson Bay, 1703–40*, ed. K.G. Davies with A.M. Johnson (London: Hudson's Bay Record Society, 1965), 131–2.

34 Letter of Richard Staunton and George Howy, Aug. 1738, *Letters from Hudson Bay, 1703–40*, ed. Davies, 266, 324. George Howy was a factor and mapmaker.

35 Letter of Richard Staunton and others, 17 Aug. 1739, *Letters from Hudson Bay, 1703–40*, ed. Davies, 304–5.

36 *Report from the Committee Appointed to Inquire into the State and Condition of the Countries Adjoining to Hudson's Bay, and of the Trade Carried on There* (1749), 14, 18–19, 22–3, 25, 30–31, 34, 40, 44.

37 *Report from the Committee*, 14; E.E. Rich, *The History of the Hudson's Bay Company 1670–1870*, vol I: *1670–1763* (2 vols, London: Hudson's Bay Record Society, 1958), 540.

voyages by more explicitly participating and advertising their own interests in natural history and the sciences. Earlier interventions by individual traders were hampered by HBC secrecy in relation to the potential value of its North American possessions, but the Company more readily embraced science following the 1769 Transit of Venus, observed by astronomer William Wales at Prince of Wales Fort.[38] Thenceforth, scientifically minded traders utilised everyday activities like gardening or recording temperatures as forms of scientific practice and knowledge production. Suzanne Zeller finds that George Simpson, as HBC governor from 1821, promoted the sciences to 'satiate both his private and his public need for acceptance and recognition'.[39] This observation can be extended to the traders working in the territory in the preceding four decades. Situating pre-Victorian traders' participation within a variety of observational and experimental activities and knowledge dissemination as a means of asserting scientific identities, demonstrates that such concerns had a longer if not entirely continuous history.

Fur traders, aware of increasing European intellectual appreciation for the role of the sciences in exploration and imperial expansion, used kitchen gardens as opportunities to cultivate and maintain scientific identities. Trader awareness of the scientific potential of gardening is evident in the detailed records kept of the success rates of assorted crops under various planting and growing conditions, and the dates of sowing, germination, transplantation and harvesting.[40] When successful, trading post gardens provided a source of familiar foods and balance to a diet rich in meat.[41] It is also important to bear in mind contemporary thought regarding cultivation's moderating effects on climate, which suggested that clearing forests and reclaiming marshland could stabilise extreme climates.[42] This cannot be discounted as a possible added impetus to encouraging cultivation at trading posts.

How do traders' interests in cultivation fit into the story of their participation in global knowledge networks, and the formation of scientific identities? The example

---

38   Binnema, *Enlightened Zeal*, 75.

39   Zeller, 'The Spirit of Bacon', 95.

40   Peter Fidler, 'General Report of Red River District by Peter Fidler, 1819 May' (HBCA B.22/e/1), ff. 7d–8v; Fidler, 'General Report of the Manetoba District for 1820', (HBCA B.51/e/1), n.p.; Carolyn Podruchny, *Making the Voyageur World: Travelers and Traders in the North American Fur Trade* (Lincoln: University of Nebraska Press, 2006), 239–40.

41   Theodore Binnema, *Common and Contested Ground: A Human and Environmental History of the Northwestern Plains* (Norman: University of Oklahoma Press, 2001), 50–51. Failure often simply reflected the fact that garden sites were chosen primarily for their proximity to trading forts and factories, rather than their growing conditions; Peter A. Russell, *How Agriculture Made Canada: Farming in the Nineteenth Century* (Montreal and Kingston: McGill-Queen's University Press, 2012), 17.

42   Jan Golinski, 'American Climate and the Civilization of Nature', in *Science and Empire in the Atlantic World*, ed. James Delbourgo and Nicholas Dew (New York: Routledge, 2008), 153–74.

of Andrew Graham is illuminating. Suzanne Zeller rightly counts Graham among the number of eighteenth-century fur traders who were also 'amateur' naturalists engaged in the collection of botanical specimens for the 'peace and rational pleasure of cultivating the mind'.[43] Graham was one of the first traders to regularly record his attempts (successful and unsuccessful) to cultivate parsley, purslane, celery, carrots and parsnips 'sown yearly at all the Settlements', and the undoing of attempts at the cultivation of barley and oats at Albany and Moose Forts, by early frosts. He recorded his own success at growing barley during a mild year (1768/9) and noted that radishes, lettuce, spinach, onions and other greens did well in similar conditions. He also noted that factories with south-facing gardens fared well 'with good management', producing salads and colewort.[44] As well as the recognition and plagiarism of his written observations cited above, Graham also sent botanical specimens to the Royal Society, the Royal Society of Edinburgh and Edinburgh Botanical Gardens.

Despite Graham's exemplary enthusiastic pursuit of the sciences in Rupert's Land, a close reading of fur trader records produces a picture of sporadic development of trading post gardens in the following decades. Peter Fidler proudly recorded that during his winters in the Athabasca in 1803–1805, they had 'good gardens […] and the Canadians followed our example'.[45] In 1808, Alexander Henry the Younger noted that the HBC had 'excellent Gardens' at Cumberland House.[46] However, Fidler registered his disappointment in 1820 that despite three years of settlement at Fort Dauphin, the HBC had failed to make 'the least attempt at any kind of cultivation', while the competing NWC produced 160 bushels of mixed vegetables with 'but little ground in cultivation'.[47]

George Simpson has been remembered as a keen promoter of scientific activity in the fur trade, his governorship representing something of a turning point in the HBC relationship with the sciences.[48] His correspondence reveals the extent of his interest in trading post gardens. Throughout the 1820s, he compared and contrasted the horticultural potential of different trading posts, articulating a picture of scientific activity across the territory and communicating to the London committee the worthy activities in which traders were engaged. Simpson emphasised the natural advantages enjoyed by some posts, reassuring the Committee of the economies that could be made. For example, he thought the conjunction of good soil and

43   Cited in Suzanne Zeller, *Inventing Canada: Early Victorian Science and the Idea of a Transcontinental Nation* ([1987] Montreal and Kingston: McGill-Queen's University Press, 2009), 192.

44   Andrew Graham, 'Bird Observations written at Severn, 1768', (Hudson's Bay Company Archives, Winnipeg (hereafter HBCA): MS E.2/5), f. 58.

45   Fidler, 'General Report of Red River District … 1819', n.p.

46   *The Journal of Alexander Henry the Younger, 1799–1814*, ed. Barry Gough (2 vols, Toronto: Champlain Society, 1988), vol. ii, 345.

47   Fidler, 'General Report of the Manetoba District for 1820', n.p.

48   Zeller, 'The Spirit of Bacon', 87.

'contiguity to the Buffalo hunting Grounds' made Peace River 'the most desirable abode in this part of the Country', and he found such fertile soil at Fort Colville House that 'excellent Gardens might be formed at a triffling [*sic*] expense'. He noted with some regret that HBC employees at Peace River had not 'sufficiently availed themselves of its natural advantages', whereas 'by a little attention they might be made to yield sufficient to lighten the consumption of animal Food materially, and guard against the dangers of Starvation'. Duncan Finlayson, supervisor of the district, was to implement plans to improve this. However, Simpson's account of a fertile landscape awaiting development overlooked the imperative of experimentation, trial and error, something traders emphasised to enhance the scientific nature of their activities. He praised NWC efforts to 'avail themselves of every advantage the Country affords' at Dunvegan, Vermilion and Fort de Pinette by maintaining extensive gardens 'with very little inconvenience and at a moderate expense'.[49] He minimised traders' achievements by emphasising the ease with which successful gardens could be established and maintained, as he praised the 'extensive gardens' worked by 20 men at Fort Vancouver, 'independent of the usual routine business of the Establishment', to produce 400 bushels of Indian corn, 1300 of wheat, 1000 of barley, 300 of peas, 100 of oats and 4000 of potatoes.[50]

It is important to remember the broader context in which fur trader participation in the gathering and dissemination of natural history knowledge of Rupert's Land operated. They effectively worked in a space about which knowledge had yet to be presented publicly and comprehensively. The first definitive work on Canadian botany would only be published in 1833 – William Jackson Hooker's *Flora Boreali-Americana*. Hooker's work made available a wealth of new information on Canadian botany, drawn from important recent field research, including that of the explorer-naturalists John Richardson and Thomas Drummond during Franklin's overland expeditions of 1819–1822 and 1825–1827, and that of David Douglas for the Royal Horticultural Society in 1823–1827. These expeditions resulted in 'knowledge of several new and many rare quadrupeds, birds, and plants', forming a key moment in the advancement of British knowledge of North America.[51] Richardson acknowledged the 'great facilities for the advancement of our pursuits' provided by trading posts, and traders' donations of avian and quadruped specimens to the expedition's collections, projecting an image of an informed and willing base of scientific participants in the North American fur

---

49 *Journal of Occurrences*, ed. Rich, 379–80; 383.

50 *Part of Dispatch from George Simpson Esqr, Governor of Ruperts Land to the Governor and Committee of the Hudson's Bay Company, London, March 1, 1829: Continued and Completed March 24 and June 5, 1829*, ed. E.E. Rich (Toronto: Champlain Society, 1947), 68–9.

51 John Franklin, *Narrative of a Second Expedition to the Shores of the Polar Sea, in the Years 1825, 1826, and 1827* (London: John Murray, 1828), 308.

trade.[52] The global knowledge disseminated in publications like those of Hooker and Richardson provide further evidence of fur traders' engagements in observing and collecting in ways that facilitated the production of large-scale, state-sponsored surveys, and thereby contributing in indirect but important ways to the generation and dissemination of knowledge at high levels.

Assembling Global and Local Knowledge

The forms of scientific participation and knowledge production described above can be thought of as 'assemblages', bearing the markers of the networks, ideas, places and processes by which they were formed.[53] The knowledge circulation in which fur traders were engaged took both written and material forms, products which provide fruitful opportunities for engagement with the concept of science as communication and process.[54]

The various currents of natural historical understanding operating in Rupert's Land overlapped and interacted as fur traders developed their own understandings by gaining indigenous knowledge and/or transmitting or translating knowledge for distant audiences. Imperial scientific practice in Rupert's Land, inevitably, responded to and was influenced by the nature of native–newcomer interactions and the local knowledge encountered by or accessible to traders. The extent to which scientific practice in other contemporary colonial spaces was indigenised – involving local experts and depending upon contact zones such as markets and ports – has recently been demonstrated, as have surveyors' engagements with indigenous geographical knowledge.[55] Tensions and interdependencies were played out in narratives of intercultural encounters, with contact involving what Sivasundaram calls 'the meeting of different vocabularies of nature'.[56] These vocabularies were transmitted to HBC offices in London, having been gathered locally by receptive traders, mediated or transformed by traders as reporters, and transmitted long distance to London. How that information was received in London is another story, but what is significant here is the fact of transmission of a form of knowledge that was 'local' in character, but 'global' in its reach and economic potential.

Indeed, fur traders' relationships with indigenous knowledge appear problematic. Despite some instances of interest in and respect for indigenous

52   John Richardson, *Fauna Boreali-Americana; or the Zoology of the Northern Parts of British America* (London: John Murray, 1829), xviii–xix.

53   David Turnbull, 'Local Knowledge and Comparative Scientific Traditions', *Knowledge and Policy* 6 (1993–4): 29–54.

54   James A. Secord, 'Knowledge in Transit', *Isis* 95 (2004): 654–72.

55   Sujit Sivasundaram, 'Sciences and the Global: Methods, Questions, and Theory', *Isis* 101 (2010): 146–58; Belyea, *Dark Storm Moving West*.

56   Sivasundaram, 'Sciences and the Global', 147.

knowledge (as detailed below), there is little evidence of traders' application of this knowledge to trading-post kitchen gardens, and the perseverance demonstrated by traders in their repeated attempts at cultivation of British crops reveals some reluctance to embrace native crops and horticultural methods. However, this does not necessarily indicate a complete disregard for indigenous knowledge. Reviewing fur trade horticulture in the context of scientific identity formation introduces the possibility that traders simply recognised in horticulture an opportunity to exercise experimentation and thereby enhance their scientific credibility. The crops available in Rupert's Land until (at least) the late eighteenth century depended on the supply of seeds from London, indicating a range of local problems from a lack of seed harvesting to a failure to learn from indigenous horticulture or adapt to native crops.[57]

Despite this, traces of trader gathering of indigenous horticultural knowledge emerge from beneath the surface of their daily activities. In 1768–1769, Andrew Graham listed native species names (a number of which were new to European science), information and specimens gathered from First Nations and other traders, and indigenous culinary and medicinal uses for various roots, berries and herbs.[58] Peter Fidler has been recognised as particularly sympathetic towards First Nations' geographical knowledge, not least for his important collection of indigenous maps.[59] Later, George Simpson noted the abundance of edible native roots and herbs in the Columbia and provided a list of their indigenous names, but this was 'thought superfluous, & not entered in [the] appendix'.[60] This remark is ambiguous. It seems unlikely that he would have gathered and transcribed 'superfluous' information, but he may have considered it so in the context of the other information contained within his report. That is, readers' expectations influenced the content of Simpson's account, de-simplifying his position from that of the 'typical' imperial agent ignorant of indigenous knowledge and customs, to a cultural intermediary operating between two worlds – the fur trade realities of food provision and the need for local knowledge, and the London committee's desire for particular kinds of economic knowledge. This may further explain the lack of evidence of trader engagement with indigenous botanical or horticultural knowledge to improve their own efforts at cultivation or understanding.

First Nations also played crucial roles in supplying traders with seed for grains, legumes and vegetables, thereby entering into the series of material exchanges

---

57   Soloway, 'The Fur Traders' Garden', 289.

58   *Andrew Graham's Observations*, ed. Williams, 78, 80–81, 84, 93, 95; Graham, 'Bird Observations', f. 55–6.

59   Byrne, *Geographies of the Romantic North*, 125–49; Ted Binnema, 'How Does a Map Mean? Old Swan's Map of 1801 and the Blackfoot World', in *From Rupert's Land to Canada*, ed. T. Binnema, G.J. Ens and R.C. Macleod (Edmonton, University of Alberta Press, 2001), 201–24.

60   *Part of Dispatch from George Simpson*, ed. Rich, 226.

across Rupert's Land and with Britain described above.[61] Fidler reported the presence of First Nations' gardens at Whitemud River, which may have formed part of the seed supply chain.[62]

Horticultural information was assembled within syntheses of broader climatic and natural history contexts, embodying a form of global knowledge in which the information itself was immobile (cultivation being subject to local contingencies), but was made mobile and given international interest through traders' capitalisation of horticultural information as a means of illustrating wider climatic and environmental patterns. This not only provided a means by which traders could demonstrate the extent of their natural historical knowledge of North America, but also reflects the interactions of local and global knowledge in fur trade correspondence and practice. As early as 1743, the HBC factor and naturalist, James Isham, provided detailed information on the effects of York Fort's location and orientation on the cultivation of peas, cabbage, turnips and 'Salletts'. He reported that crops which sprouted on open ground then 'wither['s] and come[s] to no perfection' due to the prevailing wind. He increased the currency of this relatively scant horticultural information by expanding on his remarks, demonstrating his understanding and appreciation of broader meteorological patterns and their effects on his horticultural efforts, such as the effects of a sea breeze on snowfall and permafrost.[63] Sixty years later, Fidler also used horticulture as a conduit for his broader natural historical knowledge. In 1819, he reported on the progress of cultivation in the Red River district, adding information about swarms of grasshoppers that had in the previous year destroyed the little barley remaining following a summer drought. He added that the grasshoppers appeared 'in great numbers generally about every 18 Years & come from the Southward'.[64] Isham and Fidler's descriptions of broader environmental influences on trading post gardens represent a genealogy of global thought influenced by local circumstances in Rupert's Land, but cognisant of the region's wider contextual position as a site of competing, contested and borrowed knowledges and exchanges, forerunning later conceptions of the interrelated nature of global and local climates and environments. This vision of Rupert's Land as operating within broader systems of nature and knowledge reflects the physical scope of the space in which traders operated. The initial uncertainty of the extent of Rupert's Land – defined only as

---

61   Fidler, 'General Report of the Manetoba District for 1820', n.p.; *The Writings of David Thompson, Volume I: The Travels, 1850 Version*, ed. William E. Moreau (Toronto: Champlain Society, 2009), 215.

62   Fidler, 'The Annual Report of the Manitoba District', 1821 (HBCA B.51/a/2), f. 3. On indigenous agriculture, see D. Wayne Moodie and Barry Kaye, 'Indian Agriculture in the Fur Trade Northwest', *Prairie Forum* 11 (1986): 171–83.

63   *James Isham's Observations on Hudsons Bay, 1743, and Notes and Observations on a Book Entitled* A Voyage to Hudsons Bay in the Dobbs Galley, *1749*, ed. E.E. Rich with A.M. Johnson (Toronto: Champlain Society, 1949), 217–18.

64   Fidler, 'General Report of Red River District ... 1819', ff. 6–6d.

the watershed of Hudson Bay, without any sense of its true scale – and the ever-extending reach of the HBC in North America into the nineteenth century were expressed in all their exciting potential as new in terms of knowledge, and as significant in its links to wider systems.

Conclusion

This chapter demonstrates just some of the means by which scientific identities were forged, expressed and maintained by Rupert's Land fur traders in the late eighteenth and early nineteenth centuries. Traders utilised the expanding network of trading posts as the basis for the creation of a network for the circulation of knowledge, a 'civilised' space and imperial-scientific identities. They called upon friends and colleagues to help build personal libraries; they traded seed with the First Nations with whom they were already trading furs. Correspondence not only provided news, but included book reviews and insights into the opportunities and limitations of trading life. They reported to each other and to Company bosses on the progress, successes and failures of their attempts at cultivation. They responded to the imperatives of food production (in the absence of adopting indigenous cultivation and gathering strategies) while simultaneously using that process as an opportunity for the creation and application of knowledge. These activities provided a great deal of personal and professional satisfaction and formed a keystone of scientific identity. Some traders more successfully communicated their importance than others, but the observations of most traders went largely unnoticed outside of the fur trading community, the fragility of this form of 'global' knowledge evident in the marginality of many of their observations and activities.

Even if traders' observations were not widely circulated, they remain an important testimony to the processes of transformation and mediation through which information passes. In 1744, the Irish politician and colonial governor Arthur Dobbs (1689–1765) scoffed that the HBC prevented its employees from making 'any Improvements without their Factories, unless it be a Turnip Garden'.[65] His words raise further questions about the value of isolated horticultural experiments at distant trading posts, whose residents he portrayed as so busily engaged in such quotidian tasks as collecting firewood that they had little time for scientific pursuits. His use of science as a weapon in his battle with the HBC demonstrates the burgeoning importance of scientific activities in empire and exploration, but his critique fails to incorporate the meaning of scientific practice for individual traders. Kitchen gardens, regardless of their success or failure, formed one of the means by which trader-naturalists and other 'men of science' in Company employ could flex their scientific muscles – a role which only increased in significance from c.1770. The symbolic meaning of crops all-too familiar to even the poorest Briton was

---

65   Arthur Dobbs, *An Account of the Countries Adjoining to Hudson's Bay* (London: J. Robinson, 1749), 2; Williams, 'The Hudson's Bay Company and its Critics', 149–71.

transformed by their transplantation to the northern imperial space. Dobbs' lowly turnip garden became the site of transformative processes by which everyday – even mundane – activities took on new meaning and power. The period under study in this chapter saw a consolidation in fur trader participation in the sciences, and an increasing tendency for traders to share their observations and findings with international audiences. Mirroring the processes of 'assemblage' whereby information was gathered and shared in Rupert's Land, traders' observations became integrated into some of the period's most significant natural histories of North America. Fur traders occupied an important place in the patchwork of global knowledge and imperial knowledge gathering in the late eighteenth and nineteenth centuries, with the widespread on-site network of trading posts providing practical support to British state-sponsored expeditions, and individual traders publicising their own findings internationally. If the overview of that network and the practices it engendered appears brittle or fragile, it serves as an example of the pressures and tensions under which the 'republic of letters' laboured. This case study presents some valuable counterpoints to the tidier or more secure networks which have received much more historiographical attention (like those of Joseph Banks, for example), but one which represents the cultivation of a scientific cosmopolitan identity at the margins of an emerging global knowledge economy.

PART 2
Collection and Display

Chapter 5

Sampling the South Seas: Collecting and Interrogating Scientific Specimens on Mid-Nineteenth-Century Voyages of Pacific Exploration

Sarah Louise Millar

Voyages of discovery and exploration in the eighteenth and early nineteenth century primarily focused on finding and claiming new territories. The science that was conducted during these expeditions was almost exclusively terrestrial in focus, as evidenced in the sailing instructions to the captains of all voyages from HMS *Endeavour* in 1769 to HMS *Challenger* in 1872. The naturalists who made their name on these voyages – Joseph Banks, Joseph Hooker, and Charles Darwin to name only a few – despaired of the amount of time they had to spend on board ship and were in constant verbal battles with their respective captains over the short periods of time spent on land. Perhaps as a consequence of this contemporary focus on terrestrial science, later treatment of the subject has likewise concentrated on the land-based activities of these men and has overlooked the scientific practices that took place on board ship and whilst at sea.[1]

The voyages of Captain James Cook and other early pioneering explorations around the coast of what is now Australia have received considerable scholarly attention.[2] Yet others did follow in their wake, even though the work of later

---

1   For notable exceptions see, for background, Margaret Deacon, *Scientists and the Sea, 1650–1900: A Study of Marine Science* (Aldershot: Ashgate, 1971) and Susan Schlee, *The Edge of an Unfamiliar World: A History of Oceanography* (New York: E.P. Dutton & Co. Inc., 1973); for instrumentation see, Anita McConnell, *No Sea Too Deep: The History of Oceanographic Instruments* (Bristol: Adam Hilger Ltd., 1982); for more recent work see Helen Rozwadowski, *Fathoming the Ocean* (Cambridge, MA: Harvard University Press, 2005) and Jordan Goodman, *The Rattlesnake: A Voyage of Discovery to the Coral Sea* (London: Faber and Faber, 2005).

2   For example, see, Paul Carter, *The Road to Botany Bay: An Essay in Spatial History* (London: Faber and Faber, 1987); Greg Dening, *Mr Bligh's Bad Language: Passion, Power and Theatre on the Bounty* (Cambridge: Cambridge University Press, 1992); Anne Salmond, *The Trial of the Cannibal Dog: Captain Cook in the South Seas* (London: Penguin, 2003); Bernard Smith, *European Vision and the South Pacific* (New Haven: Yale University Press, 1985).

explorers has been less intensively studied. For Jane Samson 'a curtain comes down after Cook, Vancouver and Bligh leave the stage'.[3] Likewise, for a later period, much attention has been paid to the voyage of HMS *Challenger*, which Richard Corfield argues 'single-handedly founded the sciences that we today know as oceanography and marine geology'.[4] There were, however, multiple journeys into the Pacific in the years following Cook, Vancouver, La Perouse and others, and before the scientific work of *Challenger*. Certainly, the ocean as an object of study was of increasing interest as the 1830s progressed. Michael Reidy has highlighted the work undertaken by the Ross Antarctic expedition during 1839–43, which had its main focus *on* the ocean rather than on travelling across it.[5]

Whilst Cook's first voyage to the South Seas in the eighteenth century had been predominantly astronomical in outlook and terrestrial in its operational base, the *HMS Challenger* sailed in 1872 with explicit instructions to concentrate on exploring the ocean and the ocean floor. In the century that separates the two voyages, a gradual shift occurred from the pre-dominance of terrestrial investigations to a greater focus on oceanographic studies in voyages of exploration. Here, the scientific practices undertaken on three such voyages in the 1830s and 1840s are considered, by three of the most influential countries in nineteenth-century maritime exploration: Britain, France and America. This chapter focuses specifically on the collection of scientific specimens, to help understand how an activity more commonly associated with naturalists on land was also shared and enjoyed by their counterparts on board ship, providing a window into the previously unobserved spaces below the water, that in turn helped fuel an interest in the ocean as a space of scientific study.

The collection of specimens was a popular and prevalent activity on maritime voyages throughout the eighteenth and nineteenth century, and a live or preserved specimen was the most credible way to bring home an image of what was seen abroad. Many authors have highlighted the importance of the collector in nineteenth-century science. Janet Browne argues, for example, that there has always been a great emphasis on collecting and bringing back native flora and fauna, whilst Bernard Smith has stressed the importance of collecting, measuring, drawing and painting on voyages of exploration in the eighteenth and nineteenth centuries.[6]

3   Jane Samson, *Imperial Benevolence: Making British Authority in the Pacific Islands* (Honolulu: University of Hawaii Press, 1998), 2.

4   Richard Corfield, *The Silent Landscape: Discovering the World of the Oceans in the Wake of HMS Challenger's Epic 1872 Mission to Explore the Sea Bed* (London: John Murray, 2005), xiii.

5   Michael Reidy, *Tides of History: Ocean Science and Her Majesty's Navy.* (Chicago: University of Chicago Press, 2008).

6   Janet Browne, 'Biogeography and Empire', in *Cultures of Natural History*, ed. N. Jardine J.A. Secord and E.C. Spary (Cambridge: Cambridge University Press, 1996), 305–21; Smith, *European Vision and the South Pacific.* For more recent work see, Glyn Williams, *Naturalists at Sea: From Dampier to Darwin* (New Haven: Yale University Press, 2013).

Whilst it is true to say that measurements of depth, water current or temperature were also a type of collection – numbers rather than physical objects – the two were commonly seen as separate entities, with a different intellectual motivation behind their acquisition.[7] Drawing on Susan Faye Cannon, Michael Dettelbach argues that Alexander von Humboldt was transformational in his science by focusing on the act of *measurement* rather than the act of collection.[8] Humboldt was undoubtedly influential on many contemporary natural historians – especially those working on land. From the voyages considered here, however, it is also clear that collecting physical specimens, as opposed to measurements, was still a hugely popular activity with members of the crew and the civilian naturalists alike, either to add to their own private collections, or to send back to their home country to form part of a greater public display. The act of collection from foreign lands not only offered an insight into what existed beyond the familiar European world, but reinforced the geopolitical nature of imperial maritime endeavour. Collecting, recording, and classifying species new to the European scientific fraternity stamped the authority and ownership of the traditional maritime countries on the new space.[9] In the nineteenth century this location was shifting, from the last unexplored terrestrial spaces, to the ocean depths.

France, America and Britain in the South Seas

The Pacific Ocean became of commercial interest in the late eighteenth century for a variety of reasons: on it parts were supplied, cargoes moved, new social customs introduced, and new voyage routes made available. Whilst war with Napoleonic France had ended the 'Age of Discovery' as exemplified by Cook, at the end of the war in 1815, Britain had a Pacific presence from Canton to the colonies of New South Wales, and from Van Diemen's Land across the Pacific to the North West coast of America. Samson argues that 'these developments had been a piecemeal response to circumstances rather than a formal policy of colonial expansion'.[10] Others have viewed the presence of British ships in the Pacific during this period as the 'taking out of insurance' against French influence in the region, rather than

---

7 I have explored depth recording on early-nineteenth-century British polar expeditions elsewhere. Sarah Louise Millar, 'Science at Sea: Soundings and Instrumental Knowledge in British Polar Expedition Narratives, c.1818–1848', *Journal of Historical Geography* 42 (2013): 77–87.

8 Michael Dettelbach, 'Global Physics and Aesthetic Empire: Humboldt's Physical Portrait of the Tropics', in *Visions of Empire: Voyages, Botany, and Representations of Nature*, ed. Daniel Philip Miller and Peter Hanns Reill (Cambridge: Cambridge University Press, 1996), 260.

9 See Daniel Clayton, *Islands of Truth: The Imperial Fashioning of Vancouver Island* (Vancouver: University of British Columbia Press, 2000) for more on how naming places on a map conferred a sense of ownership over the charted land.

10 Samson, *Imperial Benevolence*, 10.

a directed campaign of exploration.[11] The few American voyages into the Pacific in the early nineteenth century were less geopolitical in intention, largely ignoring the territorial, imperial and national claims other countries were engaging with in this period. Whilst America had been slow to establish itself on the international and scientific stage, it advanced rapidly as the century progressed and questions of national identity and colonial advantage were important motives behind the Pacific exploratory voyages of all counties. For Reidy, 'Empire … subtly transformed science: in the research pursued, the questions asked and the theories adopted'.[12]

Scientific interest in the South Seas also became more pronounced in the latter half of the eighteenth century and this continued into the nineteenth century. For Reidy and Rozwadowski the scientific study of the ocean was a particular priority for Britain and America, both keen to 'gain and consolidate economic and political power'.[13] At the end of the 1830s an increased interest in what lay at the southern most regions of the globe – potentially new, unclaimed land, and the commercial prospects of both land and sea – heavily influenced the decisions of the governments of France, America and Britain to send exploring expeditions into the South Seas. For France, Louis de Freycinet undertook the first maritime exploratory expedition in peacetime in 1817, on which a large collection of natural history specimens was obtained. In 1829, the corvette *Astrolabe* captained by Jules Dumont d'Urville finished a three year expedition exploring the South Seas. D'Urville's expedition had had two remits: a public one for science and hydrography, and a private one to find a site for a French penal colony and harbours to shelter French warships. Dumont d'Urville was eager to lead another exploring expedition into the Pacific in the late 1830s, and petitioned the Navy to award him a commission. D'Urville wrote, 'I received a communication in which I was told that the King himself, to whom my plan has been submitted, had welcomed it, but having learned that an American whaling ship had got very near the South Pole, he desired that a French expedition be sent in the same direction'.[14] D'Urville accepted the new terms of the expedition, taking the *Astrolabe*, and choosing Hector Jacquinot, his previous second in command, to accompany him on the *Zélée*. The corvettes left Toulon on 7 September 1837, covering vast areas of the South Pacific and navigating the edge of the Antarctic continent, before returning to France three years later on 6 November 1840.

An American Pacific exploring expedition had been mooted since the late 1820s, but gained momentum in 1836, when a campaigner for an Antarctic voyage, Jeremiah Reynolds, addressed the US House of Representatives calling for an expedition to the Pacific Ocean. For Reynolds, America had been living

---

11 David P. Miller, 'Introduction', in *Visions of Empire*, ed. Miller and Reill, 4.

12 Reidy, *Tides of History*, 292.

13 Michael Reidy and Helen Rozwadowski, 'The Spaces in Between: Science, Ocean, Empire', *Isis* 105 (2014): 340.

14 Helen Rosenman, *Two Voyages to the South Seas* (Melbourne: Melbourne University Press, 1992), 115.

too long in the wake of the British and French. Within six weeks the expedition was established and Lieutenant Charles Wilkes appointed captain.[15] Wilkes took charge of the flagship *Vincennes*, a 127-foot sloop of war carrying a crew of 190 men. The next largest vessel in the expedition was the *Peacock*, a ship previously launched for an aborted 1828 expedition. Two more ships were acquired, and two tenders included just two weeks before sailing. The six ships of the United States Exploring Expedition – or US ExEx as it became known – left Hampton Roads, Virginia, on 18 August 1838, to sail for the southern oceans. The expedition was at sea for over four years, lost one ship in bad weather and sent another home early due to its unsuitability for sailing. The *Vincennes* finally returned home on 10 June 1842.[16]

The British expedition to the southern ocean was the last to depart for the southern hemisphere. At the British Association for the Advancement of Science's eighth meeting in Newcastle in 1838, Colonel Edward Sabine – a leading figure in British hydrographic and natural philosophical science, as well as a veteran of Arctic exploration – made a case for the importance of the continued study of terrestrial magnetism, and a committee was appointed to present the resolutions of the meeting to the British Government. The importance of Canada, Ceylon, St Helena, Van Diemen's Land and Mauritius were stressed as suitable observatory stations, and a naval expedition was recommended. The proposal was welcomed by the Royal Society and an Antarctic expedition was supported under the leadership of famed polar explorer James Clark Ross, who received his commission for HMS *Erebus* on 8 April 1839.[17] The *Erebus* was a 370 tonne bomb ship, a specialised vessel ordinarily used for bombardment, able to carry 64 people. The accompanying ship, HMS *Terror*, was commanded by Commander Francis Rawden Moira Crozier. The scientific contingent of the British expedition, despite the prominence of scientific tasks in the instructions for sailing, was relatively small, with the naval officers being expected to carry out most of the work, as was

---

15   Wilkes had been part of a previous aborted voyage, and had already been sent to Europe to gather the instruments an expedition would require, meeting when he did so with many of the famous European savants, including James Clark Ross. He had not been first choice, however, and his lack of experience and relatively low rank left many in doubt as to the choice. He was, however, one of the few naval men with scientific experience.

16   Alan Gurney, *The Race to the White Continent: Voyages to the Antarctic* (London: Norton, 2000), 127. For more on the Ex. Ex. see D. Graham Burnett, 'Hydrographic Discipline among the Navigators: Charting an 'Empire of Commerce and Science' in the Nineteenth-Century Pacific', in *The Imperial Map: Cartography and the Mastery of Empire*, ed. James R. Ackerman (Chicago: University of Chicago Press, 2009), 185–260; Nathaniel Philbrick, *Sea of Glory* (London: Harper Perennial, 2005); Kathryn Yusoff, 'Climates of Sight: Mistaking Visibilities, Mirages and 'Seeing Beyond' in Antarctica', in *High Places: Cultural Geographies of Mountains and Ice*, ed. Denis Cosgrove and Veronica Della Dora (London: I.B. Tauris, 2008), 64–86.

17   Ernest S. Dodge, *The Polar Rosses: John and James Clark Ross and their Explorations* (London: Faber and Faber, 1973), 186.

the case in the American expedition. The surgeons operated as naturalists: Robert McCormick as surgeon and Joseph Dalton Hooker as assistant surgeon to the *Erebus*. In the *Terror*, John Robertson was surgeon, and David Lyall his assistant. The main work on natural history was to be conducted by McCormick, but he proved to be largely uninterested in the subject, and was happy for Hooker to lead. The ships left England in September 1839 and spent two winters in the Antarctic ice, reaching the farthest south of any of the three exploratory expeditions, before returning home on 4 September 1843.

In the voyages of these three expeditions we can trace the continued, if sometimes conflicted, desire of naturalists and crew, to probe the ocean depths at a time when, Antarctica aside, the last of the terrestrial land masses had been discovered. The ocean environment in the nineteenth century was a relatively untouched but increasingly accessible, and potentially profitable, global space for scientific investigation. As others have argued, the expansion of empire and the desire by the great sea-faring nations of the time to conquer new territory – be it terrestrial or marine – drove in turn an increased awareness of oceanographic science.[18] The means by which these oceans were both traversed and investigated in the mid nineteenth century was the sailing ship.

On Board the Expedition Ship

The global ambitions of the voyages can be contrasted with the fragile 'local' conditions that pertained on board the vessels co-opted for oceanic exploration. Recent scholarship on the ship as a mobile space of knowledge and social performance has moved it from 'the margins to the centre of geographical research'.[19] Sorrenson has argued for consideration of the ship itself as an instrument, not just the vehicle by which oceans were crossed. Key to this understanding is that who commissioned a ship, and the scientific instruments it carried, were integral to establishing that vessel's authority. These ships did not act as a mere platform for observation, but helped to shape the sort of information that was collected from it.[20] The ship was more than just a physical structure for carrying people and goods; it was an important space in its own right, integral to the success or failure of a voyage, and a key factor in the political and natural philosophical

---

18   Reidy and Rozwadowski, 'The Spaces In Between', 344.

19   William Hasty and Kimberley Peters, 'The Ship in Geography and the Geographies of Ships', *Geography Compass* 6 (2012): 660. For work on the ship as an instrument of globalising power see: Miles Ogborn, 'Writing Travels: Power, Knowledge and Ritual on the English East Indian Company's Early Voyages', *Transactions of the Institute of British Geographers* 27 (2002): 155–71; Miles Ogborn, *Global Lives: Britain and the World, 1550–1800* (Cambridge: Cambridge University Press, 2008).

20   Richard Sorrenson, 'The Ship as a Scientific Instrument in the Eighteenth Century', *Osiris* 11 (1996): 221–36.

work that it embodied and represented. In a fundamental sense, the type of vessel was important: Ross, following in the tradition of Cook and his adaptation of a collier vessel, sailed with two 'bombs', ice-strengthened and already tested in the Arctic. Ross had good cause to trust in his vessels: they broke further south through the Antarctic ice than had any vessel previously and better withstood the harsh conditions encountered than was the case of the French and American ships.

The ship operated as both 'floating laboratory' in John Beaglehole's words, but also as a field station or mobile collecting site.[21] As it moved through water, the ship experienced constantly shifting environmental conditions: temperature, humidity, the height of the waves, snow, ice and the glaring sun. All these variables had an immediate effect on how the spaces on board ship were used. These spaces were likewise integral to the types of scientific work that could be conducted on board. In wet conditions it became impossible to dissect, store, and draw specimens in the open air and they were taken below deck. Rozwadowski comments that 'multipurposeness [was] a pervasive feature of naturalist's accommodation', with those involved expected to sleep and work in the same space.[22] On HMS *Bounty*, for example, the space for the scientific contingent came from the crew's social area: on HMS *Challenger*, the scientific work appropriated space formerly used for military purposes.[23] The captain's cabin was often the best controlled environment for temperature and the roll of the ship but access to this for scientific purposes could interfere with the social command of the vessel. The spaces on board available to the men of science, in which knowledge of the great oceans could be uncovered, were limited, and largely contingent on the good will of the naval men who shared them.

The naturalists who accompanied voyages of exploration in the late eighteenth and early nineteenth century were not of a uniform type: the term 'scientist' was only coined in the 1830s, and, even then, related more to gentlemen amateurs, less to a dedicated profession. Randolph Cock argues that the beginning of the professionalisation of science at sea occurred through civilian naturalists and astronomers on polar expeditions in the 1820s, where 'open-air and ship-board laboratories promoted the training of the midshipmen, junior officers and ... seamen'.[24] As has been noted the 'act of going to sea defined practitioners of early ocean science more than a shared body of knowledge'.[25] As a result, the scientific work that was carried out at sea was largely influenced by the interests of the individual: Hooker and d'Urville shared an appetite for botany, Wilkes for

---

21   J.C. Beaglehole, *The Life of Captain James Cook* (Stanford: Stanford University Press, 1974).

22   Rozwadowski, *Fathoming the Ocean*, 186.

23   Dening, *Mr. Bligh's Bad Language*, 20.

24   Randolph Cock, 'Scientific Servicemen in the Royal Navy and the Professionalism of Science, 1816–55', in *Science and Beliefs: From Natural Philosophy to Natural Science, 1700–1900*, ed. David M. Knight, and Matthew D. Eddy (Aldershot: Ashgate, 2005), 95–111.

25   Rozwadowski, *Fathoming the Ocean*, 177.

measuring the height of waves and Ross for locating the southern magnetic pole. Whilst Ross's interests coincided with the mandate of the expedition, when this was not the case, time had to be hard fought in order to indulge individual passions.

The conflict over access to equipment, labour, time and space in which to pursue a scientific task, led to tensions from the outset between crew and the scientific corps.[26] There was a tradition in the British and French navies that scientific work would be undertaken by naval officers. Cook's first expedition had broken with this in taking a full complement of civilian scientific staff. The French captain, Louis de Freycinet, sailing in 1817, ensured the officers served as naturalists, botanists and astronomers, so a civilian corps was not required. When the American expedition was assembled in the late 1830s, they adopted this framework: Charles Wilkes wanted only navy men to be trained in using the ship's instruments. An awareness of the difficult situation they were in and the possibility for strained relations with the officers and crew was vital if the civilian scientific staff were to have the resources they required in order to pursue their work, which was often in direct opposition to the tasks undertaken to facilitate the smooth running of the ship. Helen Rozwadowski argues that 'scientists who failed to understand and negotiate the social and political dynamics on board compromised their scientific work'.[27] Nor was the split between crew and civilians the only factor that led to tensions on board. Christopher Lloyd has shown that the crew shared an elaborate hierarchy quite apart from the commissioned men and civilian officers.[28] How much time any one man was given to a particular task was often dependent on how vital that task was deemed by the Captain to completing the wider ranging goals of the expedition. In the instructions for sailing issued to all leaders of exploratory vessels in the nineteenth century, we can begin to see why the demands of civilian men of science were often put behind those of the naval staff, but also an increased focus on scientific accomplishment as a reflection of national prowess, rather than success that rested solely on the discovery and colonisation of new land. How this played out, however, was strongly conditioned by the contingent social and material realities of individual vessels and their changing itineraries.

Organisation Before Sailing

The importance of scientific endeavour and the role to be undertaken by the savants, officers and crew, was outlined to varying degrees in the expedition's sailing instructions. The Instructions for the French expedition in 1837 were issued to Dumont d'Urville on 26 August 1837, by the Minister for the Navy, Vice-Admiral Rosamel. The *Astrolabe* would travel to Australia and New Zealand

---

26   Gurney, *The Race to the White Continent*, 125.
27   Rozwadowski, *Fathoming the Ocean*, 193.
28   Christopher Lloyd, *The British Seaman 1200–1860: A Social Survey* (London: Paladin, 1970), 212.

via South America, at which point the *Zélée* would return to France with the collections and reports gathered up to that stage. The *Astrolabe* was to search for the Chatham Islands to confirm their existence or otherwise, specifically in the interests of French whaling, proving that commercial goals were to be of importance to the French expedition. The Minister concluded his instructions by stressing the importance of commerce for the voyage:

> His Majesty has in mind not only the advancement of hydrography and natural history; his royal solicitude for the interests of French trade and the development of our shipping has caused him to take a much broader perspective of the scope of your mission and the likely advantages to accrue from it. You will call at a great number of places which should be closely examined from the point of view of the resources they may be able to offer our whaling ships. You are to collect all the information appropriate to guide them in making their expeditions more productive. You will put in to ports where our trade is already established and where the passage of a French warship can have a salutary influence, into others where perhaps our manufactured goods could find markets that have been so far ignored, and on which you will be able to provide valuable information on your return.[29]

Commercial aspirations were also to the fore of the American expedition. Whaling was a key factor at this time, with the hunt for whales shifting by geography and species in the late eighteenth century and in the nineteenth century, according to market value, technological innovations, and the remaining populations. The American whaling fleet 'represented an advanced maritime guard for US imperial goals in the Pacific' at this time.[30] The vital importance of whaling to America was apparent from the sailing instructions for the US ExEx. The instructions began with orders from the Secretary to the Board of Navy Commissioners, James Kirke Paulding: 'The Congress of the United States, having in view the important interests of our commerce, embarked in the whale fisheries, and other adventures in the great southern ocean ... authorised an expedition ... exploring and surveying that sea, as well to determine the existence of all doubtful islands and shoals'.[31] While highlighting the importance of commerce to the expedition, the instructions only ordered the sailors 'to take all occasions, not incompatible with the great purposes of your undertaking, to extend the bounds of science, and promote the acquisition of knowledge': for the Americans science was

---

29 Rosenman, *Two Voyages to the South Seas*, 199.

30 David Igler, *The Great Ocean: Pacific Worlds from Captain Cook to the Gold Rush* (New York: Oxford University Press, 2013), 103.

31 Charles Wilkes, *Narrative of the United States Exploring Expedition During the Years 1838, 1839, 1840, 1841, 1842 ... 5 vols. and Atlas* (5 vols, Philadelphia: Lea and Blanchard, 1845), I: xxv.

subordinate to commerce.[32] The astronomy, terrestrial magnetism experiments and meteorology were to be undertaken by the ship's officers. For Wilkes, no special directions were thought necessary regarding the scientific tasks to be performed.

Whilst the emphasis of the instructions was on the commercial significance of the expedition, the official narrative written by Wilkes makes numerous references to the importance of the scientific work, declaring from the outset that it was the only expedition by the United States 'fitted out by national munificence for scientific objects'.[33] There were six ships in the expedition, the second largest – *Porpoise* – had a poop-cabin and a forecastle on her deck built at the personal request of Wilkes. Wilkes's flagship, the *Vincennes*, had an extra deck added for scientific work and was fitted with additional living quarters, drafting space and a preparatory room for natural historians and their collections. These adaptations made some of the vessels, such as the store-ship *Relief*, slow and ill-adapted for the voyage. In addition, every attention was paid to the materials taken on board ship to aid with navigation, scientific investigation and the encounter with new lands. Wilkes had spent time in Europe gathering the instruments he believed necessary for the voyage, and midshipman William Reynolds, in his first letter home, described having on board all the books of the French and English expeditions of the seas they were to visit.[34]

By contrast, the British geological, zoological, and botanical committees drew up contents of desiderata for the British expedition, and full instructions for the collection and preservation of specimens of the animal, vegetable and mineral kingdoms were given. The Admiralty provided the means for collecting them. The instructions given to Ross, issued by Samuel John Brooke Pechell, Lord of the Admiralty, focused on the scientific goals of the expedition, in particular those relating to magnetism. Ross, a man of science himself, was enthusiastic about the scientific aspirations, stating that the expedition aimed to 'engross the attention of the scientific men of all Europe'.[35] Ross was additionally advised to communicate frequently and openly with his sister ship and a frequent change in the observations made in the two ships was to be made in order that if any scientific discovery was made by one, it should be quickly passed to the other. The ships were not to partake in any hostile act if England was to enter into war, 'the expedition under your command being fitted and for the sole purpose of scientific discoveries'.[36] On return to England, Ross was ordered to lay a full account of the proceedings to the Board of Admiralty, which body, as was customary, also

32   Wilkes, *Narrative of the United States Exploring Expedition*, I: xxix.

33   Wilkes, *Narrative of the United States Exploring Expedition*, I: xiii.

34   Anne Hoffman Cleaver and E.J. Stann, eds *Voyage to the Southern Ocean – Letters of Lieutenant William Reynolds of the United States Exploring Expedition 1838–1842* (Annapolis: Naval Institute Press, 1988), 11.

35   James Clark Ross, *A Voyage of Discovery and Research in the Southern and Antarctic Regions During the Years 1839–1843* (2 vols, London: John Murray, 1847), I: xxvi.

36   Ross, *A Voyage of Discovery and Research*, I: xxvii.

required the logs and journals, charts, drawings and other observations from officers and crew on the voyage. The Admiralty also stated that 'You will also receive our future directions for the disposal of all such specimens of the animal, vegetable, and mineral kingdoms'.[37] Alongside the Admiralty instructions, a Report to the Council of the Royal Society concerning the scientific mandate of the expedition ran to 100 pages (although only the terrestrial magnetism section was included in the official narrative), a fact which highlights further the centrality of science to the expedition, and, specifically, the advancement of knowledge on terrestrial magnetism.

The fact that the scientific instructions for Ross's expedition were more detailed than those for the French and the American expeditions reflected a tradition of scientific investigation on British naval voyages of exploration. The British Antarctic expedition was ostensibly about pursing investigations in terrestrial magnetism, and privately about ensuring that if a southern continent did exist the British would be the first to lay claim to it. Whilst the British expedition seemed content to limit the geographic scope of the expedition in order to carry out specific, and detailed scientific tasks, the French and America voyages had much greater remits that involved covering vast distances, from the coast of Australia to the North Western coast of America, and all the islands of the Pacific in between, as well as confirming the accuracy of existing charts, to locating whales, as well as suitable harbours and refuelling stations for whaling boats. As a result of these geographically wide-ranging and lengthy voyages, the collections that were sent home, particularly to America and France, were vast in size.[38] The differences between the expeditions in terms of research priorities ran alongside a more subtle but equally valid concern shared by all three voyages, that focused on the geopolitical and commercial benefits of such journeys and that was global in both scope and ambition.

Collection and the Collector

The collection of artefacts signalled a desire from the expedition countries to stake a claim, not only on new lands, but the spaces underneath the ship – the seas and oceans the exploratory vessel spent most of its time travelling across. A collected item served as proof that what was claimed to exist had actually been seen. The processes of observing and collecting started early on all three voyages considered here. Ross recorded that they took 'daily, almost hourly, observations of various kinds, from which so large a measure of useful and important results were expected'.[39] He noted seeing large numbers of flying fish, bonito and dolphin and

---

37 Ross, *A Voyage of Discovery and Research*, I: xxviii.

38 The American expedition actually brought back over 4,000 zoological specimens, 200 of which were new species. In addition were 1,100 bird specimens and 50,000 plants as well as thousands of artefacts.

39 Ross, *A Voyage of Discovery and Research*, I: 4.

'thus early on our voyage we began the collection of natural history, by preserving as many different kinds of these creatures as we could procure, and by means of towing nets and other devices, gathered numerous curious and entirely new species of animaculae'.[40] The American expedition used a seine net to catch fish, adding what they believed to be new species in botany, conchology, zoophytes, and fossils. Wilkes wrote, 'the dredge continued to be used, and with success, and many interesting objects were obtained: among them terebratulas, chitins, corallines, sponges, many small and large crustaceans, animals and large volutes'.[41] Wilkes was keen to point out not just that specimens were collected, but that they would prove to be important for the collections at home. Off Bellingshausen Island, many new species of fish were taken, which Wilkes believed would be most sought after by the Department of Natural History in Philadelphia.

Samples from the seabed were brought up using a deep sea clamm (a sounding device invented by John Ross on his first Arctic expedition), and the traditional sounding lead. Ross described the lead bringing up black stones at one stage, which, to Ross, confirmed the volcanic origin of the sea bed. Living coral were also brought up from over 1000 feet below sea level. Ross recognised several of the species from his Arctic voyages and noted 'the extreme pressure at the greatest depth does not appear to affect these creatures'.[42] The deep sea clamm brought up green mud, sand and small stones as well as fragments of starfish and coral. Hooker made accurate drawings of many of the specimens brought up by the dredge, as well as making descriptions for the eventual publication of the *Flora Antarctica*.[43]

There was an intricate relationship between the crew and the animals they encountered. Specimens could also be foodstuffs. The means of collecting birds differed across the expeditions, and was additionally dependent on the species involved. The American expedition caught several albatross with small hooks, the birds being preserved as specimens. The British expedition took many gigantic albatross and cape pigeons with fishing lines and by means of baited hooks slung over the side of the ship, soaked in salt water. The most common way of obtaining avian specimens, however, was with the gun. The shooting of birds from on board ship and on land was an important source of new natural specimens, a welcome source of food for the crew, and for some a pleasurable pastime. Midshipman Reynolds of the *Vincennes*, stated, 'I shot a beautiful bird of the Heron kind – of white and delicate plumage, the only one of the kind that has been obtained. He makes a fine specimen for the Naturalists and had not his species been already supplied with a name, I should have had him termed the *Rinaldius*'.[44]

40   Ross, *A Voyage of Discovery and Research*, I: 16.
41   Wilkes, *Narrative of the United States Exploring Expedition*, I: 112.
42   Ross, *A Voyage of Discovery and Research*, I: 202.
43   Joseph Dalton Hooker, *Flora Antarctica* (London: Reeve Brothers, 1844).
44   Cleaver, *Voyage to the Southern Ocean*, 83.

Catching the Great Penguins. Page 159.

Sketched by Dr. Hooker.

Figure 5.1   'Catching the Great Penguins', a sketch by Dr Joseph Hooker
Source: James Clark Ross, *A Voyage of Discovery and Research Volume II* (London: John Murray, 1847). Courtesy of Special Collections, Queen's University, Belfast.

Improvements in firearms at the beginning of the nineteenth century were particularly useful to bird collectors. Instruction manuals for shooting birds at this time were 'virtually silent as to technique' – assuming the average gentleman bird enthusiast would be a proficient shot already.[45] Bird shooting was justified as a legitimate form of recreation, without the need for scientific pretensions to explain the need to shoot and collect specimens. Both Ross and Wilkes described the pursuit of particular bird species in order to obtain specimens for a collection. Ross reported that, 'numbers of the young pintado [cape petrel] were flying about, and one shot by Mr McCormick fell on board, it was the first specimen of the kind we obtained'.[46] Wilkes recorded many unsuccessful attempts to secure a petrel, one finally being shot by Mr Peale, with boats lowered onto the ocean for their retrieval.

Flightless penguins were of particular interest to all three expeditions. Ross described the difficulty in killing three large specimens that were eventually

---

45   Anne Larsen Hollerbach, 'Of Sangfroid and Sphinx Moths: Cruelty, Public Relations, and the Growth of Entomology in England, 1800–1840', *Osiris* 11 (1996): 206.

46   Ross, *A Voyage of Discovery and Research*, I: 200.

brought on board and eaten, but they proved unpopular: the flesh was dark and rank and fishy in flavour. Ross wrote: 'it was a very difficult matter to kill them, and a most cruel operation, until we resorted to hydrochloric acid of which a tablespoonful effectually accomplished the purpose in less than a minute'.[47]

Wilkes, however, wrote of the great amusement the birds' capture afforded the crew, 'it was an amazing sight to see them [the crew] associated in pairs, thus employed, and the eagerness with which the sailors attacked them with oars and boat-hooks'.[48] The animals had been first described by Georg Forster on Cook's second expedition into the southern ocean, and, according to Ross, the unpublished drawing of the creature by Forster was the only material relating to the bird then available. Ross brought the first specimen back to England: 'some of these were preserved entire in casks of strong pickle, that the physiologist and comparative anatomist might have an opportunity of thoroughly examining the structure of this wonderful creature'.[49] Whilst usually taken from land, where the penguins were slower and more ungainly, the undertaking involved the deployment of small boats and considerable effort grappling with the dangerous Antarctic sea. It was also an endeavour that brought crew and scientist staff together, albeit with different motivations: for study, food, or solely the enjoyment of grappling with a new adversary.

Personal collections were not allowed on board, however, especially when it was a crew member rather than an officer who wanted to establish them. At the beginning of the US expedition, Wilkes ordered that the crew were not to keep collections of their own. Reynolds, in his first letter home, however, wrote 'I intend to let nothing that is curious slip by me this cruise without procuring it if possible'.[50] Collecting artefacts was an important activity for the crew members, and they were unhappy with Wilkes's order. The naturalists suspected that Wilkes did not want the scientific achievements of the expedition to overshadow the naval accomplishments. Anything that was kept needed to be handed over to Wilkes at the end of the voyage: the crew blamed the presence of the civilian 'scientifics' for this measure. Reynolds, it seems, hoped to continue his collecting, presumably keeping the store secret from the captain. In one letter from home he wrote how 'everything curious is sacred to the Scientifics themselves. But I have some stones from the Southern Continent'.[51] He was, however, patently aware of the restrictions on his own collecting, adding in frustration, 'I have not collected many curiosities; the Government is so selfish as to require *all specimens* for the [Public] stock ... In the English Expedition when two specimens of each article were procured the officers were at liberty to collect for themselves'.[52]

---

47   Ross, *A Voyage of Discovery and Research*, I: 158.

48   Wilkes, *Narrative of the United States Exploring Expedition*, I: 143.

49   Ross, *A Voyage of Discovery and Research*, I: 158.

50   Cleaver, *Voyage to the Southern Ocean*, 11.

51   Cleaver, *Voyage to the Southern Ocean*, 200.

52   Cleaver, *Voyage to the Southern Ocean*, 228.

Despite the mandate to undertake scientific work on all three expeditions, there was an unwillingness of the captains to give up too much of the limited space, and time, to the scientific contingent. The American boats were greatly altered from their original purpose in order to configure the space on board in such a way that would make it more suitable for the civilian scientific contingent. Banks had attempted to do much the same on the second of Cook's expeditions to the Pacific – but upon testing the physical integrity of the vessel was shown to be severely compromised, and, as a result, the changes were scrapped. The Americans did not have the experience to draw on previous exploratory expeditions, and they made the same mistakes the British had years earlier. The American boats were not properly tested and impeded the expedition throughout its four years. Wilkes expressed his misgivings about the expedition before it sailed: 'I was well aware, from my own observations and the reports made to me, that we were anything but well equipped for such a cruise'.[53] When the boats reached Rio de Janeiro, the first major port of call, extensive repairs were needed on all ships.

Wilkes believed that it was necessary to bring the officers into closer association with the required scientific duties, with a reduction of the tasks placed under the corps of civilians, although he does claim that 'as many of these were taken as could be accommodated'.[54] The initial reception of the men of science by the crew was mixed. Some crew members welcomed their inclusion on board. Midshipman Reynolds' first opinions of the civilian men of science were good. He wrote to his sister how 'Titian Ramsey Peale, the great Naturalist is with us ... The other Scientifics are said to possess talents and much zeal in their respective pursuits. We, the ignoramuses, will no doubt take great interest in learning the origin, nature and history of many things'.[55] At the beginning of the American narrative, Wilkes expressed his own pleasure in the relations between crew and savants: 'free communications were had. It was amusing to see all entering into the naval occupation of dissecting the fish taken, and to hear scientific names banded about between Jack and his shipmates'.[56] At times, then, collecting could bring disparate factions of the ship's social structure together, and each was able, if willing, to learn from the other.

Neither did Reynolds' admiration diminish when first at sea. Commenting on their attempts to catch marine life, he recounted how,

> The Scientifics have had one chance since we sailed. On a calm day many fish were around us, and we caught them in numbers. Instead of consigning them instantly to the cooks, the Scientifics went at them with the utmost eagerness and relish, dissecting them, found out many mysterious things in the stomach

53 Wilkes, *Narrative of the United States Exploring Expedition*, I: xxii.
54 Wilkes, *Narrative of the United States Exploring Expedition*, I: xiv.
55 Cleaver, *Voyage to the Southern Ocean*, 3.
56 Wilkes, *Narrative of the United States Exploring Expedition*, I: 4.

etc., talked over many hard names, and then took drawings of the whole and the parts; all they did was Greek to us, but somewhat interesting.[57]

Reynolds showed genuine interest at such moments, and the novelty of the scientific practice to a crew member unused to sailing with civilian men of science is clear. He later wrote in the same vein how 'The Scientifics have made another haul and it is most curious to see the patient manner in which they toil, toil, seemingly for trifles ... On calm days, they drag scoop nets on board, and as the ocean is teeming with living things, they find many ... The artists copy everything from life'.[58] He also alluded to the recognition that everything of interest captured was drawn as soon as possible. Whilst actual specimens might be desirable, there was limited room on board for their storage, and the preservation was not always very good. Sketching was thus vital, as a reflection of the need to depict but also as a proof of what was originally collected, before decay or the transformation into food, took place.

If Reynolds was enthusiastic about the naturalists, other crew members were not, naming them 'clam diggers' and 'bug catchers'. They resented the additional workload, such as ferrying the 'Scientifics' from ship to shore. In the appendix to the first volume of his official narrative Wilkes appeared to address possible criticism of his treatment regarding the scientific contingent: 'to the scientific gentlemen I have only to say, that they are, and always will be considered as one of us'.[59] Wilkes admitted that the 'Scientifics' were not given the chance to do all that they had wanted but argued that they had 'messed with the ward-room officers, and received all the privileges, respect, and attention due to that rank'.[60] On the British expedition, Hooker was given a cabinet for plants, a table under the stern window, a drawer for a microscope and a locker for papers to help him with his task. Hooker was no doubt helped by Ross's keen interest in all things scientific, biological as well as physical, in comparison to Wilkes's views on the pre-eminence of the physical sciences.[61]

The civilians claimed on numerous occasions that they were not being properly treated by the captain and crew of the vessels they were on. Much of this antagonism revolved around the additional workload which the crew believed the civilians on board had caused. Reynolds described Wilkes ordering every officer on deck to help the scientific gentleman to perform their business, an additional chore to their normal duties. Reynolds stated he belonged to a division of the crew that was to help with shore excursions, 'when the Scientifics make journeys inland, *we* will accompany them for protection'.[62] The naval officers

---

57 Cleaver, *Voyage to the Southern Ocean*, 6.
58 Cleaver, *Voyage to the Southern Ocean*, 11.
59 Wilkes, *Narrative of the United States Exploring Expedition*, I: x.
60 Wilkes, *Narrative of the United States Exploring Expedition*, I: xix.
61 Gurney, *The Race to the White Continent*, 219.
62 Cleaver, *Voyage to the Southern Ocean*, 8.

were also expected to conduct the geodetic and other observations of significance to navigation. On landing at Rio de Janeiro, Reynolds wrote, 'the Naturalists, Artists, Conchologists, Geologists, and Mineralogists will have a splendid field to examine their pursuits in'.[63] He began, however, by assisting in erecting the portable houses for the instruments and fixing them in their stands: 'the experiments and observations have been going on successfully and unceasingly'.[64] The unceasing nature of the observations was to prove a trial. He described the task of checking the instruments, 'All these I observe and note, at the expiration of *every hour* in the twenty-four. When the other vessels were here the task was a light one; but now, but one officer can be spared to attend here'.[65] Whilst the crew and officers were expected to work long hours with broken sleep when the ships were at sea, the same rota on land was unwelcome. As a consequence of the understaffing of civilian naturalists, and aspirational but often unspecific scientific instructions, the ship's staff were required to perform numerous scientific duties in addition to their own workload. On a voyage with such wide ranging goals as an exploring expedition to the South Seas, requiring a considerable increase in labour, the role of the naval man and man of science became blurred.

The American naturalists became equally frustrated with the decisions of Wilkes that dictated when and where they were allowed to leave the boat and investigate onshore. Joseph Couthouy complained that whilst they collected 50,000 specimens from Rio de Janeiro (an exaggeration), what they collected from the first five uninhabited islands in the Pacific would not have filled a cigar box.[66] As with the Americans, the French sought to curb the enthusiasm of the view on land; on coming into harbour at Rio de Janeiro the French Captain d'Urville ordered that no one was allowed on shore, unless to procure supplies, in order to avoid possible extravagances related to alcohol and women. Even on board, there was conflict regarding what was to be done with the collected artefacts. Wilkes disliked the smell of dissected creatures below deck so much he forbade the practice. The cautious view captains often took of their crew directly impacted on the amount of time those on board, civilians included, could spend on land. The curtailment of this time spent onshore in many nineteenth-century voyages of exploration increased the previously less popular activity of collecting specimens from the sea rather than the more familiar artefacts from land. Even the deck of an expedition vessel, however, proved an awkward space for collecting, sketching and preserving specimens, and the below deck area even more unsatisfactory. The collections may have been made from all over the globe, but the processing of them was a profoundly local practice, contingent on the particular spaces afforded those on board at any one time.

---

63 Cleaver, *Voyage to the Southern Ocean*, 25.
64 Cleaver, *Voyage to the Southern Ocean*, 37.
65 Cleaver, *Voyage to the Southern Ocean*, 42–3.
66 Viola, *Magnificent Voyagers*, 14.

Conclusion

The expedition vessel was a complex assemblage of materials and people that together drove the types of scientific investigation that could be conducted on board ship, whilst at sea. The ship itself directly affected which geographic areas could be accessed, but also determined the scientific practices that could be performed on and below its decks. Any global knowledge that could be gathered from its predominately mobile base was only done so after a series of structural and social hurdles had been overcome.

The type of science that was performed on board ship was influenced by a number of factors: the investigations specified in the sailing instructions, the personal interests of the scientific contingent on board and the specific environmental conditions: a freshening breeze could curtail an attempt to dredge and a roughening sea prevent the continuation of a sounding event. The collection of physical specimens was an important and omnipresent activity undertaken across the ranks of ship workers. Given the amount of time which the crew, officers and civilian naturalists alike were obliged to spend on board ship, the act of collecting specimens served both as scientific practice and enjoyable pastime that could be pursued from the deck via shooting, dredging and fishing, and continued below deck through sketching, painting, mounting and stuffing of specimens. In this way the ship served both as laboratory, instrument and field site; the continually shifting conditions directly influencing the workers on board and through them, the construction of scientific knowledge at sea.

What defined science on voyages of exploration in the mid-nineteenth century, therefore, was not the practitioner, nor the act itself, but a more subtle combination of the two that importantly included the space in which it took place and the instrument used for its execution. Whilst Canon, Dettelbach and Miller, among others have argued for the importance of Humboldtian science, that focused on measurement and precision instruments as opposed to collection, it is clear from maritime voyages of exploration that collection continued to be a vibrant and important undertaking. Unlike terrestrial artefact collection that only required the physical body of the collector to hunt and gather a specimen, collecting from a ship nearly always required an intermediary – a gun, a dredge or a net – to bring the item on board. Nor did natural historical practices rest with acquiring or representing specimens. Once it had been recorded, a specimen was often eaten, in itself a type of experimental practice, that gave those involved valuable information on a potentially necessary foodstuff.

The spaces on board ship on an expeditionary voyage had multiple purposes, and in turn the individual undertaking a scientific activity was not confined to the civilian contingent. On a vessel where the captain's cabin was shared with the naturalist, and the spaces on deck used by civilians, officers and crew alike, it is unsurprising that scientific practice was not limited to any one scientific practitioner or group of such. Collecting was an integral experience in the programme of scientific investigations, and drew together individuals from across the ship's hierarchical social structure, but by doing so it also created spaces of conflict. Wilkes followed in a long tradition of expedition leaders stretching back to Cook, in wanting the officers of the ship to undertake the scientific work, and to limit the number of civilians taken on board. For Wilkes, and others like him, the ability to adapt to ship life and navigate the hierarchical structure of a naval ship was more important than any prior scientific knowledge. In addition, the natural historians were as likely to take a specimen for the pleasure of shooting or catching it as they were for drawing and preserving it, and the naval officers as likely to aid in the collection of a specimen as the men on board solely for the purpose. As such, the line between savant and crewmember or officer in the age where the professionalisation of science was just beginning to take place, was blurry and undefined. The scientific knowledge produced by the collection of artefacts of natural history on board expedition ships at this time was made by men at sea, not just men of science.

Chapter 6

Curating Global Knowledge: The Museum of Economic Botany at Kew Gardens

Caroline Cornish

The cost incurred by increasing the contents of the museum has been exceedingly small; for owing to the interest felt in these collections, it is seldom necessary to buy specimens; they are almost invariably given, and in the case of articles imported into England, often without being asked for. The majority are, however, procured by correspondence direct from the countries producing them.

Sir William Jackson Hooker 1859

When Sir William Hooker, Director of the Royal Botanic Gardens, Kew made this statement, the museum to which he referred, the Museum of Economic Botany at Kew Gardens, had been in existence for 12 years, and its collections were growing exponentially. Hooker's claim that many objects were donated 'without being asked for' is, however, somewhat disingenuous; in fact, through his constant construction and reconstruction of networks, extending across agencies of government, science, art and commerce, his museum had rapidly acquired the status of national repository for useful plants and their products. The correspondence to which he here refers is testimony to this tireless enterprise, but so are the objects themselves and it is with these, and one of these in particular, that this chapter deals.

First, however, a word on global knowledge and objects: objects have for some time now been used as primary sources in a variety of disciplines, including archaeology, art history, anthropology, material culture studies and museum studies, in attempts to trace the mechanisms and mobilities of the production of global knowledge.[1] A recurrent methodology in these studies is the 'object biography'. The notion of an object having a 'social life' emerged in the mid-1980s in the work of anthropologists Arjun Appadurai and Igor Kopytoff when important methodological guidelines were set.[2] As Kopytoff argued, 'biographies

---

1 See notes 4–7 for examples.

2 Arjun Appadurai, 'Introduction: Commodities and the Politics of Value', in *The Social Life of Things: Commodities in Cultural Perspective*, ed. Arjun Appadurai (Cambridge: Cambridge University Press, 1986), 3–63; Igor Kopytoff, 'The Cultural Biography of Things: Commoditisation as Process', in *The Social Life of Things*, ed., Appadurai, 64–91. See particularly 66–8.

of things can make salient what might otherwise remain obscure', particularly the way that objects are repeatedly 'culturally redefined and put to use' as they are produced, exchanged, and received.[3]

Applying research methods such as the object biography to studies of museum objects, archaeologists and anthropologists have been able to demonstrate the global nature of many museum collections.[4] In their exploration of the multiple meanings attributed to objects in the Pitt Rivers collection, Chris Gosden and Yvonne Marshall proposed a consideration of material objects at the stages in their 'lives' of production, exchange and consumption, and this remains particularly relevant to museums in trying to interpret their collections for twenty-first-century audiences, since 'the present significance of an object derives from the persons and events to which it is connected'.[5] Object biography has proved a useful tool in accounting for the interrelatedness of human and object agency in ways which de-centre the museum or the museum director from the narrative, thus avoiding the traps of 'parochial antiquarianism' on the one hand and hagiography on the other.[6] Similarly, object biography has allowed for an understanding of reception and meaning-making as multiple and mutable.[7]

With the advent of the 'material turn' in cultural geography over the last twenty years, there have been a number of geographers looking to objects as primary sources.[8] Such work, with its co-emphases of materiality, place and mobility, has shed new light on the role objects have played in creating and communicating global knowledge, and on their value to the researcher in reconstructing the histories and geographies of that knowledge. Historical geographers working in the field of bibliography have likewise taken a material approach to texts, focusing on the object-ness of the published word rather than on its content.[9] Like texts, objects can be read: through their inscriptions, their materiality and their associated documentation. This is particularly true of museum objects

---

3   Kopytoff, 'The Cultural Biography of Things', 67.

4   See, for example, Chris Gosden and Yvonne Marshall, 'The Cultural Biography of Objects', *World Archaeology* 31 (1999): 169–78; Stephanie Moser, *Wondrous Curiosities: Ancient Egypt at the British Museum* (Chicago: University of Chicago Press, 2006).

5   Gosden and Marshall, 'The Cultural Biography of Objects', 168–78.

6   James Secord, 'Knowledge in Transit', *Isis* 95 (2004): 654–72.

7   Moser, *Wondrous Curiosities*, 2006.

8   For the impact of the material turn on cultural geography, see particularly the 'Material Geographies' edition of *Geoforum* 35 (2004). Historical geographers who have adopted an object focus include: Jude M. Hill, 'Travelling Objects: The Wellcome Collection in Los Angeles, London and Beyond', *Cultural Geographies* 13 (2006): 340–66; Lawrence Dritsas, *Zambesi: David Livingstone and Expeditionary Science in Africa* (London and New York: I.B. Tauris, 2010); Felix Driver and Sonia Ashmore, 'The Mobile Museum: Collecting and Circulating Indian Textiles in Britain', *Victorian Studies* 52 (2010): 353–85.

9   For example, Innes M. Keighren, *Bringing Geography to Book: Ellen Semple and the Reception of Geographical Knowledge* (London: I.B. Tauris, 2010); Hayden Lorimer and Charles W.J. Withers, 'Introduction', *Geographers Biobibliographical Studies* 32 (2013): 1–5.

which are subjected to linked processes of accessioning, classifying, naming and cataloguing as part of their assimilation into museum collections. But museum objects have not only been written about, they have, quite literally, been written *on*. They bear inscriptions – numbers and letters inked or etched onto their surfaces; the labels they have been given by successive owners relating to successive epistemologies; and the packaging which contains them, not to mention that joy of joys to the researcher – the occasional note or letter pertaining to the object which has resisted transfer to the paper archive, and which remains in the box or tied to the neck of the bottle, often linking the object to its very locus of origin. Furthermore, the very materiality of objects imparts knowledge. An object's form, scale, typicality or, conversely, curiosity value, can have much to tell us as to why objects were collected, which audiences they were intended for and how they were displayed, as well as providing us with the raw material to consider their reception across the spectra of time and place. What becomes clear is the intimate relationship between objects and space, the spatial as well as the temporal contingency of meaning. Therefore in the study of objects as agents of global knowledge, we need to be mindful of where the knowledge was created, who had access to it and how it was circulated.

This chapter examines how the Museum of Economic Botany at Kew Gardens produced global scientific knowledge through its objects, in its spaces and across its networks. It demonstrates that the scientific knowledge produced in and through the Kew Museum was global in three particular aspects: firstly, those networks required to create the collection – the data for the production of knowledge – were of global reach; secondly, through a range of media, that knowledge circulated globally; and thirdly, knowledge produced at and through the Museum could have global impacts. By reconstructing the 'biography' of a particular object – a model of an Indian indigo factory – the following discussion seeks to reveal the mechanics of knowledge production at the Museum, and the spatial and temporal contingency of meaning-making. In doing so, it leans on two theoretical viewpoints, the first of which is John Pickstone's notion of 'museological science'.[10] Using the Paris Muséum National d'Histoire Naturelle as a case study, Pickstone defined what he termed 'analytical/comparative' or 'museological/diagnostic' to describe the sort of science which emerged in France in the late eighteenth and early nineteenth century. The rise of museological science was linked to the emergence of scientific specialists with command over large collections. It was new in that it was situated somewhere between the 'surface' practices of taxonomy and the model phenomena of experimentalists. Museological science was produced via analytical processes – what Pickstone termed the 'deconstruction' of specimens into elements – in order to produce classifications, or to better understand technical processes.[11] This 'deconstructive'

---

10   John V. Pickstone, 'Museological Science? The Place of the Analytical/ Comparative in Nineteenth-century Science, Technology and Medicine', *History of Science* 32 (1994): 111–38.

11   Pickstone, 'Museological Science?', 113.

process was paradoxically constructive of both museum displays and scientific diagnoses. Museological scientific knowledge was produced within the spaces of the museum, albeit in new, specialised spaces such as laboratories, which were rarely publicly accessible.

In addition, the discussion will draw on the work of those who have adopted more constructivist approaches and focused on the public spaces of the museum and the interaction between museum objects and visitors in the creation of scientific knowledge. Stephanie Moser argues that museums provide contexts for the 'visual consumption' of objects and the disciplines they represent through distinctive conventions of classification and display or 'interpretative frameworks'.[12] In the mid- to late nineteenth century, museums pre-empted university departments in establishing numerous subjects as disciplines, including Egyptology, archaeology, anthropology and, I would argue, economic botany. Furthermore, in providing these contexts or interpretative frameworks, museums in turn influenced scholarly understandings of particular fields and served to structure subsequent study by universities and others.[13] This view of knowledge production is one, therefore, in which the displays are not merely visual representations of the knowledge produced by specialists in spaces 'beyond the glass case', but are themselves active agents in the generation of new knowledge.[14]

Sites of Knowledge Creation

The Museum of Economic Botany opened in the grounds of the Royal Botanic Gardens, Kew in 1847. In 1840 the Kew estate had passed from the ownership of the royal family to the British government, and William Jackson Hooker became first Director of the new, publicly-funded Kew in 1841. The Museum was in many ways emblematic of Hooker's new order, displaying 'all kinds of useful and curious Vegetable Products' and catering to a Victorian public fascinated by what things were made of and where they came from.[15] More specifically, the Museum targeted a commercial audience: 'the merchant, the manufacturer, the physician, the chemist, the druggist, the dyer, the carpenter and cabinet-maker, and artisans of every description'.[16] The aim was to make known to British industry the range of plant raw materials available through colonial territories. However, the Museum proved a hit with a far wider constituency. Thus, in his 1857 annual report, William Hooker

---

    12 Moser, *Wondrous Curiosities*, 2006.

    13 Moser, *Wondrous Curiosities*, 1–9.

    14 Nick Merriman, *Beyond the Glass Case: The Past, the Heritage and the Public in Britain* (Leicester: Leicester University Press, 1991).

    15 William Jackson Hooker, *Museum of Economic Botany: Or, A Popular Guide to the Useful and Remarkable Vegetable Products of the Museum of the Royal Gardens of Kew* (London: Longman, Brown, Green, and Longman, 1855), 5.

    16 Hooker, *Museum of Economic Botany*, 3.

was able to claim that '[o]ne has only to see the immense numbers of people, from the prince to the peasant, who visit these Collections ... to appreciate the practical utility of these Museums'.[17] Hooker's son, Joseph, who succeeded him as Director in 1865, looked less favourably upon the presence of the *hoi polloi* in the Museums:

> The industrial class; i.e., persons in the middle and lower grades of life especially ... throng the plant houses and museums in search of general or special information. Amongst these, the mechanics and artisans are perhaps the most numerous, who, with their families, (on full days) crowd the museums to suffocation.[18]

By 1910 this popular collection had grown into four separate museums: Museum No. 1 Dicotyledons (1857), Museum No. 2 Monocotyledons (1847), Museum No. 3 Timber (1863) and Museum No. 4 British Forestry (1910). But by 1987 all four museums were closed and the collections re-housed in the Sir Joseph Banks Building at Kew where they now serve as a research, loan and outreach collection.[19]

The display principle adopted in the Museum was not that of contemporaneous museums of natural history. According to Hooker, displays were to be constituted of 'the *raw material* (and ... also the *manufactured* or *prepared article*) ... correctly named, and accompanied by some account of its origin, history, native country, etc.'.[20] Figure 6.1 shows a photograph of Case 67 taken in 1902 in Museum No. 2. Case 67 was dedicated to the kokerite palm, then known to botanists as *Maximiliana regia* (now known as *Attalea maripa*), and the plant was represented by fruits and by spectacular specimens of spadices surrounded by woody spathes (the two largest in the centre of the cabinet measure 120 cm in length).

The species had come to the attention of British botanical audiences via Alfred Russel Wallace's book *Palm Trees of the Amazon and their Uses*, but there was a more direct link between Wallace, Kew and the *Maximiliana*.[21] Before Wallace and his co-traveller Henry Walter Bates left for the Amazon in 1848 they had consulted William Hooker on plant collecting and Hooker had written them a letter of introduction to use in Brazil. Later that year they sent back palm specimens to Kew which Hooker purchased for £10 and which are still held in the Kew Herbarium and Economic Botany Collections.[22]

---

17   William Jackson Hooker, *Sir W. J. Hooker's Report on Kew Gardens, &c.* (London, 1858), 3.

18   Joseph Dalton Hooker, *Report on the Progress and Condition of the Royal Gardens of Kew for 1871* (London, 1872), 2.

19   Ray Desmond, *The History of the Royal Botanic Gardens, Kew* (London: Kew Publishing, 2007).

20   Hooker, *Museum of Economic Botany*, 4.

21   Alfred Russel Wallace, *Palm Trees of the Amazon and their Uses* (London: J. Van Voorst, 1853).

22   Sandra Knapp, Lynn Sanders and William Baker, 'Alfred Russel Wallace and the palms of the Amazon', *Palms* 46 (2002): 109–19.

Figure 6.1   Economic Museum. Hortus Kew. Fam: Palmae, *Maximiliana*
**   *regia* (Demerara) Maripa Palm**
Photograph by Johannes Lotsy 1902; © RBG, Kew.

Table 6.1   Passage from the *Official Guide to the Museums of Economic Botany* describing the contents of Case 67[23]

Case 67.

No. 140. Male and female spadices and spathes of the KOKERITE PALM, *Maximilia martiana*, Karst. (M. regia, *Mart.*), from Demerara. This magnificent palm forms a lofty smooth trunk, covered with large, terminal, pinnate leaves, sometimes 50 feet long; the petioles are persistent for some distance down the trunk. The palm produces numerous spadices from amongst the bases of the lower leaves. The spathes which enclose the spadices grow to a very large size, as may be seen from the specimens exhibited. This palm is abundant from Para to the Upper Amazon, and the sources of the Rio Negro. The fruits, of which specimens are shown, are somewhat oval-shaped, covered with a brown outer skin, which encloses a fleshy pulp said to have a pleasant, sub-acid flavour; in the centre is a hard, bony seed.

The link with Kew did not, however, extend to the publication of Wallace's book, which was funded at the author's own expense. The research for it was compiled using local knowledge of plant properties, uses and names, and for the illustrations Wallace consulted the collections of palms gathered by Richard Spruce held at the Museum of Economic Botany. Uses he observed were as follows: 'The great woody spathes are used by hunters to cook meat in, as with water in them they stand the fire well. They are also used as baskets for carrying earth, and sometimes for cradles'.[24] Wallace described the fruit of the *Maximiliana* as having a 'pleasant sub-acid flavour' and this description made its way verbatim into the Kew Museum Guide, although it is interesting that the other usages, arguably of more relevance to British manufacturers, did not.[25] On the other hand, readers were told that live *Maximiliana* plants could be seen in the Palm House at Kew, whilst preserved fruit clusters and spathes were on display in the Museum, showing Kew's strength as a botanical complex consisting of separate, yet linked facilities with complementary purposes (Table 6.1).

This strength was particularly evident in the 'museological science' produced at Kew. Sometimes a specimen intended for the Kew Museum would be sent first to the herbarium, a space inaccessible to the public, for the purpose of identification. Here the

---

23   RBGK, *Official Guide*, 52–3.

24   Wallace, *Palms of the Amazon*, 122.

25   Wallace, *Palms of the Amazon*, 121; Royal Botanic Gardens, Kew (RBGK), *Official Guide to the Museums of Economic Botany. No. 2. Monocotyledons and Cryptogams* (London: HMSO, 1895), 52. William Hooker's judgement that Wallace's work was 'more suited to the drawing room table than to the library of a botanist', has been interpreted as criticism by Michael Shermer in *In Darwin's Shadow* (Oxford: Oxford University Press, 2011), but by citing Wallace's text in the museum Hooker was clearly endorsing its status as a work of popular science.

specimen's morphology would be subjected to processes of analysis and comparison in order to determine its genus and species. From 1906 Kew's Jodrell Laboratory – another specialist space – had a permanent plant anatomist (from 1927 there was also a resident economic botanist) and specimens could be 'deconstructed', that is, examined physiologically, to determine their identity and utility according to how they functioned. Once the plant's identity was confirmed, it could then be displayed with its Latin binomial, as well as all the other information – common name (European and/or indigenous), geographical provenance, details of known uses for the plant, and the name of the donor – which the label bore with a general and commercial audience in mind.[26]

Within the museum case, the specimens were supported by photographs showing the living plant in its bio-geographical context. This information was vital to prospective planters wishing to invest in the transplantation of new species into colonial territories. Case 67 contained, in addition, an 1897 illustration from *Curtis's Botanical Magazine* in which Joseph Hooker renamed the *Maximiliana martiana* as *Scheelea kewensis*, which gives us an indication of the contested and mutable nature of botanical knowledge and the multi-layered nature of museum displays, necessitating temporal as well as spatial scrutiny. Below this were objects from a palm in the same plant family, displayed in the very process of transformation from raw materials into finished goods – from a length of palm wood into walking stick and sunshade handle. This interpretative technique – of showing an object in the various phases of its production – was what William Hooker referred to as the 'illustrative series' and objects might be supplied by manufacturers, as in this case, where they were donated by the London-based Henry Howell & Co.[27] If the plant in question was not in use in European industry, 'the manufactured article' might also be represented by ethnographic artefacts to illustrate indigenous uses. This use of indigenous objects signalled two distinct yet interconnected ideas: that the colonies were a virtually limitless source of raw materials – both plant and human – for British industry; and that indigenous practices provided the key to tapping such resources. So, by juxtaposing botanical specimens with contextual data and manufactured objects, new knowledge was created regarding the availability, scale, form, and properties of a vast range of plant raw materials from around the world. The museum was creating an interpretative framework for the public consumption of economic botany.

The Indigo Factory

In 1886 a new object went on display in Museum No. 3. This Museum was designated as a somewhat polyvalent space devoted to 'specimens of overseas timber, and large

---

26   Not all plants received were analysed in detail: often museum officers could identify plants without reference to other departments, and if the specimen had been received from and identified by a respected botanist no further checks were made.

27   RBGK, *Official Guide to the Museums of Economic Botany. No. 1. Dicotyledons and Gymnosperms* (London: HMSO, 1883), 4.

Figure 6.2   Indigo factory model
Photograph by A. McRobb; © RBG, Kew.

articles unsuited for exhibition in the … other Museums'. This display principle, based on object size, was to affect the kind of knowledge produced there. The sheer variety of the objects offered up to the public gaze leads one to suspect that it may have been encountered as something of a cabinet of curiosities. By 1886 these included wooden toys, walking sticks and 'models in cork of a ship, of the town of Fribourg in Switzerland, and of the Queen's Cottage, Kew'.[28]

At the close of the 1886 Colonial and Indian Exhibition in South Kensington, Kew Museum Curator, John Reader Jackson, had drawn up a wish-list of 'specimens' for the Museum's collection. Kew's networks with exhibition commissioners and with the Colonial and India Offices meant that the Museum frequently benefited from donations at the close of large exhibitions. Number 11 on Jackson's list read: 'Model shewing the preparation or manufacture of Indigo'. The model consisted of wooden buildings and 100 clay figures, 95 indigenous labourers, one white factory owner and four oxen (Figure 6.2).[29] The name of the modeller – Rakhal Chunder Pál – was given on the model's pedestal, marking it out as an art object, and distinguishing it from ethnographic artefacts which were typically un-attributed.

---

28   RBGK, *Official Guide to the Museums of Economic Botany. No. 3. Timbers* (London: HMSO, 1886), 69.

29   RBGK, *Official Guide to the Museums of Economic Botany. No. 3. Timbers* (London: HMSO, 1893), 83.

Life Before Kew

The model began its life in the city of Krishnanagar in West Bengal. The tradition of clay model-making in India dates to between 2500 and 1700 BC and is rooted in the production of models of deities.[30] Over time models of secular figures were also produced for the local, domestic market. By the eighteenth century, Krishnanagar was 'a contact zone of multiple presences' and the modellers reacted to the new markets represented by British trade, residency and rule.[31] One such market consisted of exhibitions both within the sub-continent and beyond, the Great Exhibition of 1851 marking the start of a significantly broader circulation of the figurines.[32]

Krishnanagar modellers belonged to the artisan class of kumhars, or potters, a caste immediately beneath that of Brahmins and writers.[33] The creator of the indigo factory figures, Rakhal Chunder Pál, was recognised as 'the best artist in miniature scenes' by the person responsible for commissioning the model. That person was Trailokya Nath Mukharji of the Indian Department of Agriculture – an agency of the British Empire with which Kew was in regular dialogue. Reporting to botanist George Watt, Mukharji was employed by the Department as an exhibition official and collector for colonial exhibitions. As he accumulated and dispatched collections to world's fairs, he compiled an 'index collection' of the economic products of India which was to provide the raw material for Watt's seven-volume opus magnum, *A Dictionary of the Economic Products of India*, a work which in many ways defined the contemporary relationship between science and empire.[34] Mukharji also authored the official publications accompanying Indian products at several international exhibitions. He was one of numerous Indian experts employed by the Department, but he was unusual in travelling overseas as an official representative of the Indian government. Mukharji was an active agent in the process of classifying Indian products and in the representation of India at

---

30   Charlotte F.H. Smith and Michelle Stevenson, 'Modelling Cultures: 19th Century Indian Clay Figures', *Museum Anthropology* 33 (2010): 39.

31   Sria Chatterjee, 'The Empire Commissioned: The Politics of Collecting in Colonial India' (paper presented at the Exploring Empire: Sir Joseph Banks, India and the 'Great Pacific Ocean' – Science, Travel, Trade & Culture 1768–1820 conference, National Maritime Museum, London, June 24, 2011).

32   Claire Wintle, 'Model Subjects: Representations of the Andaman Islands at the Colonial and Indian Exhibition, 1886', *History Workshop Journal* 67 (2009): 194–207.

33   Trailokya Nath Mukharji, *Art-Manufactures of India (Specially Compiled for the Glasgow International Exhibition 1888)* (Calcutta: Superintendant of Government Printing, 1888), 60.

34   George Watt, *A Dictionary of the Economic Products of India* (Calcutta: Government of India, Department of Revenue and Agriculture, 1885–96). Although Watt is the named author, the title page bears the qualifying statement, 'Assisted by numerous contributors', and the index which constitutes the final volume is attributed to polymath Edgar Thurston, 'assisted by T. N. Mukerji'.

home and abroad. He was, for example, a key figure in the development of trade between India and Australia.[35] The knowledge he produced as a result was as much imperial and global as it was local and national.[36]

The 1886 Colonial and Indian Exhibition aimed to show and to tell: to show via its exhibits the extent of British sovereignty overseas, and to tell of the 'development and progress' made within her territories.[37] The Indian Court figured prominently. It was sub-divided into three distinct areas: Art-Ware, Administrative and Economic. The Economic Court, where the indigo factory model was to be found, was under the stewardship of Watt and Mukharji. It was believed by the *Westminster Review* to best represent 'the progress made in India under British rule', offering us a glimpse into how the model itself may have been 'read' during its life at the exhibition.[38] It shared the space of the Court with a number of other models, and in the *Official Catalogue* clay modelling was singled out as 'an art which seems capable of attaining considerable excellence in India'.[39] The indigo factory was intended to 'make intelligible the brief account given of the process of indigo manufacture' by the specimens of indigo which were also exhibited in the Economic Court.[40] It was, therefore, adding new knowledge to the illustrative series of indigo specimens by demonstrating the 'how' of indigo production. It was deemed by one reviewer to be among the 'principal features' of the court.[41]

Indigo, Industry and Imagery

The journalist who reported on the Indian Court of the 1886 exhibition for the *Westminster Review* observed that, '[t]here are also many dyes, of which indigo is the principal, a very interesting model of an indigo factory (of which there are thousands chiefly under European management) being exhibited'.[42] Indeed, at the time of the 1886 exhibition there were 3,414 indigo factories in British India, with the majority managed by Europeans.[43] The best species of indigo dye – *Indigofera*

---

35   Cherie McKeich, 'Botanical fortunes. T N Mukharji, international exhibitions, and trade between India and Australia', *reCollections* 3 (2008): 2, accessed July 13, 2014, http://recollections.nma.gov.au/issues/vol_3_no_1/papers/botanical_fortunes_tn_mukharji.

36   Mukharji's aspirations for social and technological progress in India are explicit in his memoir, *A Visit to Europe* (Calcutta: W. Newman & Co, 1889).

37   Colonial and Indian Exhibition (C&I.E.), *Official Guide to the Colonial and Indian Exhibition* (London: William Clowes & Sons, 1886), 5.

38   Anon, 'The Colonial and Indian Exhibition', *Westminster Review* 70 (1886): 33.

39   C&I.E., *The Colonial and Indian Exhibition. Official Catalogue with Plan and Map* (London: William Clowes and Sons, Ltd, 1886), 12.

40   C&I.E., *Empire of India. Special Catalogue of Exhibits by the Government of India and Private Exhibitors. With a map* (London, 1886), 147.

41   Frank Cundall, *Reminiscences of the Colonial and Indian Exhibition* (London: W. Clowes, 1886), 21.

42   Anon, 'The Colonial and Indian Exhibition', *Westminster Review* 70 (1886): 34.

43   C&I.E., *The Colonial and Indian Exhibition*, 81.

tinctoria – is a plant native to the Indian sub-continent, and the account of Mughal chronicler Abul Fazl indicates that it was already being cultivated for its dye in the sixteenth century.[44] With the establishment of the East India Company (EIC) in 1600, plantations were introduced to the Bengal region. During the American War of Independence (1775–83), supplies of indigo from North America and the West Indies were obstructed and, in response, the EIC encouraged further planting in India and brought in planters from the West Indies who had previously managed enslaved workers. The industry peaked in 1847, but thereafter went into a steady decline, with the final blow dealt by the introduction of aniline 'Indigo Pure' in 1897. By 1914 the price of natural indigo was 50 per cent of that reached at mid-nineteenth century.[45]

Thus, by the time the Kew Museum acquired the model in 1886 the Indian indigo industry was on the wane and it is no coincidence that this was the moment chosen by the Indian Government to promote 'natural' indigo at the Colonial and Indian Exhibition in 1886, nor that the model was of particular interest to Kew after the closure of the Exhibition. Prakash Kumar relates how European growers of indigo in India had produced knowledge on indigo cultivation from a variety of sources: 'the movement of indigo, of savants, and of planters across continents and of institutions and ideologies of production'.[46] However with the arrival of synthetic indigo, planters turned to the colonial state to provide scientific answers to reducing the cost of natural indigo production.[47] The Indian Department of Revenue and Agriculture mobilised George Watt and his team to benchmark methods of indigo cultivation practised throughout British India, and Kew, by this time very much in its imperial phase under William Thiselton-Dyer, rallied its resources to produce and circulate knowledge of cost-effective methods of indigo cultivation and to promote the use of natural indigo to the British public and industries.[48]

Questions of knowledge aside, the spaces of indigo cultivation and production were also spaces of conflict between coloniser and colonised. Indian farmers were often coerced into growing indigo in place of subsistence crops and indentured labourers were kept in permanent debt, 'locked into a system akin to slavery'.[49] It is important to emphasise here that, prior to the arrival in London of the indigo factory model, there was a pre-existing, cosmopolitan discourse of indigo production which circulated widely in both popular and specialist media.[50] It seems that these spaces had long been a source of fascination for Europeans, and not only because of their commercial potential. Pictorial representations of indigo

---

44   Prakash Kumar, *Indigo Plantations and Science in Colonial India* (Cambridge: Cambridge University Press, 2012), 301.

45   Jenny Balfour-Paul, *Indigo* (London: British Museum Press, 1998), 70–85.

46   Kumar, *Indigo Plantations*, 298.

47   Kumar, *Indigo Plantations*, 299.

48   George Watt, *Pamphlet on Indigo* (Calcutta, 1890); Desmond, *History of the Royal Botanic Gardens, Kew*, 259–66.

49   Balfour-Paul, *Indigo*, 72.

50   Kumar, *Indigo Plantations*, 300.

Figure 6.3 **'Indigoterie', from Jean-Baptiste du Tertre's** *Histoire Générale des Antilles,* ***Volume 2*** **(Paris, 1667)**
Source: Wikimedia Commons/Public Domain.

production date from at least the seventeenth century and Jean-Baptiste du Tertre's 1667 illustration of enslaved workers in the French Antilles provides an early example of what became an enduring trope of such representations – that of the white owner or agent overseeing his ranks of colonised labourers (Figure 6.3).[51] In the nineteenth century, images of Indian indigo factories circulated in the popular press and bore a striking resemblance to earlier depictions. A series of illustrations in *The Graphic* in 1887 reflects the interest generated by the model at the exhibition a year earlier (Figure 6.4). It also shows how such interpretative frameworks shape subsequent thinking on a given subject: the model and previous depictions had established a way of seeing indigo production as a series of separate, consecutive processes – as an illustrative series.

Life at Kew

Once at Kew, the fact that the model was donated by the Government of India and had been commissioned for a prestigious exhibition would without doubt

---

51   Jean-Baptiste du Tertre, *Histoire Générale des Antilles Habitées par les François, Vol. 2* (Paris: T. Jolly, 1667).

CUTTING THE PLANT   IRRIGATING

LOADING THE VATS   HAND BEATING

BEATING BY MACHINERY   "RAHUT," OR PERSIAN WHEEL

PRESS AND BOILING-HOUSE   STRAINING THE INDIGO

INDIGO MANUFACTURE IN INDIA

Figure 6.4   'Indigo Manufacture in India', *The Graphic*, **4 September 1887**

Figure 6.5   Indigo factory in Museum No. 1, c.1900
Photograph by E.J. Wallis; KPI W-441;
© RBG, Kew.

Figure 6.6   Indigo factory in Museum No. 3, c.1907
Photograph by E.J. Wallis; KPI O-139;
© RBG, Kew.

have encouraged its reception as an object of artistic and technical merit. Indeed, Pál was awarded a certificate of merit by the exhibition jury which is still in his family's possession, so it is possible that visitors to Kew were aware of the model before its arrival there.[52] However, in the more spacious surroundings of Museum No. 3 its didactic value was reduced on two counts. Firstly, the indigo specimens were in Museum No. 1, a good 10-minute walk away, and secondly, the model occupied a rather isolated position amidst a collection of timbers and miscellanea (Figure 6.5). By 1907 space had been found for it on the ground floor of Museum No. 1, near the entrance (Figure 6.6), but the indigo specimens were on the third floor, once again compromising the model's didactic impact.[53]

The 1886 Exhibition has been characterised as portraying India as 'a timeless, unchanging, ancient land, dotted with jungles, natives, and village bazaars, at once geographically and temporally removed from the hectic pace of industrial life'.[54] Krishnanagar figurines have themselves been understood as belonging to an

---

52   Soumitra Das and Rabi Banerjee, 'Living Clay', *The Telegraph*, 8 June, 2008, 1. Accessed July 18, 2014 at: www.telegraphindia.com/1080608/jsp/frontpage/story_9372298.jsp.

53   RBGK, *Official Guide to the Museums of Economic Botany. No. 1. Dicotyledons. 3rd Edition, Revised and Augmented* (London: HMSO, 1907), 60–61.

54   Saloni Mathur, *India by Design: Colonial History and Cultural Display* (Berkeley: University of California Press, 2007), 10–11.

(a)

Figure 6.7   'Resistance and accommodation' demonstrated by the artist
in the detail of the model: (a) two labourers are fighting; (b)
shows a worker sleeping in the shade; (c) illustrates the artist's
attempts to individualise the workers toiling waist-high in
indigo dye

(b)

(c)

'orientalist iconography of Indian labour', depicting the Indian labourer as 'skilled yet indigent', even 'pitiful' and 'barbaric'.[55] This is, however, to underestimate the agency of the modeller in the representational process. Krishnanagar modellers were middle-caste artisans with some degree of artistic autonomy and economic agency. Commissions such as the indigo factory for the institutions of the nineteenth-century exhibitionary complex – that network of museums, exhibitions and retail outlets which formed a series of 'linked sites for the development and circulation of new disciplines ... and their discursive formations' – demonstrate in their detail a capacity for what Nick Thomas terms artistic 'resistance and accommodation' in situations of trans-cultural encounter.[56] In the Kew model, two labourers fight in the foreground while another sleeps in the shade. Moreover, the workers who are depicted beating fermenting indigo by hand in order to stimulate oxidation are all of differing physiques – to some extent they have had their humanity restored (Figures 7a–c above). But for the directors of the 1886 Exhibition and the Kew Museum, the model was a scene of economic botany in action, in which indigenous labour and plant raw materials on the one hand, and British investment and management on the other, came together in the name of improvement to create imperial wealth. To borrow a phrase later used by Halford Mackinder, the model incorporated both 'the native characteristics of the country and its people and the super-added characteristics due to British rule'.[57]

By 1927 the indigo factory was back in Museum No. 3 and had been joined by a model of a lac factory from the British Empire Exhibition at Wembley (1924–25). This pairing may have provided a new context for both models. They were certainly considered as a pair by Kew Director George Taylor, who referred to them as 'the two dioramas'.[58] The peripatetics, though, were far from over. The model survived the 'cull' which took place further to the Ashby Report of 1958, in which approximately 2,000 ethnographic objects, seen as peripheral to Kew's scientific mission, were transferred to the Horniman, Pitt Rivers, and British Museums.[59] Clearly these exhibitionary models were not perceived as ethnography

55   Lara Kriegel, *Grand Designs: Labor, Empire, and the Museum in Victorian Culture* (Durham, NC: Duke University Press, 2007), 117–20.

56   Tony Bennett, *The Birth of the Museum: History, Theory, Politics* (London and New York: Routledge, 1995), 59; Nicholas Thomas, *Entangled Objects: Exchange, Material Culture, and Colonialism in the Pacific* (Cambridge, MA: Harvard University Press, 1991), 36.

57   Halford Mackinder, 1907, cited in James Ryan, 'Visualizing Imperial Geography: Halford Mackinder and the Colonial Office Visual Instruction Committee, 1902–11', *Cultural Geographies* 1 (1994): 159.

58   Royal Botanic Gardens, Kew (RBGK), Economic Botany Collection (EBC), Indigo Factory file (unarchived); letter from George Taylor to the Museum Curator at the Birla Educational Trust, Pilani, Rajasthan, October 6, 1959.

59   The report of a visiting committee, headed by botanist Eric Ashby, recommended the immediate closure of two of the museums, closely reflecting the thoughts and wishes of George Taylor (RBGK Archives, *Report of a Visiting Group to the Royal Botanic Gardens,*

by those museums and it is ironically this very lack of epistemological determination that ensured the model's future at Kew. With the demise of Museum No. 3 in 1959, the model was returned to Museum 1. On the closure of that building in 1987, it lay in store until 2007 when it underwent a conservation exercise in preparation for a touring exhibition.[60] Since its return in 2008, it is back on display in the former Museum No. 1 which houses the ethnobotanical *Plants+People* exhibit.

Reception

Whatever the intentions of Kew Museum's curators in acquiring and displaying the model, the figures in the indigo factory have appealed to visitors to Kew in a variety of ways. Like the live shows of exotic peoples which were a popular feature of Victorian entertainment, models of 'native' workers were as often sources of wonder as they were symbols of imperialist control, and offered metropolitan audiences the opportunity to further their interests in what they understood as the natural history of humans.[61] Further, and again like displays of live peoples, displays of model peoples were also objects of fascination, perhaps prompting recollections of, and reflections on, contemporary debates ranging from science to slavery.[62] And here it is pertinent too, to revisit William Hooker's original collecting strategy, which sought to amass objects which were both 'useful and curious'.[63] 'Useful' relates to the epistemic values of eighteenth- and nineteenth-century science and to the way in which science in this era allied itself to the imperial project. By the mid-nineteenth century, the term 'curious' had taken on a decidedly unscientific hue, one which suggested the exceptional, rather than the typical specimen which was now seen as the sine qua non of scientific collections. Curiosity itself was often framed as a rather pointless impulse, the antithesis of the rational recreation that museums like the one at Kew were keen to promote. But Hooker was a committed populariser of science and knew the value of the curious in attracting visitors, and the strategy he established of collecting both useful and curious objects lived on at the Museum, though not always without contestation, long after his death. According to Stephen Greenblatt, museum visitors engage with objects in one of two ways, resonance or wonder. Resonance – 'the power of the displayed object to reach out beyond its formal boundaries to a larger world' – is the cognitive process by which the visitor, building on prior knowledge and

---

Kew [Chairman: Sir Eric Ashby] [in March 1957] [Great Britain: Ministry of Agriculture, Fisheries and Food, 1958]). Museum No. 3, where the model was at that point situated, was closed virtually overnight.

60   *Indigo: A Blue to Dye For* was shown at Birmingham, Brighton, and Southampton.

61   Sadiah Qureshi, *Peoples on Parade: Exhibitions, Empire, and Anthropology in Nineteenth-Century Britain* (Chicago, MA: University of Chicago Press, 2011).

62   Qureshi, *Peoples on Parade*, 8.

63   Hooker, *Museum of Economic Botany*, 5.

experience, comes to understand the object as representative of an idea.[64] Wonder, on the other hand – 'the power of the displayed object to stop the viewer in his or her tracks, to convey an interesting sense of uniqueness, to evoke an exalted attention' – is a sensory response to objects of high visual impact.[65] Both can be used by museums as exhibitionary strategies, but equally both are ways in which museum visitors 'receive' what they behold. Indeed, if literary evidence is to be trusted, both appear to have been influential in shaping visitors' responses to the indigo factory model.

In the 1926 novel *Adam's Breed*, set in the first years of the twentieth century, Radclyffe Hall describes the day trip of an Italian immigrant family to Kew Gardens. In Museum 1 they encounter the model:

> The museum was stuffy and very dull, two cases only were amusing. These stood by the door; they contained little people – natives with carts and oxen. The children stopped in delight before them.
>
> 'What funny clothes!' remarked Berta.
>
> Gian-Luca agreed.[66]

In Hall's account, written from a child's perspective, the attraction of the figures is their scale and their exotic otherness. They are identifiable as 'natives' because they have carts and oxen, but also because they wear 'funny' clothes. Models were strongly advocated for the teaching of geography in schools by John Scott Keltie who, in the 1880s, was appointed by the Royal Geographical Society to investigate geography teaching in Britain and abroad. During the late nineteenth and early twentieth century they were widely considered as tools with which to sculpt British imperial citizens, providing children with knowledge of Britain's overseas territories, and making 'the notion of possession meaningful'.[67] The children in Hall's novel appear to have internalised this ethos and are intrigued by, yet accepting of, the colonial subjects in the model. *Adam's Breed* certainly resonated with contemporary audiences; it was a commercial success and received critical acclaim, winning the Prix Femina and the James Tait Black Prize.[68]

---

64   Stephen Greenblatt, 'Resonance and Wonder', in *Exhibiting Cultures: The Poetics and Politics of Museum Display*, ed. Ivan Karp and Steven D. Lavine (Washington, DC and London: Smithsonian Institution, 1991), 42.

65   Greenblatt, 'Resonance and Wonder', 42.

66   Radclyffe Hall, *Adam's Breed* (London: Virago Modern Classics, 1985), 77.

67   Teresa Ploszajska, 'Constructing the Subject: Geographical Models in English Schools, 1870–1944', *Journal of Historical Geography*, 22 (1996): 395.

68   Michael J.N. Baker, *Our Three Selves: The Life of Radclyffe Hall* (New York: William Morrow, 1985), 196–7.

Some 60 years after the publication of *Adam's Breed*, the main character of the novel *Burning Bridges* talks of the indigo factory as 'one of my favourite things in London'.[69] Author Maurice Leitch here gives a description of the model which is inflected with a post-colonial unease regarding race and hierarchy, but his protagonist is, nevertheless, full of wonder for the model's scale, its detail, and the sense of 'industry' it communicates:

> a miniature world, spotlit, a piece of history fixed for eternity ... You never saw such industry. Some treading rollers, others working a great wooden press, a dozen more up to their waists in a vat of blue. There's a miniature ox-cart piled high with bundles of raw stuff, a driver hauling on the reins. Three supervisors wearing topees stand about supervising. They have identical moustaches to the workforce but their skins are much paler ... Of course, it can't all be hard graft – there's a thatched temple and ... a bench in its shade ... It's the strangest thing, but I have to tell you I could stand looking down on that little world under glass for hours on end.[70]

Oral histories of those who saw the model when it was *in situ* in Museum No. 3 pre-1959 have also revealed the visual saliency of this object and its status as a favourite amongst visitors to the Kew Museums. A visitor to Museum No. 3 in the 1950s remembers seeing the indigo factory as a child:

> It had displays that were very attractive to a young child, because it was like looking into a doll's house, really, it had these displays of what were, from memory, Chinese people, working away ... but there was something terribly exotic, somehow, about the displays as well, which made them a bit unusual, made them somewhat enticing but frightening for a small child.[71]

The fact that this visitor misremembers the figures as Chinese suggests that the model is here being encountered as the object of an orientalist gaze, though admittedly a juvenile one. The precise identity of the figures is lost amidst the overwhelming sense of 'otherness' that they convey to this viewer whose childhood was spent in suburban Kew in the immediate post-war period.

None of these eye-witnesses appears to have recalled much of the process of indigo production which the model was designed to communicate. It is clear however that the individuality conferred on the clay figures by their maker Rakhal Chunder Pál was not lost on visitors. Leitch's description in particular reveals a fascination with the diversity of poses, actions, dress and skin tones of the figures, and he picks up on the subversive message of resistance embedded in the modeller's act of individualisation: 'Oddest of all, a man at the top of a ladder looking down

---

69   Maurice Leitch, *Burning Bridges* (London: Hutchinson, 1989), 82.

70   Leitch, *Burning Bridges*, 82–3.

71   Interview conducted by the author 14 March 2011.

into a darkened room. Just a man on a ladder, but – what's he up to? What does he see? What do you make of it?'[72] Rather, it is wonder at the miniature, unknown world represented by the model, which lies at the heart of these visitors' initial engagement. Wonder, then, but also resonance. We know that late nineteenth- and early twentieth-century audiences would have been familiar with images of indigo production, whether pictorial, concrete, or textual. The model had the power to point to a world beyond the museum, and thus these visitors, building on their own prior knowledges and experiences, were able to make meaning of the model in a range of temporal and spatial contexts.

Life Beyond Kew

But knowledge of the indigo factory model was not confined to those viewing it within Kew's walls. The 'Kew' model was a replica of one produced for the Calcutta Exhibition of 1883–84, news of which had circulated widely in both official and unofficial guides and catalogues, and in press reports. Later, in 1910, the Calcutta model was loaned to Harrod's Stores for a 'small exhibition' on Indian indigo, illustrating the extent to which economic botany was subject matter for the whole of the exhibitionary complex, and to which representations of indigo production would have resonated with contemporary audiences.[73] The 'Kew' model had already acquired global recognition through circulating reports and publications whilst still on display at the Colonial and Indian Exhibition. Once at Kew, it appeared in the Kew Museum guide-book of 1893 where it was simply described as 'well-executed', emphasising its status as an art object of some standing.[74] However, prior to that it had appeared in Kew's own journal, the *Bulletin of Miscellaneous Information*, which enjoyed a global audience:

> The highly interesting and instructive model of an Indigo factory, which has now been placed in a special case in Museum No. 3, was an object of particular attention during the Exhibition; it shows the manufacture of indigo through all the various processes, from the bringing in or harvesting of the crop to the finished manufactured product.[75]

Note that the model is both interesting and instructive: instructive because it shows the processes of indigo production, it is in effect an illustrative series set

72 Leitch, *Burning Bridges*, 83.

73 RBGK Archives, MR/164 *Annual Report of the Indian Museum Industrial Section for the Year 1909–1910*, 24.

74 RBGK, *Official Guide*, 83.

75 John Reader Jackson, 'XVIII. – Notes on Articles Contributed to the Museums of the Royal Gardens, Kew, from the Colonial and Indian Exhibition, 1886', *The Bulletin of Miscellaneous Information (Royal Botanic Gardens, Kew)* 9 (1887): 18.

in a diorama; and interesting, a word that we might otherwise gloss over, surely refers us back to its visual interest, to its curiosity value.

Conclusion

By following the life of the indigo factory model through phases of production, exchange and consumption, this chapter has identified the key actors concerned in its inception and numerous lives, but further, it has defined the range of spatial and temporal locations in which the model has existed, and thus demonstrated the very global nature of the knowledge produced by and around the object. Firstly, the networks of acquisition by which the object arrived at Kew were global in their reach. Kew's global authority was underpinned by the network of colonial botanic gardens with Kew at its centre.[76] Kew's ability to request objects at the close of world's fairs was largely contingent on its involvement in such events as advisors – defining what would be required from participating nations and colonies, and as jurors – assessing entries according to their own definitions. Such involvement was achieved, not only through formal relationships with agencies such as the Colonial Office and India Office, but through less formal relationships with fellow members of the exhibitionary complex, including exhibition commissioners. Secondly knowledge produced by and around the model circulated globally. To begin with, we now know that the object itself was not merely the creation of imperial mythmakers. It originated in an indigenous Indian artistic tradition which was co-opted to serve commercial, exhibitionary and pedagogical ends by an Indian Brahmin working within the Anglo-Indian imperial framework. The Indian Department of Agriculture, the Colonial and Indian Exhibition, and the Royal Botanic Gardens at Kew were all connected through this imperial framework, and knowledge circulated between them in the form of objects, people and texts. Kew's own publications circulated across the British Empire and beyond, not only its annual reports and museum guides (in both of which the indigo factory was mentioned), but particularly its *Bulletin of Miscellaneous Information*, a monthly publication which contained the results of investigations on plants undertaken at Kew and at 'kindred institutions at home and abroad'. Its readership included officers of colonial botanic gardens, stations and research institutes, and private individuals with commercial interests in plant products.[77]

Thirdly comes the issue of global impacts. The model fit perfectly within the Kew Museum's didactic strategy, but at a point when natural indigo no longer fit the commercial imperatives of late nineteenth-century Europe. There were problems both at the supply and demand end of the distribution chain. As early as 1869, the *Illustrated London News* reported on the 'inconvenience, and often great

---

76   Desmond, *History of the Royal Botanic Gardens, Kew*, 230–43.

77   Anon., 'Index, Kew Bulletin, 1887–91', *The Bulletin of Miscellaneous Information (Royal Gardens, Kew)* (1891): 63–4.

loss' suffered by plantation owners as a result of the 'disaffection' of the Indian workers or ryots.[78] The situation had only worsened by the 1886 Exhibition; now the emphasis was on the threat of synthetic dyes. A journalist covering the exhibition for *The Times* wrote of a 'fear of some chemical indigo [which] has for years past hung over the indigo industry', this only nine years before the company, Badische Anilin Soda Fabrik, introduced the 'indigo pure' which would herald the demise of the natural product.[79] In innumerable instances, the knowledge produced in the Kew Museum and circulated thence, as in the case of rubber, cinchona and teak, had global impacts – economic, political, human and environmental – which continue to be felt today. And the fact that Kew was part of the initiative to help increase the usage of natural indigo and reduce the costs of production, was another example of Kew's involvement in networks of global knowledge. However, in the case of indigo we have to accept the limits of the knowledge produced in and by Kew – it did not succeed in improving working conditions for Indian labourers, in reversing the downward trend in demand for Indian natural indigo nor in halting the seemingly inexorable march towards synthetic dyes.

---

78   Anon., 'Indigo Culture in Bengal', *Illustrated London News*, 16 October, 1869, 286.

79   Anon., 'Colonial and Indian Exhibition: India', *The Times*, 13 October,1886, 3.

Chapter 7

'A Depot for the Productions of the Four Quarters of the Globe': Empire, Collecting and the Belfast Museum

Jonathan Jeffrey Wright[1]

For the middle classes of Belfast, the years 1830–32 were dramatic and highly charged ones. As was the case elsewhere in Britain, the reform bill's tortuous passage through parliament and the alarming spread of Asiatic cholera provided much to occupy the public's attention.[2] On 1 November 1831, however, the town's citizens were provided with a momentary distraction from these preoccupations when the Belfast Natural History Society opened its new museum on College Square North. Later to become known as the Belfast Natural History and Philosophical Society (BNHPS), this organisation was ten years old in 1831 and the opening of a museum building marked a major milestone in its development.[3] At a time when, as David Allen has put it, 'people appear to have had the utmost difficulty in conceiving of a successful corporate entity without visible substance', the Belfast Museum was a building with a message.[4] Replete with an impressive

---

1   I am grateful to the Deputy Keeper of Records, Public Records Office of Northern Ireland, the Ulster Museum and the Belfast Natural History and Philosophical Society for granting permission to use the archival material on which this chapter is based.

2   Michael Brock, *The Great Reform Act* (London: Hutchinson, 1973) remains the best account of the passing of the reform bill and its attendant drama, but more recent accounts are provided by Edward Pearce, *Reform! The Fight for the 1832 Reform Act* (London: Jonathan Cape, 2003) and Antonia Fraser, *Perilous Question: The Drama of the Great Reform Bill 1832* (London: Weidenfeld and Nicolson, 2013). For the spread of cholera to Ireland, see Gerard O'Brien, 'State Intervention and the Medical Relief of the Irish poor, 1787–1850', in *Medicine, Disease and the State in Ireland, 1650–1940*, ed. Greta Jones and Elizabeth Malcolm (Cork: Cork University Press, 1999), 202–3.

3   Established in June 1821, the society changed its name to the Belfast Natural History and Philosophical Society in August 1842. For convenience, an acronym of the later name will be used throughout the remainder of this chapter. Arthur Deane, ed., *Belfast Natural History and Philosophical Society, Centenary Volume, 1821–1921: A Review of the Activities of the Society for 100 years with Historical Notes, and Memoirs of Many Distinguished Members* (Belfast: BNHPS, 1924), 12.

4   David Elliston Allen, *The Naturalist in Britain: A Social History* (Harmondsworth: Penguin, 1978), 161.

elevated portico, styled on that found on the temple of Andronicus Cyrrhus in Athens, it was a building that conformed to the neo-classical architectural tastes favoured in Belfast and announced the BNHPS as an established presence, both literally and figuratively, in the town's civic and cultural fabric.[5]

Moreover, viewed against the backdrop of an increasingly vociferous campaign for parliamentary reform, the opening of the BNHPS's museum might be seen to take on an extra level of significance as a particularly middle class event.[6] In the English context, Samuel J.M.M. Alberti has noted, such museums, and the collections they housed, 'were the physical manifestation of civic pride, demonstrations of the sophistication of the emerging bourgeoisie writ large (and small) in material culture'.[7] Things were no different in Belfast. As one of its more recent historians has observed, the BNHPS's membership was drawn chiefly from 'wealthy middle class Presbyterian families' and represented 'a respectable professional group outside both the upper class gentry and the lower class artisans and workers'.[8] Thus, while the *Northern Whig* reported that the museum's opening attracted 'a dense mass of ladies and gentlemen, comprising a great portion of the wealth, fashion, and intelligence of Belfast', the fact that tickets to the event, although transferable, were initially granted only 'to the Members of the Natural History Society, and the Subscribers to the building', would suggest that the audience that gathered was a solidly middle class one.[9]

Whatever might be said about the makeup of the crowd that gathered on 1 November 1831, it is clear from the *Northern Whig*'s report that the opening of the museum was a noteworthy event in the civic life of Belfast.[10] But what does any of this have to do with the broader theme of global knowledge with which

---

5   'Belfast Museum', *Northern Whig*, 3 November, 1831, 2; C.E.B. Brett, *Buildings of Belfast, 1700–1914*, rev edn. (Belfast: Friar's Bush Press, 1985), 11–25; Ruth Margaret Bowman Bayles, 'Science in its Local Context: The Belfast Natural History and Philosophical Society' (PhD Diss., Queen's University Belfast, 2005), 175. For a wider discussion of the importance of museum buildings' 'physical materiality', see Sophie Forgan, 'Building the Museum: Knowledge, Conflict, and the Power of Place', *Isis* 96 (2005): 574–79, 574.

6   The campaign for parliamentary reform in Belfast is discussed in Jonathan Jeffrey Wright, *The 'Natural Leaders' and their World: Politics, Culture and Society in Belfast, c. 1801–1832* (Liverpool: Liverpool University Press, 2012), 117–35.

7   Samuel J.M.M. Alberti, 'Natural History Collections and Their Owners in Nineteenth Century Provincial England', *British Journal for the History of Science* 35 (2002): 300.

8   Bayles, 'Science in its Local Context', 79.

9   'Belfast Museum', *Northern Whig*, 3 November, 1831, 2.

10   Further reinforcing this point, the *Northern Whig*'s initial report of the opening was reprinted, the following day, in a rival paper. See 'Belfast Museum', *Belfast News-Letter*, 4 November, 1831, 4. A fuller account of the proceedings can be found in, *Address of the President of the Belfast Natural History Society, on the Opening of the Belfast Museum, 1st November, 1831; Also, the Report of the Curators on the Same Occasion: To which are Annexed, Directions for Preserving Objects of Natural History* (Belfast, 1831).

this volume is concerned? What does the BNHPS and its museum have to do with exhibition, encounter and exchange in an age of empire? As it happens, rather a lot, for the development of museums and the collection, and subsequent exhibition, of scientific and ethnographic artefacts are phenomena linked closely to the growth of empire.[11] Imperial expansion, collection and display can, indeed, be viewed as mutually constitutive processes, and particularly so in the British context. As Sarah Longair and John McAleer have noted, imperial expansion 'yielded much material for British museums' and the display of that material facilitated the development of knowledge concerning the empire and its peoples.[12]

Given the vast reach of the British Empire it is tempting to describe such knowledge as 'global', but some caution is required here. While the terms global and imperial do, at times, overlap, they are not neatly synonymous and it cannot be assumed that imperial knowledge equates to global knowledge.[13] Such qualification serves to reiterate the point, emphasised elsewhere in this volume, that the concept of 'global knowledge' is a complex and multifaceted one, which can be defined and mobilised in a variety of ways.[14] To investigate global knowledge might be to explore the ways in which foreign and exotic locations were made 'known' by the process of global expansion. It might be to reflect on new knowledge generated by the inter-cultural contacts engendered by the processes of global interconnection, or it might be to chart the dissemination of certain knowledge-types globally. In the current context, a combination of the first and second approaches is adopted. Focusing on the collections and collecting practices of the BNHPS, what follows is an attempt to shed light both on how global 'others' came to be represented in the Belfast Museum, and on the knowledge that developed from representations of these 'others'.[15]

---

11   The literature on museums and empire – and on empire and collecting more generally – is now extensive. My own thinking has been informed by Bernard Cohn, *Colonialism and its Forms of Knowledge: The British in India* (Princeton: Princeton University Press, 1996); David N. Livingstone, *Putting Science in its Place: Geographies of Scientific Knowledge* (Chicago: University of Chicago Press, 2003), 29–40; Maya Jasanoff, *Edge of Empire: Lives, Culture, and Conquest in the East, 1750–1850* (New York: Alfred A. Knopf, 2005); John M. MacKenzie, *Museums and Empire: Natural History, Human Cultures and Colonial Identities* (Manchester: Manchester University Press, 2009); and Sarah Longair and John McAleer, eds, *Curating Empire: Museums and the British Imperial Experience* (Manchester: Manchester University Press, 2012).

12   Sarah Longair and John McAleer, 'Introduction: Curating Empire: Museums and the British Imperial Experience', in *Curating Empire*, ed. Longair and McAleer, 2–3. See also MacKenzie, *Museums and Empire*, 4.

13   Miles Ogborn, *Global Lives: Britain and the World, 1550–1800* (Cambridge: Cambridge University Press, 2008), 4, 6.

14   See Diarmid A. Finnegan and Jonathan Jeffrey Wright, Introduction, this volume.

15   Here, and in the discussion that follows, the terms 'self' and 'other' are employed in the Saidian sense, for which see Edward W. Said, *Orientalism* (New York, Vintage Books, 1994), 332.

That promoters of the BNHPS's museum *thought* in global terms, and did so from an early date, is readily apparent. 'In books ... we have the descriptions of animals, plants, and minerals', declared the president of the society, on the occasion of the museum's opening, 'but the museum supplies us with the things themselves, and is a depot for the productions of the four quarters of the globe. Every land and every element contribute to its riches ... Here we can, in one view, behold the productions of the most distant and various countries'.[16] As the ensuing discussion will demonstrate, reality matched rhetoric. Rather more than a well-appointed building, what was opened in Belfast in November 1831 was a global space wherein a regional metropolitan audience encountered scientific and cultural 'others'. This global space was facilitated in no small degree by empire: the BNHPS was an imperially networked society and its members' cultivation and exploitation of imperial networks, both formal and informal, proved hugely significant in the development of its museum collection.[17] But if empire and imperial connections thus form major foci of the discussion that follows, it is important, at the outset, to acknowledge that the BNHPS's global collections were in no sense limited to material gathered within imperial contexts. To return to the point made above, the imperial and the global were not one and the same and it must be recognised that there were other, non-imperial, channels through which material reached the BNHPS. In addition to the imperial, its collections included material gathered by local naturalists and antiquarians in Belfast's immediate hinterland, and material which transcended the bounds of empire – material gathered in north America, south America and the far east by merchants, travellers and seamen whose lives and careers took them to the four corners of a world which came to be characterised, as the nineteenth century progressed, by a 'growing interconnectedness'.[18]

As might be expected, scholars of the BNHPS have long been aware of the global reach of its museum. But while it is well known that the society received donations of ethnographic artefacts and scientific specimens gathered far beyond

16   *Address of the President ... on the Opening of the Belfast Museum*, 18.

17   Here, and throughout this chapter, I am building upon an argument first sketched in brief in Jonathan Jeffrey Wright, '"The Belfast Chameleon": Ulster, Ceylon and the Imperial Life of Sir James Emerson Tennent', *Britain and the World* 6 (2013): 217–18. For the network, which has, in recent years, emerged as a critical concept in the history of empire, see Alan Lester, *Imperial Networks: Creating Identities in Nineteenth-Century South Africa and Britain* (London: Routledge, 2001); Zoe Laidlaw, *Colonial Connections 1815–1845: Patronage, the Information Revolution and Colonial Government* (Manchester: Manchester University Press, 2005); Gary B. Magee and Andrew S. Thompson, *Empire and Globalisation: Networks of People, Goods and Capital in the British world, c.1850–1914* (Cambridge: Cambridge University Press, 2010); and Barry Crosbie, *Irish Imperial Networks: Migration, Social Communication and Exchange in Nineteenth-Century India* (Cambridge: Cambridge University Press, 2012).

18   C.A. Bayly, *The Birth of the Modern World, 1780–1914: Global Connections and Comparisons* (Oxford: Blackwell, 2004), 2.

Ireland's shores, the processes by which these donations were solicited and the wider global and imperial contexts of the society have received little sustained attention.[19] Attending to the collections and collecting practices of the BNHPS will thus serve to address lacunae in the literature relating to that society, while also speaking to broader debates regarding global knowledge and empire. In particular, foregrounding the Belfast Museum as a global space, facilitated by empire, provides an opportunity to reflect on the way in which empire was experienced 'at home' in Belfast.[20] To use such language – to assert that Belfast *was* 'home', that it was a part of the metropolitan centre, rather than the colonial periphery – is to raise the question, if only implicitly, as to whether or not Ireland should be viewed as a colony.[21] Here, some subtlety is required. While the following discussion proceeds on the assumption that Belfast formed part of the empire's metropolitan centre, it is recognised that this was not the case throughout Ireland. In nineteenth-century Ireland there were those who were alienated from the British state and who opposed empire, and those, such as the middle classes of Belfast, who viewed themselves as British subjects, identifying proudly and positively with empire.[22] Put another way, Ireland was, at one and the same time, a country on which

---

19   For exceptions, see Winifred Glover, 'The Folks Back Home: Connections Between Ethnography and Folk Life', *Journal of Museum Ethnography* 9 (1997): 21–32; Winifred Glover, 'Power and Collecting: Big Men Talking', *Journal of Museum Ethnography* 15 (2003): 19–14; and Ruth Margaret Bowman Bayles, 'The Belfast Natural History Society in the Nineteenth Century: A Communication Hub', in *Belfast: The Emerging City, 1850–1914*, ed. Olwen Purdue (Dublin: Irish Academic Press, 2013), 105–24. For additional work on the BNHPS, see Deane, ed., *Centenary Volume*; Bayles, 'Science in its Local Context'; Ruth Margaret Bowman Bayles, 'Understanding Local Science: The Belfast Natural History Society in the Mid-Nineteenth Century', in *Science and Irish Culture: Why the History of Science Matters in Ireland*, ed. David Attis and Charles Mollan (Dublin: Royal Dublin Society, 2004), 139–69; and Marie Bourke, *The Story of Irish Museums 1790–2000: Culture, Identity and Education* (Cork: Cork University Press, 2011), 127–32.

20   For a seminal intervention in the debate on the way in which the empire was experienced 'at home', see Catherine Hall and Sonya O. Rose, 'Introduction: Being at Home with the Empire', in *At Home with the Empire: Metropolitan Culture and the Imperial World*, ed. Catherine Hall and Sonya O. Rose, (Cambridge: Cambridge University Press, 2006), 1–31.

21   For the long-running debate on Ireland's colonial status, see, Stephen Howe, 'Questioning the (Bad) Question: "Was Ireland a Colony?"', *Irish Historical Studies* 36 (2008): 138–52 and Terence McDonough, ed., *Was Ireland a Colony? Economics, Politics, and Culture in Nineteenth-Century Ireland* (Dublin: Irish Academic Press, 2005). Wider insights into Ireland's relationship with empire can be found in the essays collected in Kevin Kenny, ed., *Ireland and the British Empire* (Oxford: Oxford University Press, 2004) and Keith Jeffery, ed., *An Irish Empire? Aspects of Ireland and the British Empire* (Manchester: Manchester University Press, 1996).

22   My comments here are informed by Kevin Kenny, 'Ireland and the British Empire: An Introduction' and Alvin Jackson, 'Ireland, the Union, and the Empire, 1800–1960', in *Ireland and the British Empire*, ed. Kenny, 1–4 and 130–42.

some directed an imperial optic, viewing it, like the other colonies, as a source of governmental problems which needed to be solved, and a country in which some participated in empire, turning an imperial optic on others. It is this latter group that form the focus of attention here, for as has already been suggested collection and display provided a central means whereby the empire came to be 'known' and experienced. Ireland may well have been England's oldest colony, but Belfast, a workshop of the empire, formed part of Britain's metropolitan 'core' and exploring the development of its museum will reveal one aspect of its imperial culture and contribute to the development of a more nuanced understanding of the Irish imperial experience.

The BNHPS and the Belfast Museum

One of a number of provincial scientific societies that emerged in Britain in the 1820s, the BNHPS was established on 5 June 1821.[23] Its founders were a group of seven young men – Francis Archer, James Grimshaw, George C. Hyndman, James MacAdam, William M'Clure, Robert Patterson and Robert Simms – who were led by a more senior figure, Dr James Lawson Drummond, Professor of Anatomy and Physiology in the town's most prominent educational establishment, the Belfast Academical Institution.[24] The society had as its *raison d'être* the 'cultivation' (by members) and the 'diffusion' (among a wider public) of natural history, broadly defined, and it is clear that the development of a museum was viewed as an essential part of this project from an early date.[25] Indeed, the desire to establish a museum pre-dated the formation of the society, for in 1820 Drummond had urged his employers, the proprietors of the Belfast Academical Institution, to set up a museum in his *Thoughts on the Study of Natural History: And on the Importance of Attaching Museums of the Productions of Nature, to National Seminaries of Education* (1820). Such a move was necessary, he argued, in order to foster a taste for the study of natural history in Ireland, a country in which the subject was 'less encouraged, less valued, and consequently less understood, than in any part of civilized Europe'. Nothing, Drummond believed, could 'be more gratifying to curiosity, than to see, under one roof, the collected riches of the different kingdoms of Nature, fossils, and plants, and animals', and it is small wonder that the society

---

23   As Alberti notes, the natural history societies of the 1820s differed from the earlier generation of literary and philosophical societies in having 'stronger remits to establish museums'. Alberti, 'Natural History Collections and Their Owners', 298.

24   Deane, ed., *Centenary Volume*, iii–iv, 72–3. The Belfast Academical Institution was first opened in 1814 and a collegiate department was added the following year. For a standard history, see John Jamieson, *The History of the Royal Belfast Academical Institution, 1810–1960* (Belfast: W. Mullan, 1959). Wright, *The 'Natural Leaders'*, 85–8, 160–70 discusses its wider significance.

25   *Belfast Almanack*, 1830, 50.

he was instrumental in forming the following year quickly turned its attention to the development of such a collection.[26]

As early as May 1822, the BNHPS appointed a committee to consider the storage of what was already being referred to as 'the museum', and in September 1827, beginning a process which would culminate in the opening of the museum building in November 1831, the society's 'Curators' were instructed to 'draw up a plan of a House for the Society, & procure estimates ... in order that it may be taken into consideration whether there is a probability of our being enabled to raise funds sufficient for that desirable purpose'.[27] Underpinning such moves was a belief that the BNHPS had – or was, at any rate, acquiring – a collection sufficient to merit such a building. This belief was not without foundation. The report of the committee appointed to consider the issue of storage in May 1822 indicates that, by that point, the society's collections comprised botanic, conchological, mineralogical, ornithological and entomological specimens, and in subsequent years these collections grew steadily.[28] The purchase of new storage cases was required in 1824 and 1826, and when the museum was opened, in November 1831, the society's collections included 'about 2,000 specimens of minerals; a nearly complete collection of native and a considerable number of foreign shells; 200 native and foreign birds; about 3,000 insects; an extensive Hortus Siccus of indigenous, American and other exotic plants; about 200 snakes and lizards; with some skins, coins, antiquities, and miscellaneous articles'.[29] This impressive list raises two questions central to the concerns of this volume: how was this material acquired? And where, America aside, was it acquired from?

Global Collecting

Studies of the museums established by Scottish and English natural history societies in the second half of the nineteenth century have foregrounded the importance of 'local' collecting.[30] While specimens and artefacts were received from further

---

26   James Lawson Drummond, *Thoughts on the Study of Natural History; and on the Importance of Attaching Museums of the Productions of Nature, to National Seminaries of Education, Addressed to the Proprietors of the Belfast Institution* (Belfast: F.D. Finlay, 1820), 9–10, 108.

27   BNHPS Minute Book, 1821–1830, 29 May 1822 and 12 September 1827, D/3263/ AB/1, Public Records Office of Northern Ireland (hereafter cited as PRONI).

28   BNHPS Minute Book, 1821–1830, 12 June 1822.

29   BNHPS Minute Book, 1821–1830, 24 May 1824 and 18 January 1826; *Address of the President ... on the Opening of the Belfast Museum*, 9.

30   My comments in this paragraph are informed by Simon Naylor, 'The Field, the Museum and the Lecture Hall: The Spaces of Natural History in Victorian Cornwall', *Transactions of the Institute of British Geographers* 27 (2002): 500–501 and Diarmid A. Finnegan, *Natural History Societies and Civic Culture in Victorian Scotland* (London: Pickering and Chatto, 2009), 76–84.

afield, collections embracing the 'non-local' were nevertheless viewed with suspicion.[31] Were such collections not unsystematic and unscientific? Were they not throwbacks to an earlier age's cabinets of curiosities? For many, the answer to these questions was yes, and curatorial policies were adopted that prioritised the development of comprehensive collections that illustrated the natural history of the region in which a given society was located. In early nineteenth-century Belfast, however, a different dispensation prevailed.

In his *Thoughts on the Study of Natural History*, Drummond had predicted that a museum established in Belfast 'would soon become rich in specimens'. Not merely because the island of Ireland 'contains in itself a large fund of materials', but also, crucially, because 'the situation and circumstances of Belfast, as a great commercial town, are highly favourable to the acquirement of the productions of every latitude, from the shores of the North, to the plains of India'.[32] Belfast was, in short, a town with myriad links to the wider world, and when the BNHPS was established in 1821 its members did not limit themselves to collecting materials illustrative of the natural history of North-East Ulster. Nor, indeed, did they limit themselves solely to collecting specimens of natural history, narrowly defined. Rather, they made concerted efforts to exploit their town's links with the wider world, actively pursuing a policy of 'global' collecting – a policy that if focused primarily on natural history nevertheless embraced what we might today recognise as anthropology.

In the BNHPS's early days members appear to have pursued this policy instinctively, on an individual basis, but it soon became a more explicit and formalised feature of the society's activities. In May 1826, the then president of the BNHPS, Thomas Dix Hincks, observed that donations had, within the past year, been received from 'distant countries' including Norway, New Brunswick, Connecticut, Pennsylvania, Massachusetts, South America, Gibraltar, Elba, Greece, Van Diemen's Land, Madagascar and New South Wales, and that, in some instances, 'a correspondence has been commenced with private individuals or scientific societies, and an exchange of specimens commenced, from which much advantage has already been derived'.[33] Two years later, with a view to 'directing to this society the attention of our countrymen abroad' the BNHPS went so far as to prepare a printed circular, containing 'directions for the preservation of objects of Natural History and at the same time requesting their co-operation in furtherance of our views, by sending such specimens connected with our pursuits, as occur

---

31 For a detailed analysis of the non-local materials received by five provincial English museums – the Liverpool Museum and the museums of the Saffron Walden Natural History Society, the Whitby Literary and Philosophical Society, the Leeds Philosophical and Literary Society and the Natural History Society of Northumberland, Durham and Newcastle Upon Tyne – see Claire Loughney, 'Colonialism and the Development of the English Provincial Museum, 1823–1914' (PhD Diss., Newcastle University, 2006).

32 Drummond, *Thoughts on the Study of Natural History*, 121.

33 BNHPS Minute Book, 1821–1830, 24 May 1826.

in their respective places of abode'.[34] Offering detailed instructions relating to the preservation of quadrupeds, birds, fish, botanical specimens and minerals and fossils, this circular highlights the BNHPS's engagement with the natural world in all its variety, while also foregrounding its desire to develop a global collection. Directions were given for the preservations of tortoises, lizards and serpents; it was noted that scorpions, like beetles, cockroaches, bees, flies and spiders, could be killed 'by plunging them into boiling water' or 'by putting them into bottles containing any kind of strong spirits'; and it was observed, with regard to botanical specimens, that '[p]ersons residing abroad, but particularly in tropical climate, possess the means of furnishing innumerable interesting specimens in this department, since not only the plants themselves, but their flowers, leaves, fruits, seeds, seed-vessels, and woods, all supply useful materials for a museum'.[35]

Far from being viewed as problematic or unscientific, a 'global' collecting policy, that embraced both the local and the 'exotic', was thus pursued consciously by the BNHPS, and the working of that policy can be traced in two documents preserved in the society's archives – its donation book and its outgoing letter book.[36] To give just some examples, the letter-book records that in January 1828, having learned that he was due to return to Trinidad, the BNHPS contacted one James Magee, requesting that he assist in the establishment of a museum 'by forwarding, when convenient, contributions ... of whatever kind may occur to you as proper to send'.[37] From the donation book, it appears that Magee obliged, furnishing the society, at different dates, with 'fruit grown at Trinidad', a 'collection of fossils, shells, seed vessels, and corals', and a 'specimen of small tortoise'.[38] Likewise, in June 1829 the BNHPS made contact with Charles Telfair, a Belfast-born naval surgeon and administrator who had been based in Mauritius since 1812, and in October 1830 several copies of the BNHPS's directions for the preservation of objects of natural history were forwarded to John MacCormac, a Portadown-man resident in Sierra Leone, along with a letter requesting that he forward specimens of natural history and ethnological artefacts, such as 'dresses, implements of art or weapons, [which] would afford additional attraction, when ranged in the same room with the animals, birds & shells of tropical climates'.[39] Whether or not MacCormac supplied material directly is unclear, but the BNHPS did receive donations of ethnographic material from his brother Henry, a Belfast-based doctor who had himself spent several months in Sierra Leone in the 1820s, and the overture to Telfair, a founding member of the Natural History Society

---

34 BNHPS Minute Book, 1821–1830, 24 May 1828.

35 *Address of the President ... on the Opening of the Belfast Museum*, 36–44 (40 and 41 for quotes).

36 BNHPS Donation Book, 1821–1844, D/3263/J/1, PRONI; BNHPS Letter Book, 1824–1860, D/3263/BA/1, PRONI.

37 BNHPS Letter Book, 1824–1860, 56–9.

38 BNHPS Donation Book, 1821–1844, donations 348, 392 and 410.

39 BNHPS Letter Book, 1824–1860, 79–80, 93–4.

of Mauritius, established an important relationship of exchange.[40] In addition to geological specimens, Telfair supplied seeds which were cultivated in Belfast's recently established Botanic Gardens and he attempted, in 1831, to send two live Baboons – though attempted is the operative word, the unfortunate primates having died while en route to Ireland.[41]

These few examples are but the tip of an iceberg. Between 1821 and 1832, the period for which its records are most complete, the BNHPS received some 570 donations – donations which ranged from single items to extensive consignments, and which varied widely in terms of scale, geographical origin and scientific significance. Material was received from North and South America, from the West and East Indies, and from Asia, the Mediterranean and Polynesia.[42] It included geological specimens and botanical samples, animal skins and ethnological artefacts, and it ranged from the mundane to the striking to the frankly grotesque. The receipt of a specimen of 'very light wood from E. Indies, not known from what plant' seems unlikely to have caused much excitement. By contrast, the skin of an alligator, received from Charleston, or the collection of snakes and lizards preserved in spirits that was sent from New Orleans by one Ben Adair may well have set pulses racing, and the 'head of a New Zealand chief', presented to a meeting of the society in May 1829, must certainly have done so.[43]

Collecting and Imperial Power

Although it is recorded sparely in the donation book, with little elaborative comment, the BNHPS's receipt of the head of a New Zealand chief is of particular significance, insofar as it raises some of the wider issues of coercion, power and control which are related to the collection and exchange of scientific and ethnographic artefacts. Save that it was provided by Frank Archer, a founding

---

40   BNHPS Donation Book, 1821–1844, donations 530, 552, 566. See, for MacCormac and Telfair, Ian Fraser, 'Father and Son – A Tale of Two Cities', *Ulster Medical Journal* 37 (1968): 1–39 and Marc Serge Rivière, 'From Belfast to Mauritius: Charles Telfair (1778–1833), Naturalist and a Product of the Irish Enlightenment', *Eighteenth-Century Ireland* 21 (2006): 125–44.

41   BNHPS Donation Book, 1821–1844, donation 455; BNHPS Letter Book, 1824–1860, 90–91, 100–101; BNHPS Minute Book, 1821–1830, 10 March 1830. For Belfast's Botanic Garden, which was established in 1827 by the Belfast Botanical and Horticultural Society, a number of whose members were also members of the BNHPS, and which was later renamed as the Royal Botanic Garden, see *A Popular Guide to the Royal Botanic Garden of Belfast: Published for the Garden* (Belfast: W. &. G. Agnew, 1851); Bayles, 'Science in its Local Context', 145–70; and Nuala C. Johnson, *Nature Displaced, Nature Displayed: Order and Beauty in Botanical Gardens* (London: I.B. Tauris, 2011), 31–6.

42   BNHPS Donation Book, 1821–1844, passim.

43   BNHPS Donation Book, 1821–1844, donations 108, 321, 351, 428; BNHPS Minute Book, 1821–1830, 13 May 1829.

member of the BNHPS who presumably acquired it from an overseas contact, and who presented it as though it were just another object, alongside a platypus and 'a specimen of Coral from the Mauritius', little is known about the donation.[44] Neither the identity of the 'chief' – if, indeed, he was a chief – nor the means by which his head had been acquired are detailed in the archive.[45] The fact that New Zealand was, in the 1820s, one of the expanding British Empire's frontier zones, and that a lively market existed in Europe for mokomokai – tattooed Maori heads – might be seen to hint at the contexts in which it was procured, but a degree of caution is required here.[46] As recent work has demonstrated, indigenous 'go-betweens' were central to the processes whereby global knowledge was developed in the late eighteenth and early nineteenth centuries and there is evidence that some Maori, rather than being 'the gullible victims of European manipulation', acted as brokers, exploiting and catering to the demand for mokomokai.[47] More generally, some would caution against making assumptions about the power dynamics at play in collecting, pointing out that 'while obtaining objects may not be quite the result of a meeting of true minds, it is often the result of "big men talking"', a process in which '[t]he power of the collector is often equalled if not surpassed by the power of the original owners of the artefacts'.[48] This may, at times have been so, but whatever may be said about the circumstances in which the BNHPS's Maori head was acquired, the society's receipt of such an item nevertheless highlights the fact that the collecting of artefacts could be a less than benign process. Indeed, it serves to remind us that the practice in which the society participated – a practice, ultimately, of representing global 'others', both scientific and cultural, to metropolitan audiences – was one that was inextricably bound up with the expansion and exercise of imperial power.

In Edward Said's influential analysis, the ability to represent that which lay 'beyond metropolitan borders' was an ability that originated in 'the power of an imperial society', but that also contributed to that power. 'When it came to what lay beyond metropolitan Europe', Said argued,

---

44   BNHPS Minute Book, 1821–1830, 13 May 1829; Deane, ed., *Centenary Volume*, 63–4.

45   Writing in the late 1990s, Winifred Glover, curator of ethnography at the Belfast Museum's successor institution, the Ulster Museum, suggested that Archer, who had moved from Belfast to Liverpool, 'obtained the head from a sailor', and noted that it had 'since been returned to the National Art Gallery and Museum of New Zealand'. Glover, 'The Folks Back Home', 22.

46   Paul Moon, *A Savage Country: The Untold Story of New Zealand in the 1820s* (Auckland, NZ: Penguin, 2012), 24–32.

47   Simon Schaffer, Lissa Roberts, Kapil Raj and James Delbourgo, eds, *The Brokered World: Go-Betweens and Global Intelligence, 1770–1820* (Sagamore Beach, MA: Science History Publications, 2009); Moon, *A Savage Country*, 28.

48   Glover, 'Power and Collecting', 24.

the arts and the disciplines of representation – on the one hand, fiction, history and travel writing, painting; on the other, sociology, administrative or bureaucratic writing, philology, racial theory – depended on the powers of Europe to bring the non-European world into representations, the better to be able to see it, to master it, and, above all, to hold it.[49]

While museums are absent from Said's typology of 'the arts and the disciplines of representation', the applicability of his analysis to the museum is readily apparent. Thus Ruth Adams, drawing on his work, has argued that they 'illustrate in a very explicit fashion the "systematics of disciplinary order"' and can be said to have 'played a role in defining both the "core" and the "margins" of the Empire'.[50] While approaching the issue from a somewhat different perspective, Longair and McAleer have made a similar point, asserting that: 'Objects, and their collection and display, formed part of an imperial nexus'. 'Collecting and displaying were never', they contend, 'neutral activities. Knowledge, its acquisition, presentation and dissemination – key impulses driving the establishment of museums – became intertwined with the promotion of commerce and, consequently, the development of empire'.[51] The museum can thus be viewed as a central institution or, as John M. MacKenzie has it, 'a tool of empire'.[52] This was true both *in* the colonies, where museums 'often symbolised the dispossession of land and culture by whites through the rapid acquisition of specimens and artefacts', and at home in Britain, where, David N. Livingstone has noted, the museum provided a means through which 'administrators, bureaucrats, and the general public encountered the "collective improvisation" that was the British Empire'.[53]

These reflections notwithstanding, it would be somewhat tendentious to develop a Saidian analysis of the BNHPS on the basis of a single item. However, the mokomokai was by no means the only item the society received which directs our attention to the connections linking the collection of artefacts to the exercise of imperial power. During the first ten years of its existence, in addition to numerous donations illustrative of natural history, both local and global, the BNHPS received, 'Specimens of rock broken off one of the Gods in the celebrated cave of Elephanta'; 'An Indian Bow and Arrows', presumably from North America; and a consignment of Polynesian artefacts which included 'Three large pieces of cloth from Otaheite', 'A New Zealand War Club and Belt' and 'A New Zealand Mat'.[54]

---

49   Edward W. Said, *Culture and Imperialism* (London: Chatto and Windus, 1993), 119.

50   Ruth Adams, 'The V&A: Empire to Multiculturalism?', *Museum and Society* 8 (2010): 64.

51   Longair and McAleer, 'Introduction', 1–2.

52   MacKenzie, *Museums and Empire*, 7.

53   MacKenzie, *Museums and Empire*, 4; Livingstone, *Putting Science in its Place*, 32.

54   PRONI Donation Book, 1821–1844, donations 117, 147, 431; *Report of the Council of the Belfast Natural History and Philosophical Society, for the Session Ending June, 1858* (Belfast, 1858), 4.

Moreover, in the months immediately following the opening of its museum building in November 1831, it received a 'A Burmese Idol, & an Indian Manuscript, being a letter of introduction from a Rajah to a friend' from one John Irvine Esquire, while James McClean donated 'An Esquimaux canoe', Captain John M. Boyes of the 38 Madras Native Infantry forwarded 'A Burmese God' and Lieutenant Thomas Graves of the Royal Navy supplied a 'Fac Simile [sic] of writing in a book presented to the Travellers, [John and Richard] Lander, by the King of Wow Wow on the River Niger'.[55] As with the mokomokai, donated by Francis Archer, we can only speculate as to the nature of the encounters and exchanges which led to the BNHPS's receipt of these items. But what can be said is that the acquisition of such items, whether mediated by 'go-betweens' or not, entailed western interactions with non-western 'others' – interactions that were invariably complicated by issues of power, authority and control.

Complicated cross-cultural interactions, were not, of course, unique to the 'age of empire' with which this volume is concerned. It was, after all, in 1492 that Christopher Columbus 'discovered' America, and five years earlier Bartholomeu Dias had opened up a route into the Indian Ocean.[56] Nevertheless, as the European empires waxed and waned in the centuries that followed, opportunities for western/non-western encounter expanded dramatically, and particularly for the British, for Britain emerged in the eighteenth century as 'the foremost European Imperial power'.[57] Critical in facilitating this emergence was the Royal Navy, even if, as N.A.M. Rodger has argued, its 'primary function was to guard against invasion, for which purpose the bulk of the fleet was almost always kept in home waters'.[58] Those ships that were not confined to home waters enabled the expansion and interconnection of empire, not only sustaining trade and trans-imperial networks, but also facilitating global interconnection and the development of ethnographic and, particularly, scientific knowledge.[59] As Joseph M. Hodge has noted, in the years following James Cook's Pacific voyages 'the Admiralty and Navy played a central role in enlarging Britain's cartographic and hydrographic knowledge of the world' and 'became a key conduit in a worldwide operation of plant transfers that extended from the Caribbean and Central America, to the Pacific and Southeast Asia'.[60]

---

55   BNHPS Donation Book, 1821–1844, donations 545, 547, 557, 561, 564, 565.

56   Ogborn, *Global Lives*, 16.

57   G.E. Aylmer, 'Navy, State, Trade, and Empire', in *The Oxford History of the British Empire: Volume I: The Origins of Empire: British Overseas Enterprise to the Close of the Seventeenth Century*, ed. Nicholas Canny (Oxford: Oxford University Press, 1998), 479.

58   N.A.M. Rodger, 'Sea-Power and Empire, 1688–1793', in *The Oxford History of the British Empire: Volume II: The Eighteenth Century*, ed. P.J. Marshall (Oxford: Oxford University Press, 1998), 169.

59   Bayly, *Birth of the Modern World*, 128–9.

60   Joseph M. Hodge, 'Science and Empire: An Overview of the Historical Scholarship', in *Science and Empire: Knowledge and Networks of Science Across the British Empire, 1800–1970*, ed. Brett M. Bennett and Joseph M. Hodge (Basingstoke: Palgrave Macmillan, 2011), 6. For the botanical networks facilitated by empire, see Richard

Given all of this, it is scarcely surprising that those who contributed to the BNHPS in the months following the museum's opening included Thomas Graves, a serving naval officer. But Graves was just one of a number of figures linking the society to the Royal Navy. The society's minutes for 7 December 1825, for instance, detail the unanimous election, as a corresponding member, of 'Surgeon William Whyte of the R.N.', while those for 21 February 1827 record that a letter and a box of specimens had been received from another corresponding member, 'Mr Wm MGee R.N.'.[61] Moreover, in its report for 1830–31 the society's council urged members to continue their endeavours 'to obtain additional specimens for the collection', appealing in particular to 'those who have friends in the Royal Navy, on any of the foreign stations; merchants having ships going to foreign countries; and masters of vessels belonging to this port'.[62] Viewed one way, this injunction reflects the well-documented maritime connections of Belfast, a flourishing port-town whose merchants were, by the late eighteenth century, connected 'to almost every quarter of the globe'.[63] However, the passing reference to the Royal Navy assumes a greater significance when we take into account the involvement, in the BNHPS, of figures such as MGee and Whyte. If the society sought consciously to pursue a global collecting policy, it also sought to network itself within the Royal Navy, a circumstance that was not unconnected to the biographical particularities of its father-figure, James Lawson Drummond.

As noted, when the BNHPS was established in 1821 Drummond was Professor of Anatomy and Physiology at the Belfast Academical Institution. A prominent member of Belfast's intellectual elite, he held his position at the Academical Institution until 1849 – combining it, at various times, with the professorship of botany (1835–36) and presidency of the Faculty of Arts (1823 and 1831) and Faculty of Medicine (1835, 1836 and 1844) – and published several scientific works.[64] In addition to his *Thoughts on the Study of Natural History* (1820), these included *First Steps to Botany* (1823), *Letters to a Young Naturalist on the Study of Nature and Natural Theology* (1831), *First Steps to Anatomy* (1845) and *Observations on Natural Systems of Botany* (1849). All of this might seem to suggest a rarefied and scholarly life, but Drummond's was a scholarly life that had, at a formative point, been facilitated by naval service. Prior to taking his M.D. from Edinburgh University in 1814, he had completed a six-year naval apprenticeship, serving as assistant-surgeon on board the HMS *Renown* and as surgeon on HMS *San Juan*, the latter a vessel 'anchored close to the New Mole

---

Drayton, *Nature's Government: Science, Imperial Britain, and the 'Improvement' of the World* (New Haven: Yale University Press, 2000).

61   BNHPS Minute Book, 1821–1830, 7 December 1825 and 21 February 1827.

62   'Belfast Museum', *Belfast News-Letter*, 4 November, 1831, 4.

63   'Journal of the Duke of Rutland's Tour of the North of Ireland', in *Manuscripts of his Grace the Duke of Rutland, K.G., Preserved at Belvoir Castle*, vol. 3 (London: Historical Manuscripts Commission, 1894), 420–21.

64   Deane, ed., *Centenary Volume*, 72–3.

at Gibraltar', the last stronghold of Britain's Mediterranean Empire.[65] Whether or not Drummond collected specimens and artefacts during this stage of his life is unknown, but his personal experience no doubt informed the BNHPS's attempts to cultivate naval connections and it can be said that he collected knowledge, developing an expertise on the bird-life of the Mediterranean that he later shared, in anecdotal form, with William Thompson, a fellow member of the BNHPS and the author of a monumental, four volume *Natural History of Ireland* (1849–56). Thompson included some of Drummond's anecdotes in his *Natural History*, but he did not need to rely on the older man's tales, for he, too, had naval connections and had spent time in the Mediterranean. In 1841, he sailed on H.M.S. *Beacon*, a surveying ship that was tasked with charting the Aegean and that was captained by Thomas Graves – the same Thomas Graves who had, as a lieutenant, supplied the BNHPS with a facsimile specimen of native writing which the travellers John and Richard Lander had acquired while exploring the River Niger.[66]

In ways direct and indirect, obvious and subtle, the BNHPS was thus bound to empire. Imperial networks and naval contacts facilitated the gathering and transport of specimens and artefacts, and the museum that was opened in November 1831 functioned as a depot for the productions of the four quarters of a globe which the process of western and, particularly, British imperial expansion was arguably rendering 'smaller' and more complexly interconnected than at any time in human history.

Changing Practices and Making Knowledge

In the years that followed the opening of its museum, the BNHPS continued to pursue a global curatorial policy, adding, in particular, to its ethnographic collections. In June 1832, a consignment of donations, including 'a quiver with 19 poisoned arrows', 'extracts from the Koran – very rare & difficult to obtain' and 'a Mandingo's knife', was received from Gordon Augustus Thomson.[67] Gathered, among other locations, at the Cape of Good Hope and Ascension Island, this was just the first of several consignments which would be received from Thomson, a fascinating figure who had left Ireland for the West Indies in the 1820s, and who undertook a privately-financed round-the-world tour during the 1830s and

---

65   William Thompson, *The Natural History of Ireland* (4 vols, London: Reeve, Benham and Reeve, 1849–56), vol. 1, 170, 387; 'Death of Dr Drummond', *Belfast News-Letter*, 20 May, 1853, 2. For Britain's Mediterranean empire, see Linda Colley, *Captives: Britain, Empire and the World, 1600–1850* (London: Pimlico, 2003), 23–134.

66   Thompson, *The Natural History of Ireland*, vol. 1, 60, 170, 387, 423, 426; vol. 4, xvii–xviii; Deane, ed., *Centenary Volume*, 106–7. For an account of Lander's Niger expedition, see *The Niger Journal of Richard and John Lander*, ed. Robin Hallett (London: Routledge and Kegan Paul, 1965).

67   BNHPS Donation Book, 1821–1844, donation 570.

early 1840s, visiting India, Singapore, China, Japan, Australia, Tahiti, Hawaii, South America and North America. While travelling, Thomson amassed extensive collections of Polynesian and indigenous American ethnographic material – collections which he donated to the BNHPS and which comprise the bulk of the ethnographic collections of today's Ulster Museum.[68] As an independent traveller and collector, who acted both within the empire and beyond its formal bounds, Thomson might be said to illustrate the point, made earlier, that the BNHPS's global collections were not limited simply to material gathered within imperial contexts. But if he might reasonably be presented as a global actor, it would be a mistake to downplay the significance of empire and imperial connections in Thomson's biography. When he first left Ireland for the West Indies in the 1820s, Thomson did so to join his uncle, Robert Gordon, president of the council of St Vincent. He later left this colony for Africa, joining Captain Richard Meredith on HMS *Pelorus*, a vessel involved in the policing of the slave trade, and in the late 1830s, while in Brazil, he carried communications from the commander of the British navy in the Pacific, Rear Admiral Charles B. Ross.[69] None of this is to suggest that it was empire alone which facilitated Thomson's travels; patently, this was not the case. But it is to demonstrate the difficulties in separating the imperial and global, for if he was a global actor, Thomson was also an imperial actor.

Leaving Thomson aside, the importance of naval and imperial connections in facilitating the collection of ethnographic artefacts and specimens of natural history is further underlined by the substantial donations the BNHPS received from the Co. Down born naval officer Captain Francis Crozier and the Belfast-born writer, parliamentarian, traveller and colonial official, Sir James Emerson Tennent. Crozier supplied 'a very extensive collection of the skins both of marine birds, and of those belonging to several islands in the southern hemisphere', which he had collected while captaining HMS *Terror* during the Ross Antarctic Expedition.[70] Reports of this donation record that it comprised 'the whole of the collection of birds made during the voyage by our countryman, Captain Crozier', a claim that would seem to be undermined by the fact that the expedition leader, James Clark Ross, sent the natural history specimens gathered during the voyage to the British Museum.[71] This suggests that the collection Crozier donated was

68   Deane, ed., *Centenary Volume*, 110–11, 167; Winifred Glover, 'In the Wake of Captain Cook: The Travels of Gordon Augustus Thomson (1799–1886), Principal Donor of Ethnographic Objects to the Ulster Museum, Belfast', *Familia* 9 (1993): 46–61; *Illustrated Souvenir to Commemorate the Twenty-Fifth Anniversary of the Opening of the Museum and Art Gallery, Stranmillis, Belfast, in the Summer of 1929* (Belfast: Belfast Municipal Museum and Art Gallery, 1954), 11.

69   *Blackwood's Edinburgh Magazine* 27 (1830): 549; Glover, 'In the wake of Captain Cook', 46–7, 48, 56.

70   'Natural History and Philosophical Society', *Belfast News-Letter*, 26 January, 1844, 4; Deane, ed., *Centenary Volume*, 164–5; *Illustrated Souvenir*, 12.

71   'Natural History and Philosophical Society', *Belfast News-Letter*, 26 January, 1844, 4. Sarah Millar, Chapter 5, this volume.

a duplicate collection, which had been gathered independently. But whatever the case, the BNHPS recognised the collection as a valuable one, made 'under circumstances so honourable to British skill and enterprize' and donated 'in the most liberal manner'.[72]

No less valuable were the numerous donations of specimens and artefacts made by Emerson Tennent, who served as colonial secretary in Ceylon between 1845 and 1850.[73] Although not a founder of the BNHPS, Emerson Tennent was numbered among its earliest members and was well aware of the society's collecting policy.[74] As a young man he donated geological and ethnographic specimens gathered in the Mediterranean and later in life, when his career took him to Ceylon, he continued to collect on the society's behalf.[75] While travelling out to the colony in 1845, he toured the Nile Valley and acquired two mummies from Thebes and a fragment of statue from the Grand Temple of Karnak – items he later donated to the BNHPS.[76] More significantly, upon arrival in Ceylon he set about collecting with gusto, gathering, among much else, shells, insects and 'some splendid pieces of arms, in the manufacture of which the Cingalese used to excel'.[77] Much of the material he gathered was destined for Tempo Manor, his country residence in County Fermanagh, but as the annual reports of the BNHPS record, he also donated an extensive collection of material to the society that had first nurtured his interest in the natural world.[78] This included 'an extensive and most attractive' collection of entomological specimens, 'a large and beautiful collection of shells', 'specimens of the Ant Lion' and an array of ethnographic objects, including an ornate, if fearsome, Kolam dancers mask, depicting the snake demon Naga Rasa.[79]

---

72   'Natural History and Philosophical Society', *Belfast News-Letter*, 26 January, 1844, 4.

73   This paragraph includes material first presented in Wright, '"The Belfast Chameleon"', 211 and 216.

74   BNHPS Minute Book, 1821–1830, 19 July 1821.

75   BNHPS Minute Book, 1821–1830, 7 December 1825; BNHPS Donation Book, 1821–1844, donation 249.

76   *Special Meeting of the Natural History and Philosophical Society of Belfast, Held in the Music-Hall, on Wednesday, the 23d of October, 1850, Relative to two Mummies Transmitted from Thebes, by Sir James Emerson Tennent, and Unrolled in the Museum, on the 17 and 18 of the Above Month* (Belfast: Finlay, 1850), 26–8. Deane, ed., *Centenary Volume*, 18–19.

77   James Emerson Tennent to Robert James Tennent, 11 January 1847, D/1748/ G/661/212A–B, PRONI.

78   *The Langham Family Collections from Cottesbrooke Hall and Tempo Manor: Monday, 27th September, 2004 at 10.30am* (Blackrock, Co. Dublin: HOK Fine Art, 2004), 90–104.

79   *Report of the Council of the Belfast Natural History Society, for the Session Ending May, 1849* (Belfast, 1849), 3; *Report of the Council of the Belfast Natural History & Philosophical Society, for the Session Ending June, 1853* (Belfast, 1853), 4; *Report of the Council of the Belfast Natural History and Philosophical Society, for the Session Ending*

The donations by figures such as Thomson, Crozier and Emerson Tennent testify to the ongoing success of the BNHPS's collecting policy. But this success was not without its drawbacks and there are clear signs that, as the nineteenth-century wore on, the society's members began to worry about the composition of their collections and to reconsider their enthusiastic embrace of global collecting. Indeed, in its annual report for 1868, the BNHPS's council called, in a passage that bears quoting at length, for a reorientation of the collecting policy:

> For a long term of years ... the Museum has served to draw together to one centre series of objects collected at home and abroad by scientific friends connected by family ties or other associations with Belfast and the North of Ireland ... Such general, though necessarily incomplete collections of natural history objects, and specimens of art from foreign countries, have been the means, in your Museum, of conveying pleasure and information to thousands, who otherwise would never have had an opportunity of seeing them. Your Council feel deeply indebted to the kindness of those who have presented such donations, and they continue to welcome similar gifts ... But while they by no means wish to under-estimate the popular element of attraction thus created, they entertain the hope that they may be able, through future modifications, to utilise these specimens further by some more systematic arrangement. Primarily, however, they desire to complete within your walls collections illustrative of Irish Fauna and Flora, and of Irish antiquities. In other words, they feel convinced that the principal aim of a local Museum should be to gather collections illustrative of the natural history, mineral resources, and ethnological relics of the neighbourhood and country in which it exists.[80]

There are a number of points worth highlighting here. Firstly, and most obviously, there is the call to complete local collections and the assertion that the local museum should build collections reflecting 'the neighbourhood and country in which it exists'. Reinforcing Diarmid A. Finnegan's observations regarding the falling status of overseas donation in the 'reputational economy of civic society in late-Victorian Belfast', this suggests that the BNHPS was, by the late 1860s, turning away from its earlier curatorial policy.[81] If not rejecting the global entirely, it was clearly moving towards the more focused, regional approach common in the museums established by natural history societies elsewhere in Britain at this time.

Curatorial policy aside, the council's statement, with its reference to the museum's collections 'conveying pleasure and information to thousands' and its articulation of a desire to 'utilise these specimens further by some more systematic

---

June, 1854 (Belfast, 1854), 3; *Report of the Council of the Belfast Natural History and Philosophical Society, for the Session Ending June, 1857* (Belfast, 1857), 4; Deane, ed., *Centenary Volume*, 46, 167; Glover, 'The Folks Back Home', 28.

    80   *Annual Report by the Council of the Natural History and Philosophical Society. Session, 1867–8* (Belfast, 1868), 3–4.

    81   See Diarmid A. Finnegan, Chapter 3, this volume.

arrangement', serves also to raise a series of interlinked questions concerning who visited the museum, what precisely they encountered and what they were meant to learn from their visit. Beginning, firstly with the question of visitors, it appears that, in its early days, access to the Museum was primarily reserved for the solidly middle-class membership of the BNHPS, with the wider public admitted only on certain specified days.[82] However, in April 1837 a decision was taken to open the Museum for six days a week and to charge admission at 'three pence for mechanics and children, and six pence for other persons'.[83] Implicit in this decision was a recognition that a wider public should be admitted to the Museum, and a further step towards this aim was taken in 1845 with the introduction of a scheme whereby members of Belfast's working classes were admitted to the museum for a 'nominal charge' on Easter Monday. This Easter Monday scheme was one which proved to be hugely successful. In 1845, the year of its introduction, somewhere in the region of 1,000 non-members availed of the opportunity to visit the museum. This figure rose to an average of nearly 5,000 during the period 1849–53 and 'the influx of similar large numbers for many years' led the BNHPS's council to conclude, in 1869, that there was a 'marked relish amongst the artisan classes for objects connected with Natural History and Ethnology'.[84]

It thus appears to have been true that the Belfast Museum conveyed 'pleasure and information to thousands', but what was the nature of this information? How were the BNHPS's collections disposed within the museum, and what lessons were they were intended to generate? From an archival point of view, these are difficult question to answer: the BNHPS's records reveal much about the artefacts received, but have relatively little to say about the ways in which they were displayed. The evidence relating to the organisation of the museum's holdings that does survive, however, is of interest insofar as it suggests that a sharp distinction was drawn between the domestic and the foreign in the museum's displays. A visitors' guide circulated in the 1880s, for instance, records that 'British Birds' were displayed in the Thomson Room, while the 'valuable collections of the more attractive foreign birds' could be viewed in the 'Lecture Room'. Likewise, while 'Irish Antiquities' were located in the Benn Room, what might be described as overseas antiquities were housed in a separate 'Ethnology Room'.[85] Included in

---

82   The following section includes material first presented in Wright, '"The Belfast Chameleon"', 218. See also, for a discussion of public access to the museum, Bayles, 'Science in its Local Context', 183–5.

83   Deane, ed., *Centenary Volume*, 9.

84   *Report of the Council of the Belfast Natural History Society, for the Session Ending May, 1849* (Belfast, 1849), 2–3; *Report of the Council of the Belfast Natural History & Philosophical Society, for the Session Ending June, 1853* (Belfast, 1853), 5; *Annual Report by the Council of the Natural History & Philosophical Society. Session, 1868–9* (Belfast, 1869), 4.

85   [BNHPS], *Conversazione in Connection with Visit of Sir William Thomson, LL.D., F.R.S., &c., to Belfast: Museum, College Square North, 17th April, 1889. Programme.* (Belfast, 1889), 1, 2, 3, 6.

this Ethnology Room were Egyptian mummies, 'idols worshipped by various races' and 'two models of New Zealand war canoes', and its utility as a source of global knowledge was foregrounded by the museum's organisers.[86] While the visitors guide noted that its exhibits were 'illustrative of the customs and habits of various races', Samuel Alexander Stewart, the museum's curator, made the point more bluntly in his *Notes on Some of the More Interesting Objects in the Belfast Museum* (1893): the Ethnology Room served, in his words, to 'throw light on the habits and mode of life of many savage and semi-civilized races'.[87] By the 1890s, the local and the foreign were being starkly differentiated and a form of global knowledge was being promoted that cast the foreign as an 'other'. Indeed, the casual use of derogatory language – 'savage', 'semi civilized' – suggests that visitors viewing the museum's ethnographic artefacts were presented with exoticised displays which stressed the difference and foreignness, the essential 'otherness', of non-western societies.

However, while it appears to have been particularly pronounced by the 1880s and 1890s, and while it is certainly true that the focus of the BNHPS's curatorial policy had, by this period, swung from the 'global' to the 'local', it does not follow that the domestic/foreign distinction was unique to the later-nineteenth century. Quite the reverse, there are signs that similar conceptualisations were at play as early as the 1830s. Of particular relevance here is the report delivered by the BNHPS's council for the 1832–33 session, which detailed the museum's entomological holdings and collections of birds and shells, all of which were divided into distinct local and foreign categories. The entomological collections comprised 'eleven cases of foreign insects', including specimens received from Brazil, North America, the West Indies and China, alongside 'a native collection, principally made by a few of the members, and, for the most part, in our own immediate neighbourhood', which filled thirteen cases. Similarly, the society's collection of birds included 'native species' and 'foreign birds', the latter being 'much more numerous' than the former, and its shells were divided into a 'general collection', in which could be found shells gathered from South America, the Mediterranean and the Great Lakes of North America, and a 'distinct collection, exclusively devoted to native specimens'.[88] Clearly, then, the imposition of native/ foreign distinctions was not particular to the 1880s and '90s. However, in the BNHPS's early years this distinction appears to have been employed in a slightly different register, and to have been tempered somewhat by a belief that the society's collections, while encompassing the native and the foreign, nevertheless reflected the breadth and grandeur of a unified creation, rather than self/other dichotomies.

---

86   [BNHPS], *Conversazione in Connection with Visit of Sir William Thomson*, 2–3.

87   [BNHPS], *Conversazione in Connection with Visit of Sir William Thomson*, 2; Samuel Alexander Stewart, *Notes on Some of the More Interesting Objects in the Belfast Museum* (Belfast: Alexander Mayne and Boyd, 1893), 5.

88   'Belfast Museum', *Belfast News-Letter*, 7 June, 1833, 4.

Influenced by traditional natural theology, these views were articulated clearly in the Presidential address delivered by Rev. Thomas Dix Hincks in May 1826. Discussing the society's collecting policy, Hincks sought to highlight the advantages that flowed from co-operation with like-minded naturalists based elsewhere in the world. 'When the inhabitants of remote countries are thus engaged in a common pursuit', he observed, 'the investigation of the great works of the creation by which they are surrounded the true end of knowledge is attained ... Men forget the distinctions of climate and of language, they view each other as members of one vast family, and join in adoration excited by the contemplation of those words "By boundless love, and perfect wisdom formed"'.[89] Particularly striking here is Hincks' characterisation of humanity as 'one vast family' – a characterisation which cut against blunt native/foreign and self/other dichotomies, reflecting a certain cosmopolitanism of outlook. Equally striking, however, is Hincks' assertion that the 'investigation of the great works of the creation' constituted the 'true end of knowledge'. Such arguments reveal the influence of Paleyesque natural theology, an influence Hincks was by no means alone in reflecting. Addressing the BNHPS in 1834, Robert Patterson, a founding member of the society and the treasurer of the museum building fund, conceptualised the museum as 'a temple, where the humble student of nature may enter in, and behold, in all its compartments, those evidences of Almighty wisdom, with which every part of creation is replete'.[90] Likewise, in his *Letters to a Young Naturalist* (1831), William Hamilton Drummond made a number of references to William Paley's work, highly recommending his *Natural Theology* (1802) and impressing upon his notional reader that the 'chief object' of his letters was 'to impress upon your mind the importance of studying the works of nature with a continual reference to the great and Almighty God, whose offspring they are'.[91] Native/Foreign distinctions might, then, have been drawn in the early nineteenth century, but they were combined with a worldview influenced by natural theology – a worldview which was to wane as the century progressed and scientific knowledge and practice developed, but which nevertheless served, in the early years of the BNHPS's existence, to emphasise the unity of the natural world and to moderate, if only slightly, self/other dichotomies.

Turning from the museum's natural history collections, what of its ethnographic artefacts? Whatever might be said about the capacity of bird, plant and insect specimens to represent a unified creation, it might be suspected that ethnographic artefacts, which foregrounded the difference between human societies, were

---

89 BNHPS Minute Book, 1821–1830, 24 May 1826.

90 *Minute of the Proceedings of the Belfast Natural History Society, at the Museum, on Wednesday, Fourth June, 1834* (Belfast, 1834), 14.

91 James Lawson Drummond, *Letters to a Young Naturalist on the Study of Nature and Natural Theology* (London: Longman, Rees, Orme, Brown, and Green, 1831), 107, 307, 308, 324, 325 (307 and 324 for quotes). For further discussion of the impact of natural theology on the BNHPS's first generation of members see Bayles, 'Science in its Local Context', 124.

'othered' in a more overt way. But here, too, caution is required, for surviving evidence points to a desire to display and interpret ethnographic materials in a comparative manner, which stressed similarities, rather than differences. In 1852, on the occasion of the British Association for the Advancement of Science's first visit to Belfast, the BNHPS arranged an extensive exhibition of Irish antiquities, illustrative of 'the nature and extent of our ancient civilization'. Alongside the Irish objects displayed, however, there were to be found 'stone implements from Mexico, New Zealand, and the Pacific Islands; together with the tools employed in the native manufacture of flint Arrow-heads', and the exhibition's catalogue remarked on the similarities that existed between 'the arms of the New Zealanders and other uncivilized nations' and 'those found in Ireland'.[92] Drawing comparisons between ancient Irish societies and contemporary 'foreign' cultures served, of course, to make the point that Ireland had developed into a civilised, modern society – a society which ranked above the 'uncivilized nations' to be found elsewhere in the world. Nevertheless, the willingness to present the local alongside the 'foreign' points to an awareness that fruitful comparisons could be made between different human societies and cultures. This awareness was to endure until at least 1884, in which year the BNHPS, notwithstanding the worries that had emerged regarding the composition of its collections, noted that 'it should be known throughout Belfast and Ulster generally, that they would be very glad to receive and exhibit, not only specimens of ancient Irish weapons and implements, but also those of aboriginal tribes in foreign lands'. In 'a few years', it was hoped, the society would acquire 'in addition to their already most extensive and valuable collection of native antiquities, an interesting and instructive series of objects of comparative archaeology, such as the increasing interest of this subject demands'.[93]

The presentation of foreign specimens and artefacts within the Belfast Museum was thus more complicated than might first be assumed. But to read too much into this admittedly limited evidence would be unwise, for it is also possible to detect evidence of sensationalising and exoticising impulses in the museum's presentation of its overseas artefacts. The BNHPS was, for instance, well aware of the impact that striking 'foreign' objects, be they scientific or ethnographic, could have on the imaginations of those who viewed them. Indeed, when soliciting donations from overseas contacts during the 1820s, it expressed its desire to receive 'shewy' contributions, 'such as would be imposing to the casual spectator of their cabinets', and, later in the century, it adverted to the more exotic objects held in

92 *Descriptive Catalogue of the Collection of Antiquities, and Other Objects, Illustrative of Irish History, Exhibited in the Museum, Belfast, on the Occasion of the Twenty-Second Meeting of the British Association for the Advancement of Science, September, 1852. With Appendix of Antiquarian Notes* (Belfast: Archer and Sons, 1852), 1, 50 and appendix, 1.

93 *Report and proceedings of the Belfast Natural History & Philosophical Society, for the Session 1883–84* (Belfast, 1884), 32–3.

the museum when publicising its Easter Monday opening.[94] Enlivened with ornate bordering and bold print, one advertisement, dating from 1846, informed potential visitors that: 'Numerous interesting additions, from the South Sea Islands, China, and the Antarctic Regions have been made to the collection, within the last two years; of which, from their being alive, TWO JERBOAS, obtained by Sir JAMES Emerson Tennent, at the Pyramids of Egypt, may be mentioned'.[95] Likewise, an advertisement from as late as 1907 informed potential visitors that, in addition to a recently acquired collection of Irish Birds, they could view 'many examples of Birds, various Animals and Fishes, from all parts', and 'many interesting exhibits showing the customs of the races of the world', not the least of which were 'Three large STATE UMBRELLAS of ASHANTI KINGS (one 11 feet high and 24 feet in circumference.)'[96] In such adverts, the BNHPS can be said to have presented a certain type of global knowledge: a knowledge, informed by middle class notions as to what was likely to appeal to the working classes, which foregrounded the oddness, and 'otherness', of cultures that could be encountered in the wider global world. As significant as the 'otherness' of these objects, however, was the simple fact that they were placed on display. 'By accumulating, reorganizing, and reproducing information from the remotest corners of the earth', Livingstone has observed, 'the Victorian archive played its part in shaping worldwide geopolitical relations'.[97] If only implicitly, the Belfast Museum demonstrated that the 'other' could be possessed and this is significant when thinking, lastly, about the way in which the collections were experienced.

As with that concerning the way in which they were presented, the question as to how the Belfast Museum's holdings were *experienced* is a difficult one to answer. As Claire Wintle has noted, it can never be assumed that visitors interpreted museum exhibitions in the ways in which they were intended to: 'discrepancies between intended meaning and popular understanding of museum displays occurred' and 'curatorial control of audience reception' cannot be assumed.[98] That said, two points can be made. Firstly, it seems likely that there was a class dimension at play in the manner in which the museum was experienced. The experiences of those working class visitors who attended on Easter Mondays doubtless differed markedly from those of the BNHPS's middle class members, who not only had regular access to the museum, but were familiar with the ways in which the exhibitions had been acquired and with their broader scientific significance. Secondly, and developing out of this point, it seems reasonable to

94   BNHPS Letter book, 1824–60, 90–91.

95   Noel Nesbitt, *A Museum in Belfast: A History of the Ulster Museum and its Predecessors* (Belfast: Ulster Museum, 1979), 15.

96   Nesbitt, *A Museum in Belfast*, 15.

97   Livingstone, *Putting Science in its Place*, 32. See also, MacKenzie, *Museums and Empire*, 8.

98   Claire Wintle, 'Visiting the Empire at the Provincial Museum, 1900–1950', in *Curating Empire*, ed. Longair and McAleer, 37.

conclude that for the former group – the working classes – the way in which the museums exhibitions were arranged was less significant than the fact that they *were* arranged. As Mackenzie has argued, museums were 'a central part of the process of ordering the world, familiarising and naturalising the unknown as the known, bringing the remote and unfamiliar into concordance with the zone of prior knowledge, both geographically and intellectually'.[99] Gathered from the four quarters of the globe and presented according to the dictates of western knowledge, the Belfast Museum's exhibitions, whatever they were intended to illustrate, implicitly embodied imperial power and provided an opportunity to gaze on 'others' that were ordered and controlled. How much impact this had on the working classes of Belfast – how, or if, it informed attitudes to foreign 'others' – requires further investigation, but it is nevertheless clear that the Belfast Museum conveyed an important message – a message concerning the power and global reach of the British Empire.

Conclusion

Having begun with the opening of one Belfast museum, it is appropriate to close with the opening of a second, the New Municipal Museum. Located on Stranmillis Avenue, the New Municipal Museum was a direct institutional descendent of the original Belfast Museum and was formally opened on 22 October 1929.[100] Later known as the Ulster Museum, it had been many years in planning and, as with that of the original Belfast Museum, its opening was an important civic occasion. It was, one contemporary noted, 'a notable event in the capital of the Imperial Province. It marked a phase in the cultural life of the people difficult to overestimate, and brought Belfast, at one stride, not only into line with other great cities, but placed her even ahead of the majority'.[101] Whether intentional or not, the use of language here is significant. If Belfast could be characterised as the capital of Ireland's imperial province, the original Belfast Museum, the collections of which had passed over to the New Municipal Museum, could be said to have contributed to the development of imperial consciousness in that capital. While not a formal institution of empire, the museum opened by the BNHPS in 1831 was a museum which was facilitated by empire and in which empire was placed on display. In this sense, it was a space which offered a regional metropolitan audience access to Britain's global empire.

99   MacKenzie, *Museums and Empire*, 8

100   J.A.S. Stendall, 'New Municipal Museum and Art Gallery in Belfast', *North Western Naturalist*, 4 (1929): 171; Nesbitt, *A Museum in Belfast*.

101   Stendall, 'New Municipal Museum', 171.

Chapter 8

Malthus's Globalisms: Enlightenment Geographical Imaginaries in the *Essay on the Principle of Population*

Robert J. Mayhew

Malthus's Life and the Arc of Globalisation

The three decades either side of 1800 have been identified as a pivotal moment that crystallised an emergent global political-economic system of unprecedented density and complexity and that cemented Britain's centrality to that system. For Britain, Bayly notes that from the nadir defined by the loss of the North American colonies in the American War of Independence (1776–83), the British empire revived and came to control a quarter of the world's population and a third of world trade by 1820. As such, the so-called 'second' British empire in the east came to dwarf the first British empire in the west within a generation of the latter's loss.[1] And the British empire, viewed in a larger perspective, was part of a far broader process which consolidated and connected trade and political power at the global scale. Identifying this as an age of crisis, Bayly has argued that it is in these decades that we can locate the birth of our modern, global world system: 'contemporary changes were so rapid, and interconnected with each other so profoundly, that this period could reasonably be described as "the birth of the modern world" ... Modernity, then, was not only a process, but also a *period* which began at the end of the eighteenth century and has continued up to the present day in various forms'. The shock waves of an age of crisis reverberated around the world through networks forged by what Bayly terms 'archaic globalisation', but their outcome was to build a new, more highly interconnected and interdependent global system, the system of modern globalisation.[2]

1   C.A. Bayly, *Imperial Meridian: The British Empire and the World, 1780–1830* (London: Longmans, 1989), 2–5. For the historiographical construction of the notion of a 'first' and 'second' British empire see the chapters by Marshall and Bayly in Robin W. Winks, ed., *The Oxford History of the British Empire: Volume V: Historiography* (Oxford: Oxford University Press, 1999), 43–72.

2   C.A. Bayly, *The Birth of the Modern World, 1780–1914* (Oxford: Blackwell, 2004), 11, emphasis in original. See 88–9 for crisis and 41–7 for globalisation. For a similar

For all that the history and lineage of globalisation drawn by Bayly might be contested by rival versions, he makes a strong case for seeing an era of five or six decades centred around 1800 as an epochal or crisis moment, a step change in the long term history of global interconnection in trade, politics and ideas. It is interesting in this context to look at the life and ideas of Thomas Robert Malthus with a view to his engagement with this moment of step change, because his life spanned thirty-four years either side of 1800, being born as he was in 1766 and dying in 1834.[3] How did Malthus's life and work reflect and respond to the broader currents in socio-political history that Bayly has identified? While the bulk of this chapter will attend to Malthus's engagement with globalisation as framed in the several editions of his great work, *An Essay on the Principle of Population*, from its first edition of 1798 through to the last edition published in his lifetime in 1826, it is worth beginning by noting the extent to which his life and work more broadly conceived were framed by the globalising moment of modernity's birth that Bayly has depicted. Thus, for example, one of the first textual traces we have of the young Malthus is a teenage essay written around 1783, to all appearances a schoolboy's exercise, discussing the utility of imperial possessions to a nation. The essay opens by placing its remarks in the context of 'the very recent loss of our settlements in America', hence the putative dating of its contents to 1783, and this shows that Malthus's advanced schooling was already preoccupied with the collapse of the 'first' British empire. Unsurprisingly given its author's age, the essay rehearses commonplaces about colonies allowing commerce to flourish but it also attends to the debate, at least as old as Hakluyt, about whether British colonies depopulate the home nation or simply draw off excess workers, this being a prefiguration of Malthus's lifelong interest in matters demographic. The essay concludes in a tone that is supportive of the American position, arguing that it must be 'mismanagement' that led to American discontent and that independence should have been granted freely rather than 'entering into a long & painful war'.[4] If Malthus is often caricatured as a lackey of the establishment in good part thanks to Marx and the Romantics, this youthful essay reminds us that in his own era he is more accurately understood as a Whig of a reasonably radical stripe, perhaps of a Rockinghamite persuasion, as was Edmund Burke in his response to American independence, although in Malthus's case the inspiration for this position was probably the radical rationalism of his teacher, Gilbert Wakefield.[5] With Burke

thesis see Jürgen Osterhammel, *The Transformation of the World: A Global History of the Nineteenth Century* (Princeton: Princeton University Press, 2014).

3   For Malthus's life, see Patricia James, *Population Malthus: His Life and Times* (London: Routledge, 1979).

4   This essay is printed in J.M. Pullen and Trevor Hughes Parry, eds, *T.R. Malthus: The Unpublished Papers in the Collection of Kanto Gakuen University* (2 volumes, Cambridge: Cambridge University Press, 1997–2004), vol. 2, pp. 275–7.

5   See Conor Cruise O'Brien, *The Great Melody: A Thematic Biography of Edmund Burke* (London: Sinclair Stevenson, 1992). For Malthus's politics at this time see Donald

(albeit for different reasons), Malthus could consistently support American independence but revile the French Revolution a decade later.

Where Burke died in 1797, the year before Malthus's debut on the public stage as a critic of the French Revolution in his *Essay*, Malthus would live to see the full emergence of the new imperial system in the East. More than that, his livelihood was in fact dependent on that system because, from 1805 until his death, Malthus's main income came from his appointment as 'Professor of General History, Politics, Commerce and Finance' at the East India College in Haileybury.[6] The college was designed to train up teenagers with the requisite skills to make them colonial administrators in India and was thus part of the 'professionalization' of imperial governance that distinguished the second British empire from its North American predecessor. It is clear that many of the administrators trained at Haileybury viewed demographic and economic questions on the Indian subcontinent through a broadly Malthusian lens as S. Ambirajan and Mike Davis have shown for the later nineteenth century.[7] But we should be wary of ascribing this directly to Malthus's role as opposed to a more diffuse culture of Malthusianism in nineteenth-century British intellectual life, for the simple reason that the surviving evidence about Malthus's teaching at Haileybury suggests he did not proselytise his students with his own arguments.[8] On the contrary, the preserved notes we have about Malthus's work teaching history show him offering a catechism about politics and kingship from the fall of the Roman Empire to the age of Charlemagne that could have been of little direct relevance to colonial governance.[9] More important still, the only evidence we have about Malthus's teaching of economics suggests he refrained from using his own writings at all, preferring to construct his classes as discussions of the ideas embodied in Adam Smith's *Wealth of Nations* (1776). If the first view we have of Malthus as an author is his schoolboy essay on the colonies, one of the last traces preserved in the so-called 'Inverarity Manuscript' is a set of over five hundred questions about political economy that Malthus worked through with his students in 1830, all of them keyed around passages from Smith's celebrated treatise not his own work either in the *Essay* or his *Principles of Political Economy*

Winch, *Riches and Poverty: An Intellectual History of Political Economy in Britain, 1750–1834* (Cambridge: Cambridge University Press, 1996), 253–61; and Nobuhiko Nakazawa, 'Malthus's Political Views in 1798: A 'Foxite' Whig?', *History of Economics Review* 56 (2012): 14–28.

6   James, *Population Malthus*, 173.

7   S. Ambirajan, 'Malthusian Population Theory and Indian Famine Policy in the Nineteenth Century', *Population Studies* 30 (1976): 5–14, and Mike Davis, *Late Victorian Holocausts: El Niño Famines and the Making of the Third World* (London: Verso, 2002).

8   On the culture of Malthusianism in nineteenth-century Britain, see Piers Hale, *Political Descent: Malthus, Mutualism and the Politics of Evolution in Victorian England* (Chicago: University of Chicago Press, 2014) and Robert J. Mayhew, *Malthus: The Life and Legacies of an Untimely Prophet* (Cambridge, MA: Harvard University Press, 2014), 128–55.

9   Pullen and Parry, eds, *Malthus*, vol. 2, 166–211.

(1820).[10] Malthus, then, spent the second half of his life training a generation of imperial administrators who would govern a massive global empire, but however Malthusian their attitudes were to prove, that was not because of any indoctrination by Malthus himself.

Malthus's life as a writer, then, was framed by the collapse of the British Empire in North America and the emergence of new modes of professional governance in the Indian empire. Furthermore, his livelihood was secured not by the notoriety of his *Essay* but rather by the employment it led to through the East India Company. And yet this is merely to show, rather like Molière's Monsieur Jourdain who is surprised that he speaks prose, that Malthus lived and gained employment in Bayly's age of global modernity. How did Malthus actually respond to and interact with the globalising society of which he was a part? To track the answer to this question we need to attend to Malthus's writing rather than his biography. In particular, we need to look at the first (1798) and second (1803) editions of the *Essay*, for it is here that Malthus's global knowledge and what can be termed his 'global imaginary', his implicit understanding of the structure and subdivisions of the world, can be uncovered.

Imagining the Globe in the *Essay* of 1798

The critical reception of Malthus's *Essay* has not attended in detail to the global information on which it drew and the geographical imaginary it built on the basis of that information.[11] In good part this neglect may spring from two interrelated and undeniable truths. First, the *Essay* of 1798 was intentionally framed as an anti-utopian tract designed to cool revolutionary fervour at the moment of its peak in Britain and as such its role in Franco-British, rather than global, relations has rightly been to the fore. Second, the conceptual rhetoric of the *Essay* is overwhelmingly cast in quasi-mathematical terms, with references to axioms and propositions, to ratios and resultant laws of social behaviour assailing the reader in the opening two chapters. Malthus, as is well known, argued that food production could only grow in an arithmetic ratio, where population, uncontrolled, would grow geometrically, the imbalance between these ratios ensuring that population

---

10   J.M. Pullen, 'Notes from Malthus: The Inverarity Manuscript', *History of Political Economy* 13 (1981): 794–811.

11   Although see Alison Bashford, 'Malthus and Colonial History', *Journal of Australian Studies* 36 (2012): 99–110 and Alison Bashford, *Global Population: History, Geopolitics, and Life on Earth* (New York: Columbia University Press, 2014), 29ff. The publication of Alison Bashford and Joyce Chaplin, *The New Worlds of Thomas Robert Malthus* (Princeton: Princeton University Press, forthcoming) will further contribute to these inquiries. Bashford and Chaplin's focus is on Malthus as a global thinker entwined with imperial, settler and indigenous knowledges; my interest in this essay is with Malthus's geographical imaginary, not its specific grounding in imperial data.

could only be kept in balance with resources by means of either the 'positive' checks of famine, war and plague or the 'preventive' checks of self-restraint, late marriage and 'vices' such as prostitution and infanticide. Malthus's transparent indebtedness to a Newtonian approach to understanding the social world lends his work a universalising rhetoric akin to David Hume's in *A Treatise of Human Nature* (1739–40) (whose subtitle was 'an attempt to introduce the experimental method of reasoning into moral subjects'), which seems at some remove from the historical and geographical specificities of globalisation. As Malthus put it '[it is] to the constancy of the laws of nature ... we owe all the greatest, and noblest efforts of intellect. To this constancy, we owe the immortal mind of a Newton'.[12] And yet as Simon Schaffer has shown, the image of Isaac Newton as the lonely seer disconnected from the world is a myth; the real Newton relied on information from around the globe, acting as a centre of calculation for data about tidal ranges, sightings of comets and other such phenomena flowing in from the networks of Bayly's archaic globalisation via imperial settlers and long distance traders.[13] Put differently, Newton's universal laws were evidenced on the basis of global empirical data collection. The same point can be made about Malthus's social laws: no sooner had Malthus proposed the existence of Newtonian ratios and checks in the operation of demography in the first two chapters of the *Essay* than he felt obliged to confirm their reality by empirical evidence. And it is here that Malthus's *Essay* of 1798 was drawn into the ambit of global information.

It is in chapters 3 to 7 of the *Essay* that Malthus undertakes a global 'tour' to evidence the ubiquity of his principle of population, that is, the tendency of population to outstrip food supply and the checks that prevent this. This amounts to a considerable portion of the *Essay* as a whole; in a brief tract of circa 55,000 words, these five chapters amount to roughly a quarter of the total text.[14] In truth, the majority of the enquiry is in chapters 3 and 4, and it is on these chapters which form a complementary pair that we will concentrate. Chapter 3 addresses, as its thumbnail outline has it, 'the savage or hunter state' and 'the shepherd state'. These phrases immediately flag the fact that Malthus organises his understanding of the history and geography of the globe around the stadial theory of the Scottish Enlightenment literati such as Adam Ferguson and David Hume wherein all societies pass through a set of four stages, moving from hunter-gathering to pastoral (shepherd) nomadism, thence to settled agriculture

---

12   The first, 1798, edition of *An Essay on the Principle of Population* is cited from E.A. Wrigley and David Souden, eds, *The Works of Thomas Robert Malthus* (8 volumes: London: Pickering, 1986), vol. 1, 126–7. This edition is referenced as 1798 and by page number hereafter in the text.

13   Simon Schaffer, 'Newton on the Beach: The Information Order of the *Principia Mathematica*', *History of Science* 47 (2009): 243–76.

14   By a rough and ready count, I estimate these chapters to contain around 14,000 words.

and finally to the commercial societies anatomised by Adam Smith.[15] Malthus takes the framework of stadial history for granted and wants to demonstrate that geographical and historical data suggest that at all these social stages, for all their variety, the principle of population manifests itself albeit in shifting forms. In other words, Malthus immediately conjoins Newtonianism with a universalising Enlightenment theory of social development, viewing the latter through the lens of its manifestation in distinct demographic stages. Starting in what he sees as the most 'primitive' social stage, hunter-gathering, Malthus refers to North American Indians and to the peoples of the Cape in Southern Africa. In both cases, he sees evidence that population can grow rapidly where no checks exist or fresh resources are discovered, but Malthus was more impressed by, and discussed at greater length, the checks to population in such regimes, these being manifested most notably in the high rate of infant mortality which arises because women cannot tend to infants because of their 'constant and unremitting drudgery of preparing everything for the reception of their tyrannic lords' (1798, 19). As such, Malthus detects in hunter-gatherer societies a different class dynamic from that in his own age, with warriors equating to modern gentlemen and 'women, children, and aged, [equating] with the lower classes of the community in civilized states' (1798, 19).[16] Malthus, keen to emphasise that his principle of population is not just a theoretical construct, then, concludes his brief comments on the earliest stages of social development by saying that 'actual observation and experience' show that 'misery is the check that represses the superior power of population' in such hunting societies (ibid.).

Malthus then moves on to what he calls 'the next state of mankind' (1798, 20), that of shepherds or nomadic pastoralists. He initially refers to the collapse of the Roman Empire, arguing for a single cause of the narrative Edward Gibbon had written and of which Malthus had been an avid reader, seeing in 'the want of subsistence ... the goad that drove the Scythian shepherds from their native haunts' such that 'as they rolled on, the congregated bodies at length obscured the sun of Italy, and sunk the whole world in universal night' (1798, 20).[17] Moving forward in time, but unlike the paragraphs on the hunter-gatherer stage never drawing on the contemporary world, Malthus saw the same dynamic in play for the centuries after the Sack of Rome, the centuries addressed in the later volumes

---

15   The literature on stadial theory is vast: see, *inter alia*, Ronald Meek, *Social Science and the Ignoble Savage* (Cambridge: Cambridge University Press, 1976); Frederick G. Whelan, *Enlightenment Political Thought and Non-Western Societies: Sultans and Savages* (London: Routledge, 2009); Christopher Fox, Roy Porter and Robert Wokler, eds, *Inventing Human Science: Eighteenth-Century Domains* (Berkeley: University of California Press, 1995); and P.J. Marshall and Glyndwr Williams, *The Great Map of Mankind: Perceptions of New Worlds in the Age of Enlightenment* (Cambridge, MA: Harvard University Press, 1982).

16   For Malthus and gender see: Karen O'Brien, *Women and Enlightenment in Eighteenth Century Britain* (Cambridge: Cambridge University Press, 2009), 222–5.

17   Malthus read Gibbon as an undergraduate in Cambridge in 1788: see Parry and Pullen, eds, *Malthus*, vol. 1, 53.

of Gibbon's *Decline and Fall*, arguing that for 'an Alaric, an Attila or a Genghis Khan ... the true cause that set in motion the great tide of northern emigration, and that continued to propel it till it rolled at different periods, against China, Persia, Italy, and even Egypt, was a scarcity of food, a population extended beyond the means of supporting it'. (1798, 21).[18] As such, Malthus sees a different dynamic operating for shepherd societies from that active in the previous stage. If hunter-gatherers are depicted as static but as suffering a high infant mortality rate due to their internal gender dynamics, the shepherds are viewed as using emigration as an escape valve by which to release the pressure of population, but this merely shifts the form of misery that the superior power of population engenders, it does not remove it: 'independent of any vicious customs that might have prevailed amongst them [i.e. shepherd societies] with regard to women ... the commission of war is a vice, and the effect of it, misery' (1798, 22).

Chapter 4 moves on to the population dynamics of 'civilized nations' as the headnote has it (1798, 23), this category lumping together settled agriculture and commercial society for Malthus in a way it had not for most stadial theorists. The key distinction that marks a civilised nation for Malthus is the slowness of population growth: 'instead of doubling their numbers every twenty five years, they require three or four hundred years, or more' (1798, 26). The cause of this slowing rate of demographic growth is the emergence of preventive checks for the first time in human history, of 'foresight of the difficulties attending the rearing of a family' (1798, 26), and an awareness of the social implications of poverty in terms of status in a stratified society. Malthus's key exemplar of an advanced society is England, and yet even here he is at pains to point to the continued empire of positive as well as preventive checks, epidemics still holding sway periodically as is evidenced in chapter 7, while the poor in general still face an attenuated form of the problem identified in hunter-gatherer societies in that 'it has been very generally remarked by those who have attended to bills of mortality, that of the number of children who die annually, much too great a proportion belongs to those, who may be supposed unable to give their offspring proper food and attention; exposed as they are occasionally to severe distress, and confined, perhaps, to unwholesome habitations and hard labour' (1798, 29).

Malthus's global imaginary in the 1798 *Essay*, then, built out of the fourfold stadial theory of the Scottish Enlightenment, but it in fact operated with a threefold or twofold classification of the world when refracted through the lens of his principle of population. In the threefold classification, Malthus distinguished the hunter society from the shepherd society not in terms of the empire of positive checks but instead in terms of which positive checks applied most forcefully, the hunter society being driven by the binary of warriors and others leading to high infant mortality, where the shepherd society used emigration to avoid the pressure

---

18 For Gibbon's geographical imaginary and the scholarship behind it, see J.G.A. Pocock, *Barbarism and Religion: Volume 4: Barbarians, Savages and Empires* (Cambridge: Cambridge University Press, 2005).

of population and therefore experienced a positive check of continual warfare. As such, the globe was a patchwork of hunter societies, shepherd societies and modern agricultural or commercial societies, this last category being unified by the extent to which the empire of preventive checks had eroded (but not eliminated) the role of positive checks in balancing food and population. Implicit in this is the twofold classification of the world that also operated in Malthus's *Essay* in 1798 and that would become still more stark (as we shall see) in the 1803 edition: across time and space, when viewed through the lens of the principle of population, the world could be divided into societies where the positive check predominated (shepherd and hunter societies) and societies where the preventive check complemented it and thereby considerably blunted its force (settled agrarian and commercial societies).

There were two complicating cases that Malthus allowed to make the global mosaic more nuanced than this in 1798. First, colonies were discussed as a separate case in chapter 6 because, as Malthus opened, 'it has been universally remarked, that all new colonies settled in healthy countries, where there was plenty of room and food, have constantly increased with astonishing rapidity in their population' (1798, 39). As such, regardless of era from antiquity to the present and of mode of governance from Spanish tyranny and superstition to British libertarianism, colonies would experience a prolonged period before the checks that defined other societies made their aegis felt. Malthus, of course, by such an argument ignored the fact that most of the colonies he discussed had not been empty spaces before colonial occupation and that the flourishing of a colonial population often came at the expense of (more than) decimation of an indigenous population.[19] The second complicating case in the global mosaic was posed by the examples of China and India. For Malthus, 'the only true criterion of a real and permanent increase in the population' was 'the increase in the means of subsistence' (1798, 48). And yet for him China and India were examples where population had been, as he termed it, 'forced' (1798, 49), that is, over the long run these societies had grown accustomed to lower calorific intakes which allowed the population to be larger vis-à-vis the means of subsistence than was observed elsewhere. For Malthus, such regimes were characterised by the harshness of famine episodes as the populace was already weakened by its dietary norms. But the discussion of forced populations was important as it allowed for the fact that the amount of food needed for subsistence was not mathematically invariant but was at least in part a social construct that could vary over time and space.

In five short chapters of the 1798 *Essay*, then, Malthus traversed the world and built an Enlightened geographical imaginary for it on the basis of the operation of his principle of population. The binary of positive and preventive checks overlapped with that of a four stage theory of societal development he learned

19 On these questions see William Cronin, *Changes in the Land: Indians, Colonists and the Ecology of New England* (New York: Hill & Wang, 1983); and Alfred Crosby, *Ecological Imperialism: The Biological Expansion of Europe, 900–1900* (Cambridge: Cambridge University Press, 1986).

from the writings of Enlightenment scholars to create a global mosaic whose logic was rendered less stark by the acknowledgement that colonisation and societal forcing could complicate the picture. And yet in truth Malthus offered precious little evidence to support his claims at this time. The material in chapters 6 and 7 about the continued power of the positive check of epidemics in advanced civilisations drew on the best evidence available from bills of mortality in Europe and North America, with the work of Richard Price, Johann Süssmilch and Gregory King, amongst others, being cited. And yet in the main Malthus's previously-cited claims of recourse to 'actual observation and experience' were pure bluster. In chapter 3 about hunting and savage societies where Malthus made that claim to rely on observation and experience, not a single traveller, geographer or historian was cited. While the language of Gibbon permeates the chapter and we know Malthus had read Gibbon, for example, he is not directly referenced. Nor can we know who Malthus is drawing on to depict North American Indians or Hottentots at the Cape. Indeed, the only contemporary traveller directly mentioned in the text of all of these chapters combined is Antonia de Ulloa in chapter 6's discussion of the population of Lima (1798, 39).[20] Likewise for the more advanced stages of civilisation depicted in chapter 4, the only explicit references offered by Malthus are to David Hume's essay, 'Of the Populousness of Ancient Nations' (1757) and to Smith's *Wealth of Nations*. In 1798, then, Malthus's claim to empirically substantiate his argument about the universal sway of a principle of population was purely rhetorical, far more so in fact than Newton's claim to substantiate his arguments in the *Principia*. Malthus had built a demographic global imaginary in 1798, but he would only go about evidencing it in the massively expanded *Essay* of 1803.

Evidencing the Principle of Population: The *Essay* of 1803

Five years and one day separate the dating of the 'Preface' to the 1798 *Essay* from that which Malthus penned on 8 June 1803.[21] Commentators have noted many changes between the two editions – notably its massive expansion in length, the less bleak conclusions about the power of positive checks to population and the elimination of the theological elements of the project – and thereby agreed with Malthus's prefatory remark that 'in its present shape it may be considered as a new work' from that of 1798 (1803: 1: 2). And yet it can be contended that in

---

20 Malthus owned a 1752 copy of Ulloa's *Voyage Historique de l'Amerique Meridional*: see Thomas Robert Malthus, *The Malthus Library Catalogue* (Oxford; Pergamon Press, 1983), 175–6.

21 T.R. Malthus, *An Essay on the Principle of Population*, ed. Patricia James (Cambridge: Cambridge University Press, 1989), vol. 1, 3. This is a variorum edition covering the editions of the *Essay* from 1803 to 1826. I have used the 1803 elements of this edition except for numerical information offered below about the last edition. This edition is referenced as 1803 and by page number hereafter in the text.

terms of scholarly labour the single biggest change made is embodied in the work Malthus undertook to make the later edition's global evidencing of the principle of population more empirically rich and convincing. Having built an Enlightened global demographic imaginary, Malthus now wished to demonstrate its reality more convincingly and, in the process of achieving this, also adjusted details of that imaginary.

Malthus had long since been interested in historical, geographical and travel accounts, his library being well stocked with such material.[22] Although we cannot be certain when he acquired these books, we do know that from his time in Cambridge he sought, as he put it in a 1788 letter to his father, to 'get some little knowledge of general history & geography', the same letter going on to say he was reading Gibbon's *Decline and Fall* (whose impact on his depiction of pastoral societies we have already noted) and also deploying D'Anville's maps of the ancient world and Samuel Dunn's *General Atlas* to aid his historical work.[23] And yet these activities seem to have become more intense and more focussed after the publication of the 1798 *Essay* as Malthus sought to bolster the empirical credentials and the geographical coverage of his argument. It was at this stage that Malthus started to act as a centre of demographic calculation, drawing together historical, geographical and statistical data from around the globe. Indeed, the first textual evidence we have of Malthus after 1798, a letter of February 1799, shows this process commencing as Malthus asks his father, then residing in London, to seek out a long list of books, these including Pierre Wargentin's 1772 analysis of Swedish demography and parallel studies by Murat in Switzerland (1766) and by Kersseboom in Holland (1742).[24] And it is in the same context that a tour Malthus took to Scandinavia and Russia in 1799–1800 can be understood. Obviously this trip had a recreational element, but Malthus diligently collected demographic and political-economic information on an area that was still very little frequented by travellers.[25] He then used this information to offer first-hand demographic information in chapters I to III of Book 2 of the 1803 *Essay* (1803: 1: 148–80), the longest discussion focussing on compulsory military service and its operation as an effective check to population growth given its impact on average male age at marriage.[26] The travel diaries Malthus kept in Scandinavia form a pair with the books he requested from his father, both showing him keen to acquire detailed empirical data to add geographical precision to the rather impressionistic survey he had offered in 1798. The diaries, then, reflect a concern Malthus had

---

22 For evidence of which see Malthus, *Library Catalogue*.

23 Pullen and Parry, eds, *Malthus*, vol. 1, 53–4.

24 Pullen and Parry, eds, *Malthus*, vol. 1, 63–5.

25 Brian Dolan, *Exploring European Frontiers: British Travellers in the Age of Enlightenment* (Basingstoke: Palgrave Macmillan, 2000), pp. 27–112.

26 See Patricia James, ed., *The Travel Diaries of T.R. Malthus* (Cambridge: Cambridge University Press, 1966), 274–95 which has Malthus's diary and the text of the Scandinavian section of the *Essay* on facing pages for comparison.

expressed as early as 1785 with what really exists in nature in all its complexity;[27] this attention to nuance would be an abiding characteristic of his work from now on. It would also resurface in Malthus's periodic trips as a traveller – to France in 1802, 1820 and 1826, to Scotland in 1810 and 1826, to Ireland in 1817 and to Germany in 1825 – each of which resulted in manuscript travel notes attending to prices, population and polity, but none of which was as comprehensive as that for the Scandinavian trip.[28] On his return to England in 1800, Malthus penned a short tract on the high food prices then wracking the country, at the conclusion of which he advertised his new edition of the *Essay* and that it would be made 'more worthy of the public attention, by applying the principle [of population] directly and exclusively to the existing state of society, and endeavouring to illustrate the power and universality of its operation from the best authenticated accounts that we have of the state of other countries'.[29] In both of these elements, the *Essay* of 1803 would display an empirical globalism only hinted at in 1798.

We can begin to chart the changes Malthus wrought in his global tour of the principle of population for the 1803 edition quantitatively. The *Essay* itself approximately quadrupled in length, becoming a text of circa 225,000 words, but within this expansion the growth of the element that evidenced the principle globally was even more dramatic, expanding as it did from around 14,000 words in 1798 to approximately 125,000 words in 1803 and therefore from a quarter to around a half of the total length of the work. The evidence from 'the best authenticated accounts' would also expand in later editions such that by 1826 it amounted to around 135,000 words thanks to the addition of new material from later British and French censuses. A parallel story can be told of a step change in Malthus's use of textual evidence in the 1803 edition: the 1798 *Essay* had only used some eleven named sources in total of which eight were cited in the chapters which toured the earth. This expanded by an order of magnitude to 125 sources in 1803 (and reached 162 by the 1826 edition), of which 105 were cited in the global tour that Malthus undertook in books 1 and 2.[30]

Moving from the quantitative to the textual, Malthus organised his tour as the first two books of the 1803 *Essay*, the first book addressing what its heading called 'the checks to population in the less civilised Parts of the World, and in Past Times', whilst book 2 addressed checks 'in the different States of Modern Europe' (1803: 1: v–vi). We can delve into Malthus's global tour by looking at specific chapters from 1803 as they mirror the efforts of 1798. Just as chapter 3

---

27 Pullen and Parry, eds, *Malthus*, vol. 1, 41.

28 For the other travel diaries see James, ed., *The Travel Diaries*, 226–72 and Pullen and Parry, eds, *Malthus*, vol. 2, 25–55, and vol. 2, 212–41.

29 Malthus, 'An Investigation of the Present High Prices', *Works of Malthus*, ed. Wrigley and Souden, vol. 7, 18.

30 Total citation of sources is taken from Wrigley and Souden, eds, *Works of Malthus*, vol. 1, 153–9. Figures for sources cited in the global tour sections of 1798 and 1803 are my own counts.

in 1798 had started with hunting societies as the lowest stage of civilisation, so Book 1 Chapter III in 1803 started at the same point. But where in 1798 Malthus offered no citation to support his analysis, in 1803 he deployed a bevy of the Enlightened travellers of his age, starting with James Cook's depiction of the inhabitants of Tierra Del Fuego, Australia and New Zealand, before moving on to George Vancouver's comments on the natives of Tasmania and also deploying material from the *Asiatic Researches*, a periodical which collected information from the British empire in the east under the auspices of Sir William Jones (1803: 1: 25–9). The overall point Malthus makes is unchanged in 1803 – 'their countenances exhibit the extreme of wretchedness, a horrid mixture of famine and ferocity; and their extenuated and diseased figures plainly indicate the want of wholesome nourishment' (1803: 1: 25) – but it is backed by the sort of extensive citation that Malthus had promised in his comments in 1800. A similar pattern is found in Malthus's discussion of shepherd/pastoral societies in the 1803 edition. Where this discussion had occupied the second part of chapter 3 in 1798 and not offered any citations to support its claims, in 1803 it covered two whole chapters – chapters VI and VII of Book 1 – and was split into a discussion of ancient societies in Northern Europe (chapter VI) and a discussion of modern societies still in that stage of social development (chapter VII). Analysing ancient Europe, Malthus showed his classical erudition in drawing on Tacitus's *Germania* and Caesar's *De Bello Gallico* for accounts of the nomadic societies encountered by the expanding Roman empire, but the bulk of his citation came from Gibbon's *Decline and Fall* and from the highly influential stadial history of European civilisation offered by William Robertson as an introduction to his *History of Charles V* (1769). For modern pastoral societies, Malthus began again by drawing on Gibbon but then relied on modern travel accounts such as Volney's trip to Syria and Egypt, Sir John Chardin's narrative of a voyage to Persia and, most recently, Karsten Niebuhr's *Travels through Arabia* (1792).

In Book 2 a similar pattern is detected for Malthus's treatment of modern European nations, with the structure of argumentation being all but unchanged from the 1798 *Essay* but the level of evidence offered being of a wholly different order. If we take the case of checks to population in England, for example, Book 2 chapter X starts with six paragraphs taken almost verbatim from chapter 4 of the 1798 edition, a chapter that had only cited Smith and Hume, but then expands on this discussion by drawing on the returns from the first British census of 1801 as excerpted by John Rickman. Malthus also utilises the pioneering demographic work on English bills of mortality by Thomas Short and Richard Price. And as the chapter continues Malthus moves into a comparative analysis of mortality rates which benchmarks English data against the work of Süssmilch in the German states, of Jacques Necker in pre-revolutionary France, and of Wargentin (whose work we have already seen Malthus requesting his father secure a copy of in 1799) on Sweden (1803: 1: 25–66). The type of evidence Malthus used in his discussion of England points to his normal approach to analysing modern nations in Book 2 of the *Essay*. Throughout, Malthus is drawn to the statistical digests and census data that were emerging from the scientific periodicals of Enlightenment Europe. Indeed, if one breaks down Malthus's citations throughout book 2, of some thirty-seven works quoted, of the order of twenty seven of them are statistical or census material such that Malthus seeks to ground his argument in aggregate data not impressionistic anecdotes. And this also points up an important difference in terms of the types of data Malthus used in book 2 as opposed to book 1 of his global evidencing of the principle of population. For where three quarters of Malthus's citations in book 2 were quantitative, in book 1 there were some sixty-eight works cited, of which thirty-six were voyage and travel accounts, with a further thirteen being Enlightenment stadial histories such as the works of Robertson and Gibbon, and a further eleven being works from classical antiquity. As such, the evidence base for the rest of the world and the earlier stages of society was invariably more anecdotal than for modern Europe and also made the stadial assumption of a parallelism between ancient European societies and those encountered around the globe by modern travellers.

This cleavage in the types of sources Malthus used to address the population dynamics of modern Europe on the one hand and of past and geographically remote societies on the other allows us to abstract from the details of his citations in the 1803 *Essay* and ask what the global geographical imaginary of the book was and how it triangulates with that we have found in the 1798 edition. Above all, the cleavage in Malthus's sources mirrors his adoption of a stark binary between the societies addressed in book 1 and those in book 2, the former being deemed to live under the aegis of positive checks to population while the latter are in good part defined by their escape from those checks in favour of preventive checks. Indeed, Malthus made this binary explicit in his 'General Deductions' in Book 2 chapter XIII, the chapter that concluded his global survey of the principle of population: 'in comparing the state of society which has been considered in this second book with that which formed the subject of the first, I think it appears that in modern Europe the positive checks to population prevail less, and the preventive checks more, than in past times, and in the more uncivilized parts of the world' (1803: 1: 304). As such, when compared with the 1798 edition, the 1803 *Essay* encoded a binary vision of the globe far more strongly, be this by splitting the world into nations ruled by the positive or the preventive check, by dividing it into uncivilised and civilised societies, or again by partitioning the globe into societies visited by travellers and societies where the advancement of the arts and sciences had allowed quantification to replace anecdotes. Resultantly, the four stage stadial theory is less prominent in 1803 than it was in 1798, notably due to the diminished sense of a demographic difference between hunting and pastoral societies in the later version, but also due to the vast increase in levels of information offered which submerged the structural device of a stadial theory under the sheer bulk of evidence presented.

If Malthus's geographical imaginary in the 1803 *Essay* became more starkly bifurcated than in 1798, he did retain the two complicating features noted above, that is, the role of colonies and the idea of population 'forcing' in China. And yet the enormous expansion in the scope of the global treatment in 1803 made neither feature as a prominent adjustment to the binary picture of a world of preventively- and of positively-checked societies. Indeed, the question of the population dynamics of colonies became something of a footnote in 1803, only being addressed in the final chapter of book 2 as a small part of the general deductions to be drawn from the first half of the *Essay*. As such, however much colonies were discussed in Book 1, the argument that colonies displayed their own distinctive demographic dynamics was considerably weakened after 1798, Malthus generally seeing such places as simply further instances of societies ruled by positive checks. And yet the 1803 *Essay* did add two nuances to the global picture it drew of the principle of population, both of them addressing what might be termed the 'contact point' between the zone of preventive checks in Europe and the zone of positive checks still dominating the rest of the globe. In other words, from his pattern of citation and analysis, it would appear Malthus found two geographical zones of ambivalence in his otherwise twofold schema beyond those identified in 1798. First, and drawing on the more general Enlightened ambivalence about how to understand Eastern Europe, whether Russia and Siberia were modern nations ruled by preventive checks or more akin to pastoral nations ruled by positive checks was ambiguous.[31] Despite having visited Russia himself on his 1799–1800 trip, Malthus in the 1803 *Essay* still drew on travel accounts as well as statistical returns to describe this part of Europe in Book 2 chapter III, most notably relying on William Tooke's *View of the Russian Empire* (1799), a text he had also used to depict modern pastoral nations in Book 1 chapter VII. As such, where exactly one drew the boundary between Malthus's two demographic regimes which otherwise divided up the globe so neatly on his analysis was unclear on Europe's eastern periphery. The second and similar ambivalence can be found in Malthus's treatment of Southern Europe;[32] the 1803 *Essay* only treats ultramontane Europe in terms of ancient Greece and Rome (Book 1 chapters XIII and XIV). As such, Greece and Rome only appear in book 1 of the *Essay* to address the population dynamics of ancient societies, their modern demography being wholly ignored and the question of whether they are demographically 'modern' (that is, governed by the preventive check) and thereby 'European' and 'civilized' in Malthus's demographic-cum-geographic imaginary, is effectively elided. Malthus, then, addresses ancient Greece by looking at the thoughts on population to be found in Plato and Aristotle and moves on to Rome through the writings of Tacitus and the earlier eighteenth-century debate about the population of the empire between Montesquieu and Hume. For ancient Rome, Malthus's conclusion depicts it as predominantly under the aegis of positive checks: 'of the preventive check ... though its effect appears to have been very considerable in the later period of Roman History ... yet, upon the whole, its

---

31   For which see Larry Wolff, *Inventing Eastern Europe: The Map of Civilization on the Mind of the Enlightenment* (Stanford: Stanford University Press, 1994).

32   Again, there was a more general ambivalence about the depiction of Southern Europe, for which see Dolan, *Exploring European Frontiers*, 113–52.

operation seems to have been inferior to the positive checks' (1803: 1: 147). But at least Greece and Italy are addressed albeit only in antiquity; Iberia is wholly neglected and therefore amounts to one of the only parts of the world Malthus makes no attempt to discuss in his global account of the principle of population, this reflecting the more general neglect of the area in the writings of the era.[33] In Malthus's 'great map of mankind', then, modern nations were strictly coterminous with what we might term 'Western Europe', the only resultant classificatory issue being where, precisely, to place that entity's southerly and eastern limits, an issue Malthus fudged for Eastern Europe by treating it as both pastoral/ancient and as modern and ignored for Southern Europe by only dealing with the area in classical antiquity and completely neglecting Iberia.

Malthus and the Limits of Enlightened Globalism

At one level, the five years separating the first edition of the *Essay* from the 1803 incarnation that would be updated for the following quarter of a century witnessed a massive expansion of Malthus's engagement with global data. It saw the integration of Malthus's principle of population into the imperial meridian whose rise Bayly has chronicled. And yet Malthus's *Essay*, in its later versions quite as much as in its first edition, shows as powerfully the limits and evasions of Enlightenment globalism as it does its reach and scope. Above all, for all the addition of information and citation offered in 1803, Malthus's *Essay* remains firmly wedded to a highly schematic view of the societal development of civilisations across the world built out of the varieties of Scottish and English stadial theory that he discovered in Hume, Robertson, Smith and Gibbon. If in 1798 Malthus spliced settled agriculture with commercial societies to make a three stage model and this in 1803 was effectively eclipsed by a two stage model comprised of an advanced-preventive social-demographic type and of a developing-positive type, in each of these variants Malthus remained wedded to a notion of universal stages where societies spread across geographical space could be used as proxies for the dynamics of past eras in more advanced societies. And for all the rhetoric, both in terms of textual argumentation and scholarly apparatus, Malthus did not seek to test stadial theories but rather to exemplify and confirm their worth as tools for understanding demographic development by selective use of the evidence of travel accounts and general histories as well as statistical information and the digests of newly emergent censuses. In this, Malthus followed a pattern manifested more commonly amongst Enlightenment writings from Montesquieu's *Persian Letters* (1721) to Gibbon's *Decline and Fall* (1776–89) where recourse to global information was used to confirm not to interrogate an argument about

---

33 Charles Batten, *Pleasurable Instruction: Form and Convention in Eighteenth Century Travel Literature* (Berkeley: University of California Press, 1978) notes the lack of British travellers to Iberia.

social change. Malthus's great exemplar as a political economist, Adam Smith, for example, chose to ignore the evidence that 'primitive' economies were predicated on an amalgam of hunting and gathering with a strong gender division between these activities, viewing them only in terms of hunting in direct contradiction to the sources he was using.[34] For Malthus, selective citation from a vast array of travel accounts and statistical digests was deployed to confirm a pre-established binary geography of the aegis of preventive and positive checks. There was, of course, a profound (North-Western) Eurocentrism encoded in this geographical imaginary, this even as it drew upon a vast range of data being generated by indigenous peoples across the globe. And as with Adam Smith, things which did not fit this binary imaginary were elided in Malthus's argument, most notably how to frame Eastern and Southern Europe as the contact zones between the two demographic-cum-social regimes.

Malthus's global vision of 1803 (or indeed of the final edition of 1826), then, was another world and yet the same as that of 1798; for all the accession of evidence and for all the expansion of the global account of the operation of the principle of population, the limitations of the empirical spirit of Enlightenment globalism were just as plainly in evidence. Indeed, it might even be argued that the 1803 edition and its later incarnations operated within an even more rigid binary than its 1798 predecessor. Evidence, geographical, historical and statistical, continued to be adduced to support a way of imagining the social world rather than to question it, and that imagining was squarely grounded in the Enlightenment stadialism Malthus had imbibed in his youth. In this continued adherence to Enlightenment stadialism, Bayly shows Malthus was by no means alone, and that it 'was not really dethroned until the early Victorian Biblical revival'.[35] More generally, if the three decades either side of 1800 that encompass Malthus's life were, on Bayly's account, a coherent moment of globalisation, it was only after that time that intellectual argumentation more sensitive to geographical difference emerged in English intellectual life in the form of a comparative method in history and anthropology.[36] For all that projects such as William Jones's *Asiatic Researches*, on which we have seen that Malthus drew, moved in this direction, it was only in the Victorian era, then, that the essentially universal and static models of the Enlightenment came to be questioned by more dynamic models that allowed for divergent and geographically different processes of societal evolution. In this transition Malthus has been accorded a key role, George Stocking suggesting that 'among the various discontinuities that separate nineteenth-century evolutionary

---

34   See Christian Marouby, 'Adam Smith and the Anthropology of the Enlightenment: The 'Ethnographic' Sources of Economic Progress', in *The Anthropology of the Enlightenment*, ed. Larry Wolff and Marco Cipolloni (Stanford: Stanford University Press, 2007), 85–102.

35   Bayly, *Imperial Meridian*, 152.

36   John Burrow, *Evolution and Society: A Study in Victorian Social Theory* (Cambridge: Cambridge University Press, 1966).

progressivism from its eighteenth-century precursors, none was more powerfully disjunctive than that introduced by Thomas Malthus'.[37] However much Malthus's work may have contributed to this transition towards more geographically- and anthropologically-sensitive understandings of global geographic difference, his own global vision most assuredly belongs to an Enlightened worldview from prior to that watershed.

---

37   George Stocking, *Victorian Anthropology* (New York: Free Press, 1987), 220.

PART 3
Circulation and Translation

Chapter 9

Brokering Knowledge in an Age of Mis-Recognition and Ignorance; or Displaying Haiti to the Masses

Karen N. Salt

Introduction

In 1870, *Ballou's Monthly Magazine*, a Boston area popular magazine and pictorial digest, carried an article on the last emperor of Haiti, Faustin I (1849–1859) titled 'An Emperor's Toothpick'.[1] It included information about Faustin I's life and reign, referring to him near the beginning of the piece as 'a tyrant that his people deposed and sent wandering in the world, a dark specimen among the uncrowned vagabonds'.[2] Although characterised as despotic, Faustin I is also sketched as a richly pompous ruler who managed to charm his people into tolerating his tyrannical rule. Within a few lines, though, the article shifts from describing Faustin I's policies to tracing the objects associated with his imperial empire.

The most important object, according to the article's writer, was Faustin I himself. Care was taken to link the emperor – a 'dark specimen' – to other items from his kingdom, imparting to the magazine's literate and urbane readership knowledge about the 'dark' world of Haiti. Curiously, the writer's opening rant on the strangeness and almost burlesque persona of Haiti's imperial sovereign is not followed through, in tone, within the rest of the article. This is surprising given that the object identified in the title of the piece is of a transitory, throwaway nature – something decidedly unworthy of an emperor.

And perhaps that is its point. If the article followed the pattern of public discourse and knowledge about Faustin I, it should have presented him as a fool in imperial robes; a buffoon pretending to have airs and graces; or a befuddled and simian-faced ruler wearing glass baubles on his hands and surrounding himself with crass emblems of fortune and power. Instead of a toothpick, the article catalogues the material reality of the emperor's treasures, focusing explicitly on one of these, something that the writer identifies as the 'imperial dagger'.

Neither a trinket, nor a transitory object, this dagger is powerful and awe-inspiring in its grandeur. The writer is at pains to stress that it is also a dangerous

---

1   'An Emperor's Toothpick', *Ballou's Monthly Magazine* 32 (1870): 507.
2   'An Emperor's Toothpick', 507.

tool that hopefully does not have any blood on its hilt. It is identified as an object 'worthy the high position it held as protector of the august potentate of the Antilles'.[3] This dagger, in the text of the article and in the accompanying illustration, is depicted as a thing of beauty. Called 'exquisite', it is an imperial object that is fully portrayed as a rare and significant piece of 'imperial authority'.

The writer is at pains to distinguish Faustin I's imperial objects from those thought to inhabit the world of Henrí Christophe, one of the leaders of the Haitian Revolution and Haiti's last emperor before Faustin I's reign. Although befriended by British notables, Christophe is described in the article as a pretender who wore as an imperial uniform only 'a cocked hat and a pair of spurs'.[4] Faustin I's imperial dagger symbolises how much Haiti, and the Atlantic world, had changed as Haiti's brilliance and power by the 1870s correlated with the recognition accorded it, as the USA had recently recognised Haiti as a nation in 1862. The text of the article stresses that the world now recognises the possibility of 'colored rulers'.

Although the article is not completely free of language that alludes to perceptions of violence and disease often associated with Haiti, it marks a significant turning point in the knowledge of and presentation about the nation. It is an example of the writings that responded to the cultural objects and images related to Faustin I's imperial reign that, I argue, he purposefully circulated throughout the Atlantic world in order to trouble and correct pejorative ideas and ill-formed knowledge about Haiti.

This chapter takes up the ways Faustin I and his brokers, or go-betweens, positioned knowledge of and about Haiti in various uncharted cultural spaces in the 1850s. In so doing, it creates an archive of knowledge and discovery that, although representing various cases and 'display sites', suggests a concentrated effort to re-position ideas about Haiti in the Atlantic world. It does this through an examination of Haiti's participation at the 1853 Exhibition for the Industry of All Nations held in New York; writings by one of Haiti's commercial agents to challenge pejorative assumptions about the nation and its people in the Atlantic world; and the circulating images, including an album of daguerreotypes and lithographs of Haitian imperial royalty, that sought to educate various publics about the vitality of Haiti. In examining these documents, this chapter builds upon the work of Sibylle Fischer who notes, in her unravelling of Haiti's disavowal by Atlantic metropoles and their colonial Caribbean peripheries, that 'there are layers of signification in the cultural records that cannot be grasped as long as we pay attention only to events [or objects] and causality [or collecting] in the strict sense'.[5]

In what follows, I argue that disparate and fragmentary archival materials can be used to understand intangible processes, such as prestige, power and even international recognition. Although the archive contains individual and often disconnected pieces, it is my contention that they reveal, as Neil Safier stresses,

---

3   'An Emperor's Toothpick', 507.

4   'An Emperor's Toothpick', 507.

5   Sibylle Fischer, *Modernity Disavowed: Haiti and the Cultures of Slavery in the Age of Revolution* (Durham, NC: Duke University Press, 2004), 2.

a pathway 'linking individuals, objects, and impulses between sites that are often taken for granted [or ignored] in the Atlantic system of knowledge production and that frequently lie outside the purview of metropolitan institutions and imperial capitals'.[6] In considering how this knowledge about Haiti moved, this chapter extends Safier's contention that objects have a mobility that extend beyond centres and peripheries, and notes that knowledge, itself, constitutes a vital and mobile entity. By focusing on mid-nineteenth-century Haiti, this chapter responds to Safier's challenge to critics to deal with and carefully consider a range of sites for the production of knowledge and power. It also illuminates the role that cultural knowledge played in a tumultuous time in Atlantic imperial and racial history.

In turning to Haiti, this chapter offers up a site for analysis that is still relegated to the periphery of historical understandings of politics, modernity and ideas, even though it was the second republic to form in the Americas and the only nation to emerge from a slave revolution. Frequently disavowed and stereotyped, Haitian officials have resisted reductive assumptions about the nation since its founding in the early 1800s. By mid-century, its battle for recognition, power and place was well and truly underway.

As the USA struggled with the cultural upheaval brought on by the enactment of the Fugitive Slave Law of 1850 and the British Empire grappled with the fallout from the violent uprising in India in 1857, Haiti – the small Caribbean nation – was in the midst of a revolt of its own. This revolution did not draw blood or force out a monarch; although it definitely, at times, carried a desperate, violent tension. Instead of physical violence, this revolt used cultural knowledge as its sword on an international stage in the battle over Haiti's future – something that remains an issue in our present considerations of Haiti.

Although Haiti is no longer sidelined to the margins of history, newspapers and pundits tend to proclaim Haiti's 'failure' as a nation, lament its political corruption or pity its poverty.[7] Since the 2010 earthquake, Haiti has either dominated the news as a site in need for charitable giving, or as a site for capitalist expansion, a turn that David Harvey has labelled 'the new imperialism'.[8] In many ways, these new ways of talking about Haiti contain old knowledge about its purported backwardness, its violence and its absurdity, often typecasting it as a horror.[9] Haitians have countered these assumptions and declarations for more than two centuries. As J. Michael

---

6   Neil Safier, 'Global Knowledge on the Move: Itineraries, Amerindian Narratives, and Deep Histories of Science', *Isis* 101 (2010): 138.

7   A wide range of examples exists on this. Two recent ones are David Brook's *New York Times* Op.-Ed article, 'The Underlying Tragedy', January 14, 2010, and the USA television broadcast on CBN of evangelist Pat Robertson's comments after the 2010 earthquake in Haiti in which he attributes the earthquake as retribution from the devil due to a purported pact entered into by the rebels during the Haitian Revolution.

8   See David Harvey, *The New Imperialism* (Oxford: Oxford University Press, 2003).

9   For more on this, see Michel-Rolph Trouillot, *Silencing the Past: Power and the Production of History* (Boston: Beacon Press, 1995) and Fischer, *Modernity Disavowed.*

Dash contends, although Haiti has been defined and described by others for a very long time, 'the first nation to abolish slavery in the Western hemisphere as well as the second nation to achieve independence in the Americas' always engaged in crafting its own knowledge and cultural thought.[10] Instead of identifying as the Atlantic's anathema, many of Haiti's leaders, especially its last emperor, Faustin I, championed Haiti as the Queen of the Antilles and used various material objects to transmit this knowledge.

In gathering this neglected and understudied repository of cultural knowledge, this chapter makes the case for the role of material, especially fragmentary and disjointed material, in generating alternative histories and competing logics and knowledge about maligned or disavowed groups. This is a timely examination as it adds to what Diane Coole and Samantha Frost describe as a new scholarly moment that draws together 'scattered but insistent demands for more materialist modes of analysis and … new ways of thinking about matter and processes of materialization'.[11] As such, it argues for matter's vibrancy, its politicisation and its contribution to the production of knowledge.

'Object' Lessons

Faustin I may have been a power-wielding leader, but he did not begin life with those powers on display. Faustin Élie-Soulouque was born enslaved in the French-controlled colony of Saint-Domingue in 1782 to parents purportedly from Africa. Although he fought on the side of the enslaved and freeborn rebels during the Haitian Revolution, he is not known as a revolutionary fighter or rebel leader. Those accolades and legacies belong, instead, to the early leaders, such as Toussaint Louverture, Jean-Jacques Dessalines and Henrí Christophe, who would steer Saint-Domingue through twelve years of violence and intrigue before emerging victorious as the newly independent nation of Haiti in 1804.

Soulouque, as he was then known, did benefit from his revolutionary training. After independence, he transferred his fighting skills into an unremarkable, but respectable, military career. He would serve under various leaders during calm and tumultuous decades of Haiti's political history. By the time that Haitian President Jean-Pierre Boyer was ousted in 1843, Soulouque held the highest military office in the nation, commanding the presidential guard.[12] His future trajectory changed

---

10 See J. Michael Dash, 'Nineteenth-Century Haiti and the Archipelago of the Americas: Anténor Firmin's Letters from St. Thomas', *Research in African Literatures* 35 (2004): 45.

11 Diane Coole and Samantha Frost, 'Introducing the New Materialisms', in *New Materialisms: Ontology, Agency, and Politics*, ed. Diane Coole and Samantha Frost (Durham, NC: Duke University Press, 2010), 2.

12 Scattered scholarly works on Haiti tend to cover Faustin's history briefly, but two in-depth pieces provide more focused coverage, albeit fairly negative in tone: John

in 1847 after a drawn out fight for political control of the government saw him suggested as a potential compromise for the office of the president. He was not anyone's first choice, but would do, offering plotting politicos the puppet that they so desperately sought. Or so they thought.

His presidency began quietly. Soon, though, he changed, quelling dissent, violently routing out adversaries and aggressively controlling all public discourse about him and his policies. In short, he grew to crave and love the power of his position. He then went the next step and ensured that he would rule for life with absolute control. In 1849, President Faustin Élie-Soulouque became Faustin I, the last emperor of Haiti, after a questionable 'public' petition called for his ascension to the throne and the legislative bodies approved the decision.

Tyrannical, violent and often paranoid, Faustin I morphed from a puppet official into a dangerous ruler well remembered for his brutality and his attempts to stifle any opposition to his rule. But that is not all that he did. He would helm a cultural and political revolution that sought to mark out Haiti's presence in the Atlantic world. Through these efforts, he forever altered the cultural geography of knowledge about Haiti and its powerful history.

Imperial 'Displays'

In 1849, just after he made himself emperor, Faustin I created a noble class in Haiti, complete with coats of arms and ranks. He then began a series of disastrous attempts to invade and re-conquer Santo Domingo (an area now known as the Dominican Republic that had previously been unified under an earlier Haitian leader). He would face stiff opposition to these actions from Dominican rebels and American, English and French politicians. The latter three had imperial designs of their own in the region that Faustin I's actions disrupted primarily because he demanded that Haiti be treated as an equal and modern imperial Atlantic nation-state. Displaying those attributes to his critics must have played some part in Faustin I's decision to participate in the New York-based Exhibition of the Industry of All Nations that took place in 1853 (hereafter referred to as the Fair).

The impetus for the Fair began in London during the Great Exhibition of the Works of Industry of All Nations of 1851.[13] After its resounding success, its exhibitors and promoters sought additional outlets for the goods and other material

---

E. Baur, 'Faustin Soulouque, Emperor of Haiti His Character and His Reign', *The Americas* 6 (1949): 131–66 and Jennie J. Brandon, 'Faustin Soulouque – President and Emperor of Haiti', *Negro History Bulletin* 15 (1951): 34–7.

13   Scholarship on the Fair is extensive. Two recent examples of the current positioning of the exhibition can be found in Jeffrey Auerbach, *The Great Exhibition of 1851: A Nation on Display* (New Haven: Yale University Press, 1999) and Jeffrey Auerbach and Peter H. Hoffenberg, eds, *Britain, the Empire, and the World at the Great Exhibition of 1851* (Aldershot: Ashgate, 2008).

objects that had gone on display in London. Visitors to the exhibition's site in Hyde Park were entranced by the architecture, technological wonder and industry of Britain, its empire and other regions and nations around the world. They would encounter more than just an interest in further commercial glory. Dubbed the Crystal Palace Exhibition for the centrality of the cast plate glass building at its heart, this gathering, the first official world's fair of industry and culture set the tone, and the bar, for future international exhibitions. It would be here that the seed for a similar event situated in the USA took hold.[14]

It began, most tangibly, in the efforts by the USA Commissioner to the London Crystal Palace, Edward Riddle, who found backers and obtained permission from the New York Board of Aldermen (for a small fee) to use Reservoir Park, an area near the New York Public Library's main building and the private-public park near 6th Street and West 42nd in Manhattan (now known as Bryant Park). Although Riddle would fade away and turn the project over to other investors, including poet and journalist William Cullen Bryant and the nephew of novelist Catherine Maria Sedgwick, interest in the Fair gained momentum and support.[15] By the time of its completion, the Fair covered nearly 5 acres of space with its domes, galleries and towers. Its centrepiece, the New York Crystal Palace, directly imitated London's grand Crystal Palace structure, but heralded a decidedly less European vision of the world.

According to Charles Hirschfeld, 'the exhibition flung a challenge to aristocratic Europe even as it deferred to Old World tradition. It bespoke a radical utopian vision as well as hard-driven ideals, a universal human fellowship and the petty calculations of merchants'.[16] An 1853 article published in *The Illustrated Magazine of Art* argued that the Fair was essential viewing as it 'affords us opportunities to learn the present condition of the arts and progress of the race, such as we have never enjoyed before. It is the Normal School of Art for the nation'.[17] The author goes on to assert that Americans, especially, 'need its lessons'.[18]

Robert C. Post suggests that the exhibition's leadership emphasised that this educational setting would provide the masses with knowledge about:

> speedy avenues and modes of transportation and communication, on rapid
> methods of exploiting natural resources, on keeping track of time, measure and

---

14   For more on the Fair, see Horace Greeley, *Art and Industry: As Represented in the Exhibition at the Crystal Palace, New York, 1853–4* (New York: J.S. Redfield, 1853) and William C. Richards, *A Day in the New York Crystal Palace* (New York: G.P. Putnam & Co., 1853).

15   For more on this view of the Fair, see Robert C. Post, 'Reflections of American Science and Technology at the New York Crystal Palace Exhibition', *Journal of American Studies* 17 (1983): 337–56.

16   Charles Hirschfeld, 'America on Exhibition: The New York Crystal Palace', *American Quarterly* 9 (1957): 101.

17   'The American Crystal Palace', *The Illustrated Magazine of Art* 2 (1853): 258.

18   'The American Crystal Palace', 258.

quantity, on protecting the safety of persons and possessions, and on mechanizing the production of consumer goods and the whole range of operations relating to farming, food and textile processing, and clothing manufacture.[19]

Unsurprisingly, the Fair's 'educational setting' brought together the leading nations of the world, featuring exhibits and objects from Italy, Great Britain, Germany, France, Austria, and Holland. Together, nearly 2,700 exhibitors provided these foreign contributions.[20] One of those nations, ensconced in the northwest corner of the Fair's building along with Italy, Austria and Holland, was Haiti.[21]

Haiti's presence at the Fair and its participation in an event meant to educate the public about the industry and culture of the world may seem surprising given the pejorative articulations that circulated in the Atlantic world about Faustin I and his imperial desires. While French caricaturists, such as Charles Amédée de Noé (who went by the pseudonym Cham) and Honoré Daumier, produced a substantive body of derogatory images of Faustin I (a topic that I will return to below), the Fair offered a decidedly different portrait of Haiti and its ruler.[22] Fair documents repeatedly note the presence of Haiti (identified as Hayti) in maps, official guides and other materials prepared for the voracious public intent on consuming everything about the Fair.

The sheer fact that Haiti was included in a gathering whose sole goal was to offer an educational setting to the public that articulated the 'progress of the [human] race', signaled a marked shift, at least by some parties, as to the knowledge and import of Haiti and its culture. By linking it with other presumably modern, cultured and civilised nations and regions, Fair officials cast Haiti before its insatiable public as something other than the Atlantic world's *bête noire*. Instead, it appeared as an equal member of the family of Atlantic nation-states. As opposed to being isolated to a display of 'primitive' or purportedly less civilised peoples, Haiti's contributions to the Fair were given equal footing with other European sites.

At present, a full history of Haiti's objects at the Fair has yet to be written, including how the government was approached to participate and why they were positioned in the particular quadrant in the Crystal Palace. Contemporaneous writings on the Fair suggest that Faustin I was personally involved in the process, undoubtedly responding to the circular sent by the USA Secretary of State to 'American consular and diplomatic officials directing them to give all legitimate though limited aid to the [Fair] Association's European agents in their efforts

---

19   Post, 'Reflections of American Science and Technology', 341.

20   'The American Crystal Palace', 254.

21   Richards, *A Day in the New York Crystal Palace*, 26.

22   For more on this, see Cham, *Soulouque et sa Cour; Caricatures* (Paris: au bureau du journal *Le Charivari*, 1850) and Elizabeth C. Childs, 'Big Trouble: Daumier, Gargantua, and the Censorship of Political Caricature', *Art Journal* 51 (1992): 26–37.

to secure foreign exhibits'.[23] How Haiti would be able to use this targeting of 'European agents' for its own positioning is not fully clear. That it would be able to get its foreign objects to the Fair is undisputed. If Haitian diplomats, government representatives and business leaders did use consular networks and the circulation of tracts and missives to gain entry to the Fair this would be a major coup, as USA congressional discussions in the 1850s concerning Haiti tended to censor Faustin I's imperial desires regarding the Dominican Republic while at the same time, labelling the nation as inherently dangerous (especially to the legal system of slavery in the USA). To many USA politicians, Haiti, no matter its self-described majesty, remained troublingly black.[24]

For these and other reasons, the Haitian objects that circulated at the Fair – identified in one Fair document as including samples of mahogany – represent a different image of Haiti. While some USA politicians had difficulties seeing Haiti as modern and industrious, others were able to accept a different view. For the latter, Haiti was, for all intents and purposes, a nation deserving of being on display. Even with this success, and perhaps because of it, Faustin I faced numerous criticisms of his power and his imperial designs and responded to them by circulating knowledge within a variety of other materials – including official and unofficial governmental pamphlets on race.

Pamphlets of Racial Protest

As Faustin I engaged in his ill-fated invasions of the Dominican Republic and the USA tried to influence the outcome through manipulation and shadowy deals, Faustin I found himself in the midst of another fight – over the race of Haitians. Although Saint-Domingue historically had a very intricate system of colour difference and gradation, power in post-independent Haiti tended to be held between two elite groups of mulattos and blacks, with the peasants residing far below both groups.[25]

While these differences in colour and status provoked divisions and an entrenched way of soliciting support and dissension amongst the elite and the military, many politicians from outside of the Caribbean tended to regard Haiti as simply a black nation. This articulation of black Haiti, reinforced by the constitution declaring all Haitians (regardless of colour) as black, morphed into far

---

23   Hirschfeld, 'America on Exhibition: The New York Crystal Palace', 104.

24   For a wide range of approaches and details concerning the invasion, see 'Soulouque and the Dominicans', *Democrat's Review* 30 (Feb 1852): 137, Britannicus, *The Dominican Republic and the Emperor Soulouque* (Philadelphia: T.K. Collins, Jr., 1852) and Rayford Logan, *Haiti and the Dominican Republic* (New York: Oxford University Press, 1968).

25   For more on this, see David Nicholls, *From Dessalines to Duvalier: Race, Colour and National Independence in Haiti* (New Brunswick, NJ: Rutgers University Press, 1996).

more pejorative considerations about Haiti than just the nation's colour, as Haiti's blackness became tangled up in Atlantic world ideas about the nation's political aptitude and displays of power. For example, fears of black Haiti and Haitians prompted a number of USA politicians in the early- to mid-nineteenth-century to fight against any form of recognition of Haiti's independence, arguing that formal acceptance would bring black diplomats into Washington's political culture. If that was not worrying enough, their fantasies of racial dis-ease prompted apocalyptic imaginings of cross-racial interaction at diplomatic functions. One example of why that could never happen can be found in the 1862 debates in the USA Congress about recognising Haiti's independence. During a heated exchange in the House of Representatives, Samuel Cox, a Democrat from Ohio, declared why he objected to full recognition and diplomatic exchange with Haiti:

> Objection? Gracious heavens! What innocency! Objection to receiving a black man on equality with the white men of this country? Every objection which instinct, race, prejudice, and institutions make. I have been taught in the history of this country that these Commonwealths and this Union were made for white men; that this Government is a Government of white men; that the men who made it never intended, by anything they did, to place the black race upon an equality with the white. The reasons for these wise precautions I have not now the time to discuss. They are climatic, ethnological, economical, and social.[26]

Cox suspected that this national equality hid something far more sinister: individual and political equality across races. He believed that recognising Haiti was just the first step to extending African Americans' recognition and political rights. These racio-political considerations were clearly a step too far for some politicians, prompting some to use them as the rationale for not recognising Haiti's independence. This racialisation and demonisation of Haiti would not go unchallenged by Faustin I.

In the 1850s, Faustin I would instigate the circulation of a set of pamphlets on Haiti. At the centre of all of them was one curious question: Was Haiti a real nation? This question was posed, on the one hand, in response to attempts by the USA government (and other empires) to establish naval bases within the important Caribbean global shipping lane that went between Haiti and Cuba, en route to Colón. On the other hand, the question was meant to subvert the logic of those who questioned the legitimacy of Haiti's existence using arguments that offered convoluted formulas about who could and could not be and act as political agents, such as those expressed by Cox, above. Written by Faustin I's Commercial Agent to the USA, B.C. Clark, and published in Boston, these pamphlets tell the heroic story of Haiti's national history and its brilliance.

---

26   *The Congressional Globe*, 37th Congress, 2nd Session, June 2, 1862, p. 2502. *A Century of Lawmaking for a New Nation: US Congressional Documents and Debates, 1774–1875*, Library of Congress.

Clark's life is a bit of a mystery. Records indicate that he was a successful businessman and prolific writer on Haiti, sending letters to USA politicians and publishing essays as both an independent businessman and then in his guise as an official commercial representative of Faustin I's Haitian government. The series that interests us in this chapter are two works published one after the other, *A Plea for Hayti, with a Glance at Her Relations with France, England and the United States for the Last Sixty Years* (1853) and *Remarks Upon United States Intervention in Hayti with Comments Upon the Correspondence Connected With It* (1853).[27]

Just as a representative of a state legislature might use a speech on the chamber's floor to fight for or pull apart alternative points of view and ideas, Clark used these essays to present new knowledge about Haiti for an intrigued readership. While the essays provide an answer to the question of the race of Haitians and Dominicans (and what that might mean for foreign intervention into Haitian affairs), both essays focus primarily on Haiti's imperial legitimacy, as can be seen in this passage from *Remarks*:

> No one will deny that the Republics of South America have been, since the day of their existence, vastly more unfortunate in relation to internal dissensions, *'fearful atrocities'*, and *'bloody tragedies'*, than Hayti has been during the same period; but no one asks or desires that they shall be blotted out from the Map of Nations. No public agent of humanity is sent to either of these countries,—no Commissioner writes in relation to them, that their *'destruction can scarcely be considered a cause of grief, and their epitaph will have no claim to be written with a pen dipped in tears;'* and yet, this is the sentiment communicated by the Commissioner to the Hon [USA] Secretary of State, in relation to Hayti, and the communication is published throughout the United States.[28]

Clark felt that the US public had received distorted information that allowed them to formulate an image of Haiti from limited or discriminatory knowledge. His essays, therefore, would correct this error and provide important evidence for the public's consideration. This evidence included correspondence between diplomats and other parties, official decrees supplied by Faustin I and snippets from established and respected researchers and critics. It would be in the last category that Clark would act as mediator, translating material for readers in the USA from the original French written by Haitian historian, Thomas Madiou. In curating these sources, Clark set the case for his 'plea for Hayti' – asking, ultimately, for

---

27   B.C. Clark, *A Plea for Hayti, with a Glance at Her Relations with France, England, and the United States, for the Last Sixty Years* (Boston, MA: Eastburn's Press, 1853) and B.C. Clark, *Remarks Upon United States Intervention in Hayti, with Comments Upon the Correspondence Connected with It* (Boston, MA: Eastburn's Press, 1853).

28   Clark, *Remarks Upon United States Intervention in Hayti*, 16.

Haiti to be recognised, and 'placed upon the same footing as other nations'.[29] For Clark, not recognising Haiti politically while profiting from interactions with it economically was a moral wrong.

Clark's essays, though, are not anti-slavery tracts internationally-focused, but intent on attacking the system of slavery in the USA. While he takes care to underline the hypocrisy of those who fail to note the similarities between slavery in the USA and slavery in the Caribbean, his main critique is for those USA politicians and businesses who choose to valorise the Latin American republics or the rise of the Dominican Republic while engaging in extensive financial transactions with non-recognised Haiti. 'Hayti', he argues, 'is the only nation having extensive commercial relations with us, that has not been recognized by the government of the United States'.[30] Clark felt that recognising the South American republics was short-sighted and potentially crippling, given the resources available in Haiti and the attributes of its people.

Haitians, he argued, are tranquil and peaceful, with travellers frequently remarking about Haitians' kindness. Although Clark presents the history of Haiti from its indigenous people to its earliest colonial days as Saint-Domingue and its revolutionary rulers in the pages of his pamphlets and essays, he sets aside the bulk of his pages to informing his reader about the life and rule of Faustin I. He asserts that current Atlantic rulers isolate and ostracise Faustin I, even though they have no knowledge of his true character or position within Haiti:

> The present Chief [Faustin I] has no … [foreign] friend, nor does he require any; his strength is at home; it is not too much to say, that there is not a town, village or hamlet in Hayti, however distant from the Capital that does not spontaneously and joyfully claim to honor him as a Chief, and to love him as a man; and there are strong reasons for this, for in matters which are generally deemed subordinate, but which in relation to Hayti are vital, the Emperor has shown a wisdom beyond all precedent in his own country.[31]

This wisdom, I argue, extended beyond the borders of Haiti and included the intentional and opportunistic circulation of materials by and about Haiti that helped to correct pejorative ideas about the nation and its ruler. Although the full extent of Faustin I's network remains to be mapped, the proliferation of materials, the dates of circulation and their reach suggest a more coordinated effort than mere happenstance. Regardless of intent, these objects and materials represent an intriguing archive of knowledge regarding Haiti and its last emperor, Faustin I – an archive that includes images that he commissioned that offered photographic proof of his power.

---

29   Clark, *A Plea for Hayti*, 3.
30   Clark, *A Plea for Hayti*, 4.
31   Clark, *A Plea for Hayti*, 46–7.

Imaging Empire

An 1856 issue of *The Illustrated London News* (hereafter referred to as *ILN*) contains a curious image. A weekly British newspaper committed to offering its readership visual depictions of newsworthy events, this particular issue does not disappoint. In it, readers were presented with two nearly half-page and intricately detailed images of the last emperor of Haiti, Faustin I and his wife, Empress Adelina. Although the illustrations correspond to the article, 'His Imperial Majesty Faustin, Emperor of Hayti', the images' visual iconography seems to contradict the textual narrative chronicling the rise of Faustin I and his world.[32] Similar in tone to the caricatures of Faustin I mentioned above, the text of the *ILN* article casts Haiti as the Atlantic world's dark-skinned – and distempered – foundling, an aggressive and bumbling entity engaged in a comedy of errors as a nation-state on the international stage.

The image makes no such case. It conveys the imperial majesty of the emperor – and his empress, Adelina – displaying them, in separate illustrations, in all of their regal glory. Both Faustin I and Empress Adelina exude power on the page in these detailed full-body portraits that show them in full monarchal regalia, complete with their ermine coronation robes and crowns. In the case of Faustin I, the additional objects within the visual field reinforce his imperial status; namely, his eagle-crested sceptre and necklaces adorned with medals, possibly the Orders of the Cross for military and civil service to the nation. Although the illustrations are saturated with layers of imperial power and French (and possibly African) referents of sovereignty and kingship, the textual narrative takes pains to represent the emperor and empress as grand simpletons burlesquing empire.

After recounting the history of Haiti from Columbus's intrepid march from Europe to Hispaniola through to the present day, the author takes on Haiti's current imperial turn. Haiti's emperor, the author claims, 'exceeds the production of any writer of fiction'.[33] The entire imperial pageantry was a farce; 'the best burlesque upon an empire that the most fertile brain ... could have imagined'.[34] According to the text's author, and many other foreign chroniclers of Haitian history, free, independent, sovereign, and imperial Haiti was not just unthinkable; it was ridiculous. And Faustin I, the last emperor of Haiti, was the despotic ringleader of this parody of customs and imperial power. Unfortunately, twentieth and twenty-first century scholars have done little to challenge this pejorative view.

John E Baur probably best captures the dismissive critical tone toward Faustin I in a mid-twentieth-century essay in *The Americas* in which he announces that 'the nations of the free New World have produced but seven sovereigns; all have been deposed. Some were ridiculous and cruel, but Haiti's third and last monarchical attempt was the most ludicrous', producing a 'drama of strangely mingled comedy

---

32   'His Imperial Majesty Faustin, Emperor of Hayti', *The Illustrated London News* 16 February 1856, 186.

33   'His Imperial Majesty Faustin, Emperor of Hayti', 186.

34   'His Imperial Majesty Faustin, Emperor of Hayti', 186.

and tragedy'.[35] Repeatedly described as 'stupid', 'naïve', or slow, Baur gives ample space to Faustin I's savagery, noting the efforts that the emperor took to quell dissent, to control the opposition and to silence those wanting more liberal policies and reforms. With views pivoting between Faustin I's ridiculousness and his violence, Baur ably captures Faustin I's current reputation.

Historiographical accounts that trace the emperor's life remain mixed. Often labelled as tyrannical, he would also be described as an inept ruler, something that is surprising given the many battles that he would find himself fighting and the more generic battles for recognition that would bring him, and his brokers, into contact with other agents throughout the Atlantic world.[36] Laurent Dubois notes 'if Soulouque was usually dismissed by external critics as stupid and inept, it was ... in part precisely because he proved rather stubborn in the face of outside pressures, granting few concessions to foreign governments'.[37] It is perhaps this stubbornness – and the danger associated with him – that informed thoughts about the emperor found in the *ILN* article and in other documents, such as the caricatures of Cham and Daumier.

Faustin I first appeared in images by Daumier in 1850. Elizabeth Childs asserts that Faustin I was never the intended target and argues that Daumier's caricatures of Faustin I were in fact 'clever surrogates' that attempted to re-focus domestic critiques of French President Louis-Napoléon onto an exotic, and therefore acceptable, locale.[38] Instead, Daumier's satire 'casts Louis-Napoléon in blackface in order to criticize the French government's recent repression'.[39] This was a racist insult, to be sure, but Childs stresses that it was directed at the French president, not at Soulouque.

That may have been one of the direct impulses for Daumier's creation of the images, but they correspond too tellingly with derogatory images that already circulated about Faustin I. In fact, they perform a concentrated echo of the thoughts of mid-nineteenth-century diplomats, such as Robert M. Walsh, a USA agent sent to Haiti in 1851, who described Faustin I as 'stout and short and very black with an unpleasant expression and a carriage that does not grace a throne'. He finishes by noting that Faustin is 'ignorant in the extreme'.[40] According to Walsh, Faustin I's inward deficiency manifested in his outward countenance, a line of thought familiar to those who study nineteenth-century notions of colour,

---

35   Baur, 'Faustin Soulouque, Emperor of Haiti', 131.

36   See above source material by Dash for more on this.

37   Laurent Dubois, *Haiti: The Aftershocks of History* (New York: Metropolitan Books, 2011), 146.

38   Childs, 'Big Trouble: Daumier, Gargantua and the Censorship of Political Caricature', 36.

39   Childs, 'Big Trouble', 36.

40   In William R. Manning, ed., *Diplomatic Correspondence of the United States, Inter-American Affairs, 1831–1860*, Vol. VI (Washington: Carnegie Endowment for International Peace, 1925), 92–3.

race, and intellectual acuity.[41] Faustin I was not just the wrong colour; he was the wrong colour for a ruler. It is this view that makes the illustrations of Faustin I and Empress Adelina in the *ILN* so curious.

Instead of Faustin I being presented as a fool, too ignorant and uncivilised to exist outside of the parodic world that his mimicry had created, the images present a formidable ruler – a sovereign of power; an emperor clearly in charge of his own iconography, determined to present certain 'truths' about his nation-state. Instead of presenting him as a ruler of a dark, dangerous space, overflowing with black savages, the *ILN* illustrates a powerful entity whose imperial desires echo those of other Atlantic nation-state rulers. Although his sovereignty would be familiar, his claims went beyond those exuded by other circum-Atlantic countries because they emerged from and were sustained by a complex notion of blackness.

Rather than emerge from the imagination of an illustrator employed by the *ILN*, the images of Faustin I and Empress Adelina were copied from a critically important and under-studied album of daguerreotypes and lithographs, titled *Album Imperiale d'Haïti* (translated as *The Imperial Album of Haiti*).[42] It is a rich and significant resource as it contains an array of imperial objects associated with Faustin I's reign, the grandeur of his royal family, and the nobility created once he was crowned. In total, there are 12 daguerreotypes and lithographs in the collection, including images of Emperor Faustin I and his wife, Empress Adelina. In addition to the royal court, there are community scenes, such as Faustin I's coronation ceremony, that captures the view of the actual placement of the crown on Faustin I's head from a vantage point amongst the crowd. There are also rare portraits of Faustin I's family. In addition to an image of Empress Adelina is perhaps the only known representational image of Princess Olive (their daughter).

Caricaturists in France tended to capture the Princess as a stooped figure shuffling through dance lessons and other cultural teachings without grace or style.[43] The Princess that looks back at the viewer in the album has wide, open eyes that appear to scrutinise the intention of all who claim to know her or her thoughts. Bejewelled and dressed in an impressive gown, Princess Olive is, without a doubt, beautiful. That this image, and others of the royal court, is included in the collection suggests that the album, as an object, is an important symbol of the power of Haiti and its ruler.

---

41   For more on this, see Bruce Dain, *A Hideous Monster of the Mind: American Race Theory in the Early Republic* (Cambridge, MA: Harvard University Press, 2002) and Mark M. Smith, *How Race is Made: Slavery, Segregation, and the Senses* (Chapel Hill: University of North Carolina Press, 2006).

42   I first encountered this album after the British Library acquired one of the original published editions a few years ago. I thank Carole Holden for allowing me to carefully examine the edition.

43   For examples of this, see Cham, *Soulouque et sa Cour; Caricatures.*

Figure 9.1   Title page of album
Source: Album Imperiale d'Haïti, title page (1852). Image courtesy of the British Library.

In the album, Faustin I is presented as a force of imperial power. His portrait shows him standing on a raised dais, before a regal throne. The throne, itself, is surrounded by a curtain of fabric that is adorned with additional representations of Haiti and its ruler's reign, such as his crest. He looks, in tone and stature, as powerful as any European monarch. The album also captures further details

FAUSTIN 1er

EMPEREUR D'HAITI

Figure 9.2   Portrait of Faustin I
Source: Album Imperiale d'Haïti, Plate 3 (1852). Image courtesy of the British Library.

L' IMPÉRATRICE ADELINA.

Figure 9.3   Portrait of Empress Adelina
Source: Album Imperiale d'Haïti, Plate 4 (1852). Image courtesy of the British Library.

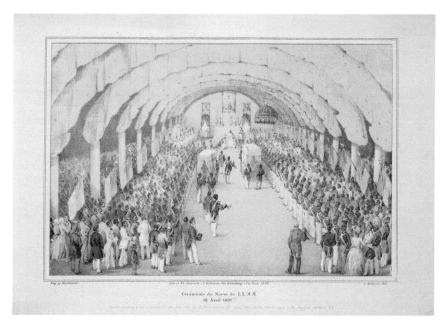

Figure 9.4   Faustin I's coronation ceremony
Source: Album Imperiale d'Haïti, Plate 1 (1852). Image courtesy of the British Library.

that the *ILN* illustration leaves out, such as the name of the daguerreotypist and lithographer responsible for the image. Their names appear at the base of the print. The image is attributed to one A Hartmann, a daguerreotypist based in New York, and made into a lithograph by lithographer Th. Lacombe, whose base of operations was in the heart of Manhattan on Broadway, one of the oldest avenues in the city. Lacombe, it seems, worked closely near the site of the 1853 Fair. The closeness in the publication date of the album and the opening of the exhibition suggest that these two sites of display were connected.

Exactly what may have transpired to bring the album to fruition remains to be further charted, but the images within the album, those staged and observed, had to have been taken by the creative artists attached to it at the time of the coronation of Faustin I, an event that occurred in Haiti in 1852. The album, as an object of its times, is an important archive for those interested in the material culture of people of African descent. Due to the high production quality of the album, and the association of Hartmann and other Broadway-based lithographers associated with it, *The Imperial Album of Haiti* is of significant import to the study of mid-nineteenth-century visual culture of people of African descent and those interested in the earliest uses of photography to communicate aspects of black power. Yet, for all of its importance as an example of early black images of power, it is also

S.A.I.Madame Olive,
Fille de L.L.M.M.

P.A. Ott Lithographer

Print of Th. Lacombe 3M Broadway NY

Figure 9.5   Portrait of Princess Olive
Source: Album Imperiale d'Haïti, Plate 5 (1852). Image courtesy of the British Library.

an important collection of images that challenged negative ideas that proliferated about Faustin I, his reign and his people.

Questions remain about its reach and its purpose. Did the album appear amongst the objects at the Fair? Extensive evidence suggests that Faustin I commissioned the portraits and the album, but why? Could he have used the album as his way to counter-attack the images circulating by Daumier and Cham that depicted his family as buffoonish, ill-cultured simpletons? If that is the case, how did the images reach the other side of the Atlantic?

As stated earlier, an *ILN* illustrator copied (with some morphological alteration) two of the images from the album, those of Faustin I and Empress Adelina, to accompany a negative article about Faustin I. Although the illustrations are not signed, it may be possible to learn more about the illustrator at the *ILN* and his or her knowledge of, or ties to, Haiti. Perhaps this illustrator travelled to the exhibition space at the Fair, saw the album on display and quickly copied the impressive images only to unveil them, years later, as the incongruous illustrations for a derogatory article on Faustin I. What cannot be disputed is the fact that these objects – created, championed and crafted by a range of actors officially and unofficially associated with Faustin I's reign – offered knowledge about Haiti that challenged negative ideas about Haiti that circulated in the Atlantic world.

Conclusion

The imperial dagger presented at the beginning of this chapter is a curious object. Unlike the imperial album of daguerreotypes and lithographs, the material and presence of Haiti at the Fair or the writings by official Haitian representatives, such as Clark, this object appears disconnected to Faustin I. Or was it? A concentrated search through databases of mid-to-late-nineteenth-century USA magazine and newspaper articles reveals the publication and illustration of more Haitian imperial objects connected to Faustin I and his rule, including an engraving of his seal, insignia and crown, and further details about the manufacturer of these items, such as the fact that the firm provided jewels and other items to distinguished royals around the world. One particular article, from an 1852 issue of *Gleason's Pictorial Drawing-Room Companion*, discusses these items and contains no derogatory language about Haiti or Faustin I.[44] In fact, Faustin I's coronation is described as having outshone that of Napoleon I's royal celebrations.

This positive narrative and its accompanying engraving of Faustin I's regalia match each other in tone and depiction. Why? How did the illustrator of *Gleason's* produce or obtain the engraving? Did other collections and objects about Faustin I's imperial sovereignty circulate, much like the album? And if they did, what else

---

44   'Crown, Seal, Etc., of Faustin I', *Gleason's Pictorial Drawing-Room Companion* 2 (1852): 364.

did this circuit produce? New social and cultural networks? New political allies? New knowledge?

I address some of these questions in my larger research project on Haitian recognition, power and political future in the nineteenth century. Clearly, many questions remain to be answered about the cultural knowledge of Haiti and other self-identified 'black' spaces in the nineteenth-century Atlantic world – such as Liberia and Sierra Leone. Although this chapter focused on Haiti and offers more questions than definitive answers, it does begin the arduous work of curating the archive of cultural knowledge and material objects that offered counter-narratives to Haiti's negative image in the Atlantic world. What I have offered is more than just a forgotten narrative of imperialism. This is, in truth, a story about power and who can wield it and in what form. It would be through these objects that various publics learned new ideas about Haiti, the culture of its people and the stature of its last emperor, Faustin I. Through them, they encountered material evidence of the sovereignty of nations far beyond European and American shores and began to grapple with the many spaces of knowledge-production in the world.

Chapter 10

'Throughout Bihar and Beyond': Dublin University Mission and the Structuring of a 'Global' Medical Practice

Sarah Hunter

Introduction

The Dublin University Mission (DUM) arrived in the town of Hazaribagh in the Chota Nagpur Division of western Bengal in January 1892, with a clear strategy to draw upon and utilise the collective knowledge of its graduate missionaries. Contemporary missionary wisdom emphasised the importance of third level education in order to justify financial support and ensure victory in the foreign mission field.[1] Initial missionary encounters with local populations were facilitated by an apparent need for western teaching and healthcare, and it was hoped that the provision of this practical, medical assistance would help lead the local population to Christianity.[2] However, upon arrival in Hazaribagh, the Dublin University missionaries recognised the need to adapt the philanthropic approach to evangelising they had devised while preparing for mission in the comfort of the college rooms and parochial houses of Victorian Dublin. For the mission to succeed and appeal to the local population, it needed not only to transcend geographical and social boundaries, but to engage directly with individuals and traditional practices. Within the wider apparatus of the colonial project, the Mission utilised medicine as a mechanism to penetrate spatial and social barriers. As will be demonstrated in this chapter though, it was the assimilation of local social customs and medical practices with the Irish missionaries' medicine, and the subsequent alteration in the way in which the missionaries' medical knowledge was practiced, which ensured the sustainability of the Mission in Chota Nagpur.

---

1 For further information on university brotherhood missions see Daniel O' Connor et al., *Three Centuries of Mission: The United Society for the Propagation of the Gospel 1701–2000* (London: Continuum, 2000), 62–5.

2 Raj Sekhar Basu, 'Healing the Sick and the Destitute: Protestant Missionaries and Medical Missions in 19th and 20th Century Travancore', in *Medical Encounters in British India*, ed. Deepak Kumar and Raj Sekhar Basu (New Delhi: Oxford University Press, 2013), 189.

In concentrating on interactions between Irish practitioners and the Indian populous, this chapter moves beyond traditional interpretations of the diffusion of ideas from west to east, arguing rather that human exchange, interaction and interdependence, even in remote imperial spaces, created a product which could be described as organically global.[3] In identifying a need to teach and heal the local community and implementing a paternalistic programme, the DUM undoubtedly echoed existing colonial policy. At the same time, though, the Mission incorporated local custom in their medical practice, creating what Chris Bayly has termed a hybrid healthcare practice, a 'dual system of scientific endeavour'.[4] Pradip Kumar Bose has termed the transition of western medicine from the west to the east as 're-situation', but concedes that this 're-situation' inevitably led to 're-formulation', arguing that '[f]or any kind of "science", perfect transplantation in a different culture is impossible'.[5] As the missionaries and Indian practitioners engaged with transnational networks of knowledge, this chapter is informed by the work of Kapil Raj, and acknowledges the emergence of 'global' medicine as a consequence of the movement of people and ideas over space and time. In refuting diffusionist theories, Raj argues that South Asia was an 'active' partner, albeit an 'unequal' one, in the 'emerging world order of knowledge'.[6] In recognising the complex formations and interpretations of global knowledge, this chapter is influenced by Sujit Sivasundaram's argument concerning the fragmented and diverse character of global science. For Sivasundaram, the process of globalisation brings together 'the local and the international', each informing the other and the two becoming intertwined. Reconstructing the activities of the DUM will bring to light one aspect of this phenomenon, for, as will become clear, the DUM's missionaries reformulated their medical practice as a result of interaction with the particular culture and customs of a remote region of Chota Nagpur.[7]

This chapter will begin by outlining the ways in which the Dublin University medical missionaries utilised medicine as a mechanism to overcome spatial and social barriers. In particular, the medical missionaries identified a lack of female healthcare in the region and so employed Irish female practitioners to enter segregated female spheres. Secondly, the chapter discusses how the Mission acknowledged the need to actively engage with local social customs by employing Indian female practitioners in order to develop their medical practice

---

3   Christopher A. Bayly, *The Birth of the Modern World, 1780–1914: Global Connections and Comparisons* (Oxford: Blackwell, 2004), 1.

4   Bayly, *The Birth of the Modern World*, 319.

5   Pradip Kumar Bose, *Health and Society in Bengal: A Selection from Late 19th-Century Bengali Periodicals* (Calcutta: Sage, 2006), 10.

6   Kapil Raj, *Relocating Modern Science: Circulation and the Construction of Scientific Knowledge in South Asia and Europe, Seventeenth to Nineteenth Centuries* (Delhi: Permanent Black, 2006), 13.

7   Sujit Sivasundaram, 'Sciences and the Global: On Methods, Questions, and Theory', *ISIS* 101 (2010): 155–6.

in Hazaribagh. Finally, the chapter will analyse how climate, geography and social norms facilitated the development of a transnational network of expertise exchange, by concentrating upon those female medical practitioners who worked for the Mission, attending and delivering medical courses in Dublin, London, Delhi and Lucknow.

Dublin, Ireland to Hazaribagh, India: Medicine as a 'Tool'

The DUM was established in 1891 in response to an appeal from within the divinity school at Trinity College, Dublin to create a foreign missionary brotherhood on a par with the Universities' Mission to Central Africa, the Cambridge Mission to Delhi and the Oxford Mission to Calcutta.[8] The initiative was financially supported by the Society for the Propagation of the Gospel (SPG) under the stipulation that it be willing to establish itself in any location where the SPG saw a need, providing there was 'promise of useful and hopeful work' – that is, work that was both practical and spiritual.[9] Upon being appointed the inaugural bishop of Chota Nagpur, then west Bengal, Jabez Whitely appealed to the SPG for a brotherhood mission to be sent to the region, identifying Hazaribagh as the district within the diocese which was in most need of Christian intervention.[10] The concurrent calls for university brotherhoods and the subsequent cooperation between Trinity College's divinity school and the SPG was beneficial for both parties; neither the divinity school nor the SPG had enough funds to entirely support a company of missionaries, though both fervently believed in the educational expertise which could be harnessed within universities for the benefit of Christian mission.

In mid-December 1891 six Irish missionaries left Ireland for India, arriving in the town of Hazaribagh on 22 January 1892, having travelled by boat, train, cart and finally on foot.[11] Information on the geographical and sociological composition of the region, gleaned from contemporary annals, serves to highlight its isolation. Hazaribagh district comprised an area of 7,021 square miles in the north-eastern division of Chota Nagpur, with a population calculated as 1,164,321 in the 1891 census.[12] Set amidst jungle at an elevation of 2,000 feet, Hazaribagh

---

8   The Universities Mission to Central Africa was founded c. 1857, Cambridge University Mission to Delhi in 1877 and the Oxford University Mission to Calcutta in 1879.

9   *Dublin University Mission to Chota Nagpore Working Under the Society for the Propagation of the Gospel in Foreign Parts, Second Annual Report for the year 1891* (Dublin: University Press, 1892), 4. Hereafter the DUM annual reports will be referred to in their series and year number only.

10   For further information on Bishop Whitley's call for a missionary brotherhood see the S.P.G. publication, *Church Work in India: Being Sketches of the Work of the SPG in India, Burma, and Ceylon* (Westminster: Society for the Propagation of the Gospel, 1910), 13.

11   *Second Annual Report for the year 1891*, 8.

12   William Stevenson Meyer et al., *Imperial Gazetteer of India* 13 (Oxford: Clarendon Press, 1908–1931): 85, accessed April 1 2013 http://dsal.uchicago.edu/reference/gazetteer/

lay on Bengal's geographical frontier, despite its relative proximity to the coastal capital of Calcutta, approximately 240 miles to the east. As such, Hazaribagh was separated socially and economically from the remainder of Bengal Province. Although there had been a British presence in Hazaribagh since 1780, the majority of this community left in 1874 due to an outbreak of enteric fever.[13] The local population thus had little interaction with the European community living in India, or with European medical ideas or practices. Under such circumstances, suspicion and fear was the natural response when the DUM arrived, having transcended geographical barriers, desiring to engage with Hazaribagh's population.

Immediately upon arrival, the Mission identified medicine as the means by which they could interact with the local population. As Ian Catanach and Daniel Headrick have argued, medicine had been used as a 'tool of empire' since the eighteenth century, an apparently passive and benign means by which imperial powers could enter and control urban, rural and familial spaces.[14] Taking a similar view, the DUM viewed medicine as a mechanism by which it could, under the pretence of healing the physical body, transform the spiritual body. By the end of the nineteenth century, medical missions were considered imperative for the continuance of Christian mission abroad. Indeed, the DUM acknowledged that supporters of foreign mission were increasingly of the belief 'that no Mission is fully equipped unless it ministers largely to the bodily diseases and suffering of the people whose spiritual wants it seeks to supply'.[15] Thus, in its report for the year 1890, the DUM placed an appeal for financial aid directly to the alumni of Trinity's medical school, emphasising 'this opportunity of aiding so directly and so appropriately the work, of relieving, by the results of their science, the bodily needs of those to whose souls the healing medicine of the Gospel is, at the same time administered'.[16]

The six original members of the Mission comprised five graduates (one a fully trained doctor, Kenneth Kennedy) and a nurse, Francis Hassard, who completed the contingent as the first member of the Ladies' Auxiliary. Upon graduating in 1891, Kennedy was awarded a monetary prize from Trinity College for securing the distinction of top place in his final examinations, with the single stipulation

pager.html?objectid=DS405.1.I34_V13_091.gif. Figures for 1891 census can be found in the S.P.G. publication, *Historical Sketches, Missionary Series, No. II Chôta Nagpore* (Westminster: Society for the Propagation of the Gospel, 1891), 3.

13   Meyer, *Imperial Gazetteer of India*, 13, 99. The majority of the European personnel stationed in Hazaribagh were officers who commanded an Indian infantry unit stationed in the town.

14   Ian Catanach, 'Plague and the Tensions of Empire in India, 1896–1916', in *Imperial Medicine and Indigenous Societies*, ed. David Arnold (Manchester: Manchester University Press, 1988), 150. Daniel R. Headrick first expressed this idea in his *The Tools of Empire, Technology and European Imperialism in the Nineteenth Century* (Oxford: Oxford University Press, 1981).

15   *Sixth Annual Report for the Year 1895* (Dublin: University Press, 1896), 21.

16   *First Annual report for the Year 1890* (Dublin: University Press, 1891), 7.

that the award be used to support a medical project. Kennedy used his award to establish a dispensary upon arrival in Hazaribagh. As such, even before leaving Ireland, the practical work of the DUM was directed towards utilising Kennedy's professional knowledge and financial backing. Within a month of arrival in Hazaribagh, Kennedy and Hassard established a makeshift dispensary in the abandoned military barracks which the Government of Bengal had granted to the Mission for use as a temporary mission headquarters. The missionaries worked with interpreters in order to overcome the multiple language barriers, a situation described by Hassard as akin to the Curse of Babel.[17] By the end of February 1892, barely a month after the original contingent had arrived in Hazaribagh, Kennedy wrote to the Mission's Dublin Committee requesting more medical personnel and making a particular call for female practitioners.[18] According to the DUM's reports, the local population was immediately attracted to the medical care provided by the Mission and 'both the Doctor and Nurse found work enough to tax their health and strength to the uttermost'.[19] This reflected recognition among the local population of the potential usefulness of the missionaries' medical knowledge, if not the missionaries' spiritual message – an interest the Mission responded to in May 1892 when it opened a permanent dispensary in the town of Hazaribagh.

Medicine should not be viewed simply as a tool utilised by the missionaries to access the population of Hazaribagh, but the missionaries were aware their medicine offered them 'a ready entrance to many a village where they would not otherwise have been welcomed'.[20] By 1893, the head of the Mission, Eyre Chatterton remarked that there was 'evidence of its [medical work's] influence … thirty miles away from here [Hazaribagh]'.[21] Kennedy noted in his report of the same year that patients attended the mission dispensary in Hazaribagh from many great distances, suggesting that an awareness of the missionaries' medical knowledge and expertise was emanating throughout the district. As a subtle colonial practice, medicine became the mechanism to access the villages of Ichâk and Petarbar. In its annual report for 1893, the Mission expressed its hope that the Ichâk dispensary, situated nine miles north of Hazaribagh, 'will prove to be an important opening in this large centre'.[22] The village of Petarbar, fifty miles south of Hazaribagh, was chosen as a field for mission not solely due to the missionaries' desire to work among the Santal population in the town but also in acknowledgement that it was 'a most suitable place' as it was 'cut off from all medical aid'.[23]

The employment of additional medical practitioners enabled the Mission to extend its field of work. Upon the appointment of Dr John Hearn in 1893,

---

17 *Fifth Annual Report for the Year 1894* (Dublin: University Press, 1895), 34.
18 *Second Annual Report for the Year 1891*, 2.
19 *Third Annual Report for the Year 1892* (Dublin: University Press, 1893), 5.
20 *Fourth Annual Report for the Year 1893* (Dublin: University Press, 1894), 6.
21 *Fourth Annual Report for the Year 1893*, 6.
22 *Fourth Annual Report for the Year 1893*, 10.
23 *Seventh Annual Report for the Year 1896* (Dublin: University Press, 1897), 16.

Kennedy was freed from daily practice at the Hazaribagh dispensary and hospital. This enabled him to concentrate upon touring clinics in the wider district. He often travelled for weeks, even months, living in a tent while administering medical treatment. In February 1895, he began a tradition which was to continue as an annual event. Each year, the medical missionaries attended a mela at the Surjkund Sulphur Springs – thought to heal bodily ailments – near Bagodar, twenty five miles from Hazaribagh.[24] The provision of medicine at these melas proved so popular that it was reported that locals would bring empty bottles each year so as to be filled with the missionaries' medicine.[25] However, those who availed of the medicine continued to proceed to the Springs for divine healing. The local populations intertwined a ready appreciation for western medical remedies in physical matters, while critiquing the missionaries' divine message and relying on indigenous beliefs and rituals for spiritual restoration.[26] As such, in melding different practice systems, this custom provided an indication that the missionaries needed to adjust their medical practice to allow for local accommodations. By using their medical knowledge to gain a welcome in geographical areas which had not before been explored or culturally integrated into the British Empire in India, the DUM subjugated land and a people group into the colonial sphere, although this was perhaps a subconscious process as the main aim of the Mission was the introduction of Christianity to the region through philanthropic means.

In addition to geographical boundaries, the missionaries utilised their medical knowledge to transcend spaces which were defined by social barriers. When nurse Sydney Richardson arrived in Hazaribagh in autumn 1892, her presence at the dispensary at Hazaribagh enabled Frances Hassard to spend more time in private female living quarters or zenanas, attending to poorer women in the bazaar and inviting women into her own bungalow for medical treatment. Medical knowledge was the means by which entry to these restricted spaces was allowed, and it was hoped that this would lead to spiritual conversation. Indeed, the nurses noted that they 'found no difficulty in visiting any Zenana' – but acknowledged that they were often treated as 'novelties' rather than necessities.[27] Nurse Frances Finch White, appointed by the Ladies' Auxiliary in 1893, spent a large proportion of her work touring the district of Hazaribagh with teacher Laura Wickham, to administer medicine and the Bible to female residents of district villages. But while Wickham noted that villagers who had attended the Mission's main hospital in Hazaribagh or the district dispensaries often 'gave us a friendly introduction to their fellow

24   *Sixth Annual Report for the Year 1895*, 18–19.

25   *Tenth Annual Report for the Year 1899* (Dublin: University Press, 1900), 16.

26   While it is recognised that the term 'western medicine' inevitably arose through engagement with ancient medical systems, I use the term here for the purpose of identifying it as definitive from indigenous medical practices. For further discussion on 'western' medicine see Maria Elshakry, 'When Science Became Western: Historiographical Reflections', *Isis*, 101 (2010): 98–109.

27   *Fifth Annual Report for the Year 1894*, 37.

villagers', many approached the two female missionaries with caution, and older women frequently forbade young girls from interacting with the missionaries.[28] In 1900, Frances Hassard noted that on a district tour to the north-western region of Hazaribagh district, 'the women were so afraid of me'.[29] While medicine went some way towards infiltrating spatial and social barriers, a certain suspicion directed toward the missionaries remained. Fear of the western missionaries and their foreign practices were valid concerns. Persistent prudence on the part of the local population made the missionaries realise that, in order to attract more patients to the Mission's medical institutions and ultimately to Christianity, they needed to alter their practice and to allow for, and incorporate, the cultural and social mores of the region.

The Practice of Medical Knowledge

As Deepak Kumar and Raj Sekhar Basu have noted, in seeking to understand the idiosyncrasies of the human body, medical knowledge by its very nature needs to be fluid. It is 'constantly incorporating new features' and therefore cannot simply be diffused 'from a powerful centre': it is influenced by multiple interactions and exchanges and inevitably altered in the process.[30] Pratik Chakrabarti, likewise, recognises that science was not 'diffused' from west to east but altered by geography and society. In Chakrabarti's words, medical knowledge is 'defined in the milieu of the recipient culture'.[31] In order to survive as a discipline over time and space, medical knowledge and the practice of that knowledge has had to combine differing interpretations of the body, the need to heal the body and how to heal the body. Rather than simply practicing their own medicine in a different locale, the DUM medical missionaries' practice was altered over time to allow for translations of health and interpretations of medicine as practised in a different geographical space. This evolution of medical practice, created through a dialectical exchange of western medical knowledge and eastern practice, suggests that while undoubtedly the Irish missionaries left Ireland with the view to 'diffuse' their own knowledge among the local community, this was impossible without recognising and incorporating local knowledge and practice methods.

From initial encounter, the Irish missionaries saw a need to work with and employ Indian personnel, recognising that sustainability lay with the local population. A Mrs Digby of 20 Morehampton Road, Dublin agreed to fund a three-year scholarship available to a male student from Hazaribagh district to

---

28   *Seventh Annual Report for the Year 1896*, 28.

29   *Eleventh Annual Report for the Year 1900* (Dublin: University Press, 1901), 38.

30   Deepak Kumar and Raj Sekhar Basu, 'Introduction', in *Medical Encounters*, ed. Kumar and Basu, 8.

31   Pratik Chakrabarti, *Western Science in Modern India: Metropolitan Methods, Colonial Practices* (Delhi: Permanent Black, 2004), 12.

attend the Medical Missionary Training College in Agra. The training of locals in medical education was immediately hailed as an avenue through which 'valuable assistance may be procured in this very important branch of the Mission'.[32] By 1893, the administration of the Digby Fund was slightly altered in order to finance the medical education of a number of students. The Digby Pupils, as these students affectionately become known, proved indispensable to both the medical work of the Mission and the Mission itself. The students were to be taught at Hazaribagh by the Mission doctors and were only sent to medical colleges further afield for examination. This was beneficial for the Mission as, while training, these local students could work in the Mission's dispensaries and hospital, supplementing the numbers of medical personnel which the Mission greatly needed. In 1894 the Digby Fund, as it became known, began training three male pupils.

In 1897, Shanti Prakast became the first Digby Pupil to be sent to Temple Medical College in Patna for official examination.[33] He passed and returned to Hazaribagh to take up employment at the Mission as a fully qualified and trained compounder. Masih Charan Pankaj followed, passing his examination at the Temple Medical College in 1902.[34] Pankaj qualified as a hospital assistant and became central to the Mission's medical work, single-handedly managing district dispensaries throughout his years working with the Mission. By 1902, the value placed upon the local medical practitioner was widely recorded in the Mission literature: 'Throughout the past year the medical work lay more than ever before in the hands of our Hospital Assistants and Nurses, and the result is a tribute of the regard in which their treatment and attention are held'.[35] In thus conveying gratitude to the Indian practitioners, the DUM acknowledged the imperative need for Indian personnel. Christarit Khacchap, who had also trained at Temple Medical College, Patna, joined the Mission as a hospital assistant in 1896, as did Komal Tirki. Ten years later, in 1906, Tirki single-handedly ran the district dispensary of Chitarpur, his name becoming 'a household word in the Damūdar Valley, whence most of the patients there are drawn'.[36] Additional Indian personnel were to join the medical staff, primarily recruited among locals who had received training from the medical practitioners themselves.

Medical care for Indian male patients began to falter from 1910, primarily due to competition for better paid jobs in municipal medical institutions and a want of a European male doctor to train those sponsored by the Digby Fund. Indeed by 1913, as a result of the failure to attract a sufficient number of Irish male doctors, the male medical work was firmly in the hands of the Indian workers, headed by Tirki and Pankaj. Dr Mary Iles, the only doctor working for the Mission at the time, dedicated a limited amount of time to teaching the compounding class, while Masih Charan

32   *Second Annual Report for the Year 1891*, 8–9.
33   *Eighth Annual Report for the Year 1897* (Dublin: University Press, 1898), 20.
34   *Thirteenth Annual Report for the Year 1902* (Dublin: University Press, 1903), 23.
35   *Thirteenth Annual Report for the Year 1902*, 21.
36   *Sixteenth Annual Report for the Year 1905* (Dublin: University Press, 1906), 29–30.

taught the dressers. In 1914 the Digby Class was terminated due to the strain placed upon those teaching the courses, though the Fund itself was to be maintained and re-distributed among the medical needs of the Mission. A percentage was reserved in the hope of sending a male pupil to be trained as sub-assistant surgeon in Patna Medical College in the future.[37] Eleven years later, in 1925, Prabhu Dayal Hemron, whose older brother Anand had held the position of compounder at Ichâk, was sent to Patna Medical College for teaching and examination.[38] Hemron qualified as a sub-assistant surgeon and returned to Hazaribagh in March 1929, taking up his position as the only male practitioner working for the Mission.[39] Medical training, reflective of the move toward a focus on female medical care, was to concentrate on training female students from 1914 onwards. As such, the structuring of the missionaries' medical practice centred upon the need to provide medicine administered by female practitioners within separate social spheres.

The November 1892 edition of the *Dublin University Missionary Magazine* reported that 'the amount of useless and unnecessary suffering which a lady doctor could prevent is enormous. She could have a hospital full in no time'.[40] Kennedy acknowledged that the local community were attracted to the Mission's medical institutions in part due to the work of Nurse Hassard, noting that the presence of a female nurse was 'something quite different to the treatment they are accustomed to, and a spectacle they could hardly understand to see a *mem sahib* do such work for them'.[41] Although it needs to be recognised that there was an acknowledgement of the relevance of ordinary missionary work, such as visiting and preaching, stress was placed in particular upon the need for medical personnel.[42] Indian women, constrained by social boundaries arising from Indian cultural mores, were often restricted from receiving healthcare from male doctors and with no female practitioner working in the Hazaribagh municipal hospital, little provision was in place for half of Hazaribagh's population. As such, the missionaries' medical practice began to gravitate towards the Indian women of the district.

Although the DUM employed Irish nurses in order to attract a significant number of female patients from the district to the Mission, it needed to provide adequate facilities for the nursing of these patients. Frances Hassard and Sydney Richardson both commented on the fact that they were never entirely left alone with the women when they visited the zenanas. Male members of the family were omnipresent and all questions concerning female health were dictated through the men rather than directly to or from the women themselves, thereby hindering medical practice. Hassard recorded how she often relayed the medical ailments of purdah patients

---

37 *Twenty-Fifth Annual Report for the Year 1914* (Dublin: University Press, 1915), 51.

38 *Thirty-Sixth Annual Report for the Year 1925* (Wexford: The Free Press, 1926), 27.

39 *Fortieth Annual Report for the Year 1929* (Dublin: University Press, 1930), 46.

40 Reverend T.A. MacMurrough-Murphy 'The work of the Ladies' Auxiliary "Helpers in Christ Jesus"', *The Dublin University Missionary Magazine* 1 (1892): 36.

41 *Fourth Annual Report for the Year 1892*, 14.

42 *Second Annual Report for the Year 1891*, 2.

to Kennedy and Hearn, only for the male doctors to examine the patients verbally and then prescribe treatment.[43] What, essentially, the Mission felt it needed was an opportunity to work directly with the purdah patients by employing a female doctor and providing secluded space within the mission hospital and dispensary where purdah patients could attend in privacy. As such, a number of Brahmin and other upper-caste women attended the dispensary as early as 1896. Although numbers of purdah patients constantly remained small, the Irish missionaries valued these patients and viewed their attendance as highly significant. 'Considering the strictness of their purda system', the Mission's eleventh annual report noted 'we regard this as perhaps the greatest mark of confidence in us that has yet been expressed by our Hindu neighbours'.[44] In 1900, Richardson reported that of 137 in-patients at the Women's Hospital in Hazaribagh, twenty-five were Muslim and twenty-three belonged to higher castes of Hindus. The following year, of the 113 patients at the Hospital, upper-caste attendees included six Brahmin and two Rajput patients. The year 1902 saw three Rajput and one Brahmin attend the Hospital from a total of 125 female in-patients. 'I see a great many changes in this part of the "unchangeable East" since my arrival ten years ago', Richardson remarked of these patients. 'This is a foretaste, I hope, of still greater changes in years to come'.[45]

A symbol of the Mission's success in providing healthcare for female patients of Hazaribagh district is found in the fact that more female patients attended the Mission's hospitals and dispensaries than the local municipal hospital. Separate figures for female patients attending the Mission's medical institutions were first recorded in 1904, when the number of female patients attending the Mission's hospital and dispensary in Hazaribagh alone totalled 4,417.[46] This was more than twice the female attendance at the municipal hospital for that year, which was recorded as 2,035.[47] This reflected general patient preference for the Mission's medical institutions over those provided by the local municipality. In the same year, 1904, the Dublin University medical missionaries attended to 37,041 patients in their hospital and dispensary in Hazaribagh and four district dispensaries located at Ichâk, Petarbar, Chitarpur and Ranchi.[48] This is in comparison to the municipal medical institutions which treated 29,019 patients in the municipal dispensary and hospital at Hazaribagh and four municipally-funded district dispensaries.[49]

The female missionaries even attracted greater numbers of patients than the local Dufferin nurse. Lady Dufferin's enterprise – The National Association for

43 *Tenth Annual Report for the Year 1899*, 30.
44 *Seventh Annual Report for the Year 1896*, 19.
45 *Thirteenth Annual Report for the Year 1902*, 29.
46 *Fifteenth Annual Report for the Year 1904* (Dublin: University Press, 1905), 33.
47 Colonel S.H. Browne, *Triennial Report on the Working of the Charitable Dispensaries Under the Government of Bengal and the Calcutta Medical Institutions, for the Years 1902, 1903, and 1904* (Calcutta: Bengal Secretariat Book Depot, 1905), xxii–xxiii.
48 *Fifteenth Annual Report for the Year 1904*, 33–9.
49 Browne, *Triennial Report ... for the Years 1902, 1903, and 1904*, xxxvii.

Supplying Female Medical Aid to the Women of India – was a large nationwide organisation aimed at providing medical care to Indian women by campaigning for female wards in hospitals, and by training and employing Indian nurses and doctors. The Association, more widely known by the name of its funding body the Dufferin Fund, had placed a female medical attendant, Shornolata Mitter, a graduate of Campbell Medical School in Calcutta, in charge of the Dufferin dispensary in Hazaribagh in 1895. However, the Dufferin Fund practice did not grow significantly, eventually closing in 1911 in response to the work being carried out for female patients by the DUM, which, it acknowledged, 'provides medical assistance to females of all classes, including purdah women, on a much larger scale'.[50] As such, the DUM was the prime healthcare provider for women in the Hazaribagh district.

Although great advancements had been made in the first fifteen years since the medics of the Dublin University Mission and the Dufferin Fund had arrived in Hazaribagh, work was restricted by the lack of a fully qualified female doctor, despite continual calls for such. When Dr Eva Jellett became the first female doctor to be accepted by the DUM in 1905, it became possible to fulfil a long-held goal and to open a fully female staffed hospital for women in Hazaribagh. In 1912, St Columba's Hospital for Women was opened. It was recognised early in the planning stage of the hospital that the female medical work itself would not be sustainable without a guaranteed nursing workforce. Although Irish nurses were accepted by the Ladies' Auxiliary almost year on year, the number had never been sufficient enough to fully staff a hospital and maintain the district dispensaries. There was also a recognition that this stream of Irish nurses would decrease in the wake of Florence Nightingale's campaign to improve nursing standards and recognise nursing as a valuable career within Britain and Ireland. Consequently, a nurses' training centre was included in the plans for St Columba's Hospital, becoming the only nursing training school in the newly created province of Bihar and Orissa. Furthermore, local employees were increasingly viewed among missionary circles as imperative to the longevity of the Mission. The medical missionary Irene Barnes who worked for the Church of Ireland Zenana Missionary Society, noted of Indian nurses: 'no better agents can be found … their influence and conduct must tell more powerfully with their countrywomen than that of the foreigner'.[51]

The Mission had acknowledged the value of employing Indian women before the instigation of the Nurses' School. In 1904, Hulās Tirki, a hospital assistant trained in St Catherine's Hospital, Cawnpore, was employed by the Mission. The opening of St Columba's Hospital in 1913 provided an opportunity for the Mission

---

50   *Twenty-Sixth Annual Report of the Bengal Branch of the National Association for Supplying Female Medical Aid to the Women of India. 1911* (Calcutta: Bengal Secretariat Press, 1912), 33–4.

51   Irene H. Barnes, *Between Life and Death: The Story of C.E.Z.M.S. Medical Missions in India, China, and Ceylon* (London: Marshal Brothers and Church of England Zenana Missionary Society, 1901), 292.

itself to train greater numbers of Indian women as nurses and medical healthcare workers for the first time. The 1914 Mission report first recorded an extensive list of not only Indian staff, but also Indian female students involved in medical care in the region. Mohini and Dyamani were listed as staff nurses while Mockta, Jaikumari and Tabitha were classed as junior nurses under training. Probationer nurses included Juliani and Silwanti, and Ujala was listed as the compounder under training.[52] By 1925, Mohini taught the probationer class – implying that the DUM not only saw the necessity of training Indian women as nurses but also as nursing teachers.[53]

In 1912, a visit by Sir Charles Bayley, the first Lieutenant-Governor of Bihar and Orissa, to St Columba's Hospital and Nurses Training School publicly conveyed appreciation on the part of the colonial government for the medical work of the DUM:

> The new building will give a great extension to your sphere of useful work; and I am very glad indeed to hear that you contemplate making a special feature of training Indian nurses who will bring help to many who are prevented by distance or other causes from placing themselves in your hands.[54]

In this inauguration speech, Sir Bayley recognised that Indian female practitioners were needed to extend the geographical and social parameters within which 'western' medicine was practiced. Thus in recognising a need for Indian practitioners to extend healthcare, the idea of western medicine as a mechanism of western control was altered, contributing to the evolution of something 'global' in its place. The medical knowledge and practice of the Irish medical missionaries and Indian practitioners was informed as much by their experiences working in Chota Nagpur, as by those networks which they communicated through in belonging to a global community of knowledge.

Networks of Knowledge: Agra to Vellore via Dublin–Delhi–London–Lucknow

The Irish missionaries' medicine was cultivated by their experiences in Chota Nagpur. The climate, geography and customs of the region impacted upon the way these missionaries practised their medicine in addition to improving their medical knowledge, especially of tropical medicine and diseases common to the region. In addition, both the Irish and Indian practitioners extended their knowledge and practice of medicine by attending and delivering courses in Dublin, London and throughout India.

The district of Hazaribagh, due to its large expanse, low population density and temperate climate, experienced a greater standard of natural health than other regions

---

52 *Twenty-Fifth Annual Report for the Year 1914*, 62–3.
53 *Thirty-Sixth Annual Report for the Year 1925*, 53.
54 *The Dublin University Missionary Magazine* 8 (1914): 76.

of Bengal, receiving results of 'good' and 'very good' in public health and sanitation tests.[55] However, malaria, cholera and dysentery remained the major illnesses which the medical practitioners came in contact with. The medical personnel of the DUM recognised the need for additional training in medical specialities and research most suited to the Hazaribagh region, particularly tropical disease and maternal care. This enabled the practitioners to learn additional skills relating to diseases not native to the western countries in which they had trained. Hearn, for instance, recorded that he and his colleagues encountered tropical illnesses and diseases 'which we have often read about, but never come across, at home', and consequently felt that 'we have had to become medical students all over again'. 'It has enabled us to gain a slight insight into the manners and customs of the natives', he continued, 'Thus we have learnt something of their ideas regarding the causes of disease and also a little about the methods of treatment adopted by the native physicians'. Such encounters and interactions were not, however, without their difficulties, for 'the native method of treatment is in most cases directly opposite to the European'.[56] As such, in recognising that the missionaries were informed by their experience working in India, Hearn's reports suggest that western practitioners did not always aim to 'diffuse' their own knowledge, but saw the colonial landscape as an opportunity to develop their medical knowledge and practice. While Hearn's report does reflect typical colonial dynamics in portraying his western medical practice as superior to local traditional practices, he nevertheless conveys an appreciation of the need to learn more about tropical diseases so as to effectively minister to the local population. Thus Hearn's report negates conventional 'diffusion' arguments of the spread of western science from west to east.[57]

In July 1897, Hearn returned to Ireland on leave and while at home took a course in bacteriology at Trinity College's medical school.[58] Before travelling to Chota Nagpur in 1905, Dr Jellett attended an additional course of lectures on tropical diseases at the London Hospital for Tropical Diseases.[59] Nurses Alice Roe, Margaret Burn, Emily Poole and Meta Young were all sent to St John's Maternity Home, Battersea for additional training.[60] Obtaining knowledge in female specialities was thought to be imperative among the medical practitioners, not only in order to provide obstetric and gynaecological services, but additionally to learn how to extend western midwifery care to patients. In 1899, Hassard recorded that she had personally attended to twenty-one cases in the last two months of that year

---

55   J.G. Pilcher, *Triennial Report on the Working of the Charitable Dispensaries Under the Government of Bengal for the year 1890, 1891, and 1892* (Calcutta: Bengal Secretariat Press, 1893), lxxix.

56   *Fifth Annual Report for the Year 1894*, 30–33.

57   George Basalla, 'The Spread of Western Science', *Science* 156 (1967): 611–22.

58   *Eighth Annual Report for the Year 1897*, 8.

59   *Sixteenth Annual Report for the Year 1905*, 96.

60   *Eight Annual Report for the Year 1897*, 63; *Twelfth Annual Report for the Year 1901* (Dublin: University Press, 1902), 76; *Thirteenth Annual Report for the Year 1902*, 75.

alone, and that more services were required for female patients. This, she noted, was imperative in order to overcome local superstition and to further interaction between the local population and western scientific knowledge systems.[61]

The female Indian medical students whom the Irish missionaries taught were also to interact with differing systems of medical practice. As with the male Digby Pupils, these female Indian medical students received training from the European medical staff at Hazaribagh and were examined in official qualifications further afield. Once fully trained in general nursing at St Columba's, the Indian nurses were often sent for further midwifery training at St Stephen's Hospital in Delhi, which was run by the Cambridge University Mission. Qualified nurses were also sent to the Lady Kinniard Hospital, Lucknow, St Catherine's Hospital, Cawnpore and the Duchess of Teck Hospital, Patna. Under the stipulation that they work for the Mission for a year on returning, the Indian nurses were paid expenses for undertaking additional courses. However, the SPG recorded that the medical students tended to stay with the Mission longer than the required year as they 'would rather work in the Mission to which they and their countrymen owed so much'.[62] While experience of nursing and living elsewhere in India broadened each nurse's professional and personal practice, they nevertheless desired to utilise their skills within the district in which the DUM worked – Hazaribagh.

Both Jaikumari and Mockta were sent to Delhi for additional training after qualifying in general nursing in Hazaribagh. Upon returning, Jaikumari was sent to work in the district dispensary in Singhani. Initially working alongside nurse Emily Henry, she ran the dispensary single-handed when Henry left in 1923. Mockta was sent to Chitarpur, taking responsibility as the sole female nurse in Chitarpur following the withdrawal of Emily Martin in 1919. In these roles, as the only representative of western scientific knowledge in remote district dispensaries, did these Indian nurses alter or 're-form' 'western' medicine as something 'global'? This is, ultimately, an impossible question to answer given an absence of provincial records and personal sources. But what can be said is that nurses like Mockta and Jaikumari stood as an example, demonstrating that female professional training and employment was attainable as an alternative to the traditional roles of tribal or indeed Hindu or Muslim women. That such an example was influential is suggested by the continuous growth in female medical students taught by the Mission.

In 1922, a complete list of Indian female nursing staff was published in the annual report and this continued until 1929. In 1922, the Indian nursing staff numbered sixteen, but this had risen to twenty-eight by 1928.[63] Local appreciation for the training of female medical students solidified its acceptability as a necessity

61 *Tenth Annual Report for the Year 1899*, 30.

62 Society for the Propagation of the Gospel, *Here and There with the S.P.G. in India* (Westminster: Society for the Propagation of the Gospel, 1905), 81.

63 *Thirty-Third Annual Report for the Year 1922* (Dublin: University Press, 1923), 30 and *Thirty-Ninth Annual Report for the Year 1928* (Dublin: University Press, 1929), 50.

in the Hazaribagh region. In 1923, the Rani Sahiba of Ramgarh inspected St Columba's Hospital and promised Rs. 40,000 towards building a nurses' home. Completed in 1926, this nurses' home could accommodate twenty-two Indian nurses, a European sister, a compounder and a housekeeper.[64] Almost immediately, the accommodation at the nurses' home was filled – the nursing staff increased from twenty in 1924 to twenty-four in 1926.[65] This nurses' home increased the intake of female nurses as it provided a secure environment in which nurses could live and study, adhering to Indian female social customs. Nurse Jessie Perrie wrote of this donation: 'The new building will not only represent the public spirit of the donors, but will be a concrete expression of their appreciation of the nursing done by Indian women'.[66]

In December 1926 the nurses' quarters were officially opened by the Rani Sahiba, at a purdah party – that is, a party held in seclusion from male attendees. This conveyed not only the acceptance of western healthcare in rural Chota Nagpur, but equally the importance placed upon female education and employment within the social precincts of Indian society. Recognition of the medical work of the Mission as having to conform to traditional roles within Indian or more specifically Hazaribagh society continued to ring true for decades to come. Jean Butt, in an article entitled 'Ward nursing – British versus Indian methods', published by the SPG, wrote of her experience nursing at St Columba's Hospital in 1964, commenting:

> When I first came to India I thought that the idea of having relatives to stay with the patients was a bad one. Now I'm not so sure. Even though the wards do look dirty and untidy at least one can be sure that if anything is amiss with the patient whilst you are with somebody else, the relative will call you.[67]

As the Irish nurses 're-formulated' their practice to accommodate customs and practices prevalent in Hazaribagh, so the Indian nurses 're-formulated' the missionaries' medicine, by engaging, practicing and ultimately altering it themselves.

In 1939, the DUM employed their first fully qualified Indian female doctor – Dr Binodini Porh. Five years later, Dr Porh took up the position as Head of the medical work and of the Ladies' Auxiliary. During 1948–49, Dr Porh took a post-graduate course in children's diseases at the Harcourt Street Children's Hospital and the Eye and Ear Hospital, Dublin, and in 1960 she travelled to Vellore to

---

64   *Thirty-Sixth Annual Report for the Year 1925*, 50.

65   See *Thirty-Fifth Annual Report for the Year 1924* (Dublin: University Press, 1925), 38 and *Thirty-Seventh Annual Report for the Year 1926* (Dublin University Press, 1927), 45.

66   *Thirty-Sixth Annual Report for the Year 1925*, 54.

67   United Society for the Propagation of the Gospel, *Annual Report from Diocese of Chota Nagpur, 1964* (1965).

carry out further postgraduate study. By the 1970s it was recognised impractical to send Irish missionaries to Chota Nagpur, though it was noted that 'exchanges still exists' such as Dr Pushpa Dass' coming to Dublin in 1978 to undertake postgraduate study. By 1984, St Columba's Nurses' Training School had eighty-four nurses in training, solely in the hands of Indian practitioners. The Mission increasingly saw a need to invest in training Indian members of staff. As early as 1905 the Mission noted that 'it becomes more evident that this must, after all, form the backbone of the Mission work … [and] the material for further advance', and by the 1960s the Mission, still retaining a residue of colonialism, observed that St Columba's Hospital 'sends it graduates round the district, State and Country and some to other countries as well, so we are helping India in one of her most essential services'.[68] The Chairman's Report for the years 1973–77 reported that nurses trained at St Columba's worked in America, Canada, Germany and all over India. Writing on the centenary of the DUM in 1992, Kenneth Kearon, then Chairman of the Mission, noted the success of St Columba's Hospital by stating that 'from its excellent Nursing School graduates have served the needs of others in hospitals throughout Bihar and beyond'.[69]

Conclusion

Despite records of high patient attendance and low mortality rates, the appeal of the DUM's western medicine lay not solely in the medicine itself, for if medical knowledge alone were sufficient, the majority of patients would have flocked to the municipal dispensary in the town of Hazaribagh.[70] Nor can its success be attributed to a Christian approach to administering medicine, for it was acknowledged by both the Mission and other sources that religious conversion was, perhaps ironically, not the Mission's strongpoint.[71] Rather, the appeal of the medical aspect of the Mission can be explained by the manner in which the medicine was administered and then altered in reaction to geography, climate and social customs, in employing local people in the hospital and allowing local interruption. This hybrid medical practice was moulded from two differing sources and was informed by geography

---

68   *Sixteenth Annual Report for the Year 1905*, 21; *St Columba's Hospital, Hazaribagh, a History from 1892 to 1963* (Hazaribagh, Gupta Press, 1964), 8.

69   Kenneth Kearon, *St Columba's Hospital, Hazaribagh, 1892–1992* [Record in the Representative Church Body Library, Dublin].

70   Oonagh Walsh has calculated that there were no postoperative deaths during the first five years of the Mission's medical work. Oonagh Walsh, 'The Dublin University Mission Society, 1890–1905', *History of Education* 24 (1995): 66.

71   H. Pakenham Walsh recorded in 1898: 'I do not think we can look for conversions, unless in exceptional cases, in the school'. *Ninth Annual Report for the Year 1898* (Dublin: University press, 1899), 18; The Imperial Gazetteer observed 'The Dublin University Mission, established at Hazāribāgh in 1892, has not been very successful in making conversions'. Meyer, *Imperial Gazetteer of India*, 13, 90.

and social traditions, implying that the missionaries' medical knowledge was not 'diffused' from west to east but that the practice of this knowledge was altered upon contact with a different environment.

This chapter has outlined how knowledge of western medicine enabled the medical missionaries of the DUM to penetrate regions of Bengal which were separated, due to geography as well as social and cultural mores, from interaction with other communities. Medicine enabled the Irish practitioners to transcend the spatial boundaries established by late nineteenth-century conceptions of useable space. Although medicine acted as a means by which the medical missionaries could practically travel the district in safety and enter notionally private spaces, much animosity directed toward the missionaries still prevailed, thereby negating the success of the diffusion of western ideas and practices from west to east. The missionaries were obliged to look towards the local population to recruit medical students and nurses who knew the language, culture and geography of the region and who, it was thought, might face less prejudice from the local population. Through reconstructing the way in which they practiced their medical knowledge and working, in particular, with Indian female practitioners sensitive to local traditions and medical practices, the missionaries of the DUM ensured the development and longevity of a newly developed medical practice in the region.

Chapter 11

Historical Geographies of Textual Circulation: David Livingstone's *Missionary Travels* in France and Germany

Louise C. Henderson

Introduction

David Livingstone's *Missionary Travels and Researches in South Africa* was a Victorian publishing phenomenon.[1] First published by London-based John Murray in 1857, the text had been consciously constructed to appeal to a wide range of audiences, within and beyond the scientific and geographical community.[2] This widespread appeal helped to ensure the work enjoyed commercial and critical success in Britain on a scale which was unprecedented for a travel narrative. Pre-publication subscription sales numbered 12,000 ensuring the initial print run sold out immediately while sales of the guinea edition quickly reached 30,000. A subsequent abridged 'popular account' aimed at non-specialist readers sold a further 10,000 copies.[3] However, *Missionary Travels* itself was only part of a broader network of print dedicated to constructing, communicating and debating the 'discoveries' which Livingstone outlined in his narrative. The official account of the explorer's sixteen years in Africa gave rise to a remarkable number of articles, reviews, extracts and sketches in the British periodical and newspaper

---

1   David Livingstone, *Missionary Travels and Researches in South Africa* (London: John Murray, 1857).

2   Louise C. Henderson, '"Everyone will die laughing": John Murray and the Publication of David Livingstone's *Missionary Travels*', Livingstone Online, Welcome Trust Centre for the History of Medicine at UCL. Available at http://www.livingtsoneonline.ucl. ac.uk/companion.php?id=HIST2; Louise C. Henderson, 'Geography, Travel and Publishing in Mid-Nineteenth Century Britain' (Unpublished PhD Diss., Royal Holloway, University of London, 2012); Justin D. Livingstone, 'The Meaning and Making of *Missionary Travels*: The Sedentary and Itinerant Discourses of a Victorian Bestseller', *Studies in Travel Writing* 15 (2011): 267–92; Justin D. Livingstone, '*Missionary Travels*, Missionary Travails: David Livingstone and the Victorian Publishing Industry', in *David Livingstone: Man, Myth and Legacy*, ed. Sarah Worden (Edinburgh: National Museums Scotland, 2012), 32–51.

3   David Livingstone, *A Popular Account of Missionary Travels and Researches in South Africa* (London: John Murray, 1861).

press, not to mention a number of alternative accounts which sought to benefit from the public's fascination with Livingstone.[4]

It was not uncommon, of course, for print industries to develop around nineteenth-century expeditions for as Driver notes, 'from the point of view of metropolitan science and culture, exploration without writing and publication was no exploration at all'.[5] However, the volume of commercial print which Livingstone's *Missionary Travels* inspired in the late 1850s and 1860s was certainly unusually large. Yet, while Livingstone scholars have noted that *Missionary Travels* and the broader network of print which promoted and sustained the text were crucial to constructing and communicating Livingstone's reputation both within and beyond the scientific community, it is striking that the impact of *Missionary Travels* has almost always been considered in relation to British audiences alone.[6] I have recently written elsewhere about the circulation of *Missionary Travels* in America but the extent to which the publication of *Missionary Travels* was an international or even global literary event is a question which has barely been touched upon.[7] This lack of understanding in relation to the best known nineteenth-century travel narrative points to a broader gap in knowledge regarding the international circulation of nineteenth-century travel and exploration literature (and the ideas therein). This is problematic, particularly in cases where texts crossed language barriers for, as Rupke notes,

> Translators relocate books, taking these away from the intellectual control of authors, repossessing the texts, possibly in the service of very different purposes than those for which they were originally intended. Such alterations of meaning can be effected by new, additional prefaces, by footnote commentary, by other additions such as illustrations, by omissions, and most fundamentally, by the very act of cultural relocation.[8]

Translation then regularly involves transformation of both textual meaning and of the physical construction of books as printed objects. Elshakry's recent book *Reading Darwin in Arabic* emphasises 'the creative tensions involved in the negotiation of meaning across languages and locales' to suggest that 'What

---

    4   Henderson, '"Everyone will die laughing": John Murray and the Publication of David Livingstone's *Missionary Travels*'; Henderson, 'Geography, Travel and Publishing in Mid-Nineteenth Century Britain'; Louise C. Henderson, 'David Livingstone's *Missionary Travels* in Britain and America: Exploring the Wider Circulation of a Victorian Travel Narrative', *Scottish Geographical Journal* 129 (2013): 179–93.

    5   Felix Driver, '*Missionary Travels:* Livingstone, Africa and the Book', *Scottish Geographical Journal* 129 (2013): 167.

    6   Although see Claire Pettitt, *Dr Livingstone, I Presume? Missionaries, Journalists, Explorers and Empire* (London: Profile Books, 2007).

    7   Henderson, 'David Livingstone's *Missionary Travels* in Britain and America'.

    8   Nicolaas Rupke, 'Translation Studies in the History of Science: the Example of *Vestiges*', *British Journal for the History of Science* 33 (2000): 210.

gets read is movable and transmissible. And yet it also helps to define specific communities, with distinct epistemic, didactic and literary styles, often with their own linguistic and conceptual vocabularies'.[9] In this light, failing to follow Livingstone's text into foreign markets results in a partial understanding of the nineteenth-century impact of *Missionary Travels* (and the ideas therein) at best.

This chapter examines recent work in the history and geography of science and history and geography of the book to offer a rationale for exploring the wider circulation of travel and exploration texts more generally. Thereafter, the chapter explores for the first time how Livingstone's text crossed both national *and* linguistic boundaries by examining French- and German-language editions of *Missionary Travels*. It examines authorised translations as well as unauthorised foreign-language works which sought to repackage the ideas expressed in *Missionary Travels* in alternative forms. In so doing, attention is drawn to the mediating influence of foreign publishers, authors, translators and reviewers and to regular challenges to the legal frameworks that attempted to regulate the circulation of texts in this period. By analysing foreign-language editions of *Missionary Travels*, it is shown that translation could transform cultural meaning and texts as printed objects. In the process, we gain a better understanding of the impact and reach of nineteenth-century travel narratives by exploring the full extent of the journeys they took beyond the publishing house. Although the place of Livingstone's texts and their derivatives in an emerging 'global' print culture is beyond the scope of this chapter, examining the transformative effects, and fundamental challenges, of linguistic translation offers a critical perspective relevant to that larger project.

Historical Geographies of Knowledge

Almost two decades ago, Thrift, Driver and Livingstone proclaimed that 'If it were necessary to choose the most vibrant and exciting areas of research in the social sciences and humanities today, then surely the study of science as a social construction would figure large'.[10] The subsequent conversations between geographers, historians and sociologists of science have ensured that this is an area of research that continues to thrive, with scholars gathering evidence from diverse times and locales to argue that the production of scientific knowledge is a situated practice, 'a social practice earthed in concrete historical and geographical circumstances'.[11] In so doing, such scholarship has challenged traditional

---

9   Marwa Elshakry, *Reading Darwin in Arabic, 1860–1950* (Chicago: University of Chicago Press, 2013), 5.

10   Nigel Thrift, Felix Driver and David N. Livingstone, 'Editorial', *Environment and Planning D: Society and Space* 13 (1995): 1.

11   David N. Livingstone, *Putting Science in its Place: Geographies of Scientific Knowledge* (Chicago: University of Chicago Press, 2003), 180.

understandings of scientific knowledge as untouched by the 'messiness of local circumstances'.[12] Indeed Livingstone, Withers and other geographers have convincingly argued that historical 'geographies of science' can complement the work of historians of science and provide meaningful insights into how and why certain claims to knowledge gather prominence in certain spaces.[13] Crucial to this endeavour is an attention to the way that knowledge travels in and between different spaces at a number of scales, for transmission regularly results in transformation, facilitating the production of new forms of knowing in the process. In the wake of this work, geographers and others concerned with the historical development of scientific knowledges have explored a diverse range of knowledge-making venues, from the laboratory, lecture hall and field to the public house and ship's deck.[14] Such work tends to challenge rigid distinctions between notions of production and consumption, by considering how knowledge is made and re-made as it travels in and across different sites. Thinking about how knowledge was produced in and moved between different sites reveals that, in the words of Livingstone,

> scientific knowledge is not just about how and where the worlds of natural objects or material artefacts are experienced, nor about how the rendezvous between human culture and nature is stage-managed. It is also about the encounter with scientific texts.[15]

---

12   David N. Livingstone, *Science, Space and Hermeneutics: Hettner Lecture 2001* (Heidelberg: University of Heidelberg, 2001), 10; See also, Charles W.J. Withers, *Geography, Science and National Identity: Scotland Since 1520* (Cambridge: Cambridge University Press, 2001), 1–29; David N. Livingstone, 'The Spaces of Knowledge: Contributions Towards a Historical Geography of Science', *Environment and Planning D: Society and Space* 13 (1995): 5–35; David N. Livingstone, *Putting Science in its Place*; David N. Livingstone, 'Text, Talk and Testimony: Geographical Reflections on Scientific Habits. An Afterword', *British Journal for the History of Science* 38 (2005): 95; Simon Naylor, 'Introduction: Historical Geographies of Science – Places, Contexts, Cartographies', *British Journal for the History of Science* 38 (2005): 1–12; Steven Shapin, 'Placing the View from Nowhere: Historical and Sociological Problems in the Location of Science', *Transactions of the Institute of British Geographers* 23 (1998): 5–12.

13   See especially, Livingstone, 'The Spaces of Knowledge: Contributions Towards a Historical Geography of Science', and *Putting Science in its Place*.

14   Naylor, 'Introduction: Historical Geographies of Science – Places, Contexts, Cartographies'; Dorinda Outram, 'New Spaces in Natural History', in *Cultures of Natural History*, ed. Nicholas Jardine, James A. Secord and Emma C. Spary (Cambridge: Cambridge University Press, 1996), 249–65.

15   David N. Livingstone, 'Science, Text and Space: Thoughts on the Geography of Reading', *Transactions of the Institute of British Geographers* 30 (2005): 391; Charles W.J. Withers, *Placing the Enlightenment: Thinking Geographically about the Age of Reason* (Chicago: University of Chicago Press, 2007).

By such reckoning, if we re-examine the conditions surrounding the production, circulation and consumption of scientific texts, we can better understand how knowledge is made and remade in particular historical and geographical circumstances. Indeed if the 1990s were characterised by a flourishing interest in the social (and spatial) construction of science, we could argue that the period since has been marked by a growing awareness of the need to investigate the production and mobilisation of knowledge through print and print industries.[16]

Historical Geographies of the Book

Recent work by historians of science concerned with the connections between print and knowledge production in the nineteenth century has emphasised the importance of place and location in shaping and re-shaping knowledge as it moves through different spaces at a variety of scales. That there is a geography as well as a history of the book has long been recognised, as for example in Febvre and Martin's 1958 work, *The Coming of the Book*.[17] In this work, the geography of the book is interpreted through the diffusion of the printed press and the resulting printed materials over space. However, as Withers and Ogborn have recently suggested, this account

> is, in essence, a story of the diffusion of something (or some things) already made. As a result it has little to say about how we might construct an account of the technologies of making books (printed or otherwise), of those involved in producing those books, of the books themselves and crucially, of their readers which would show how their diverse geographies can illuminate how those people and things actually come to be as they are.[18]

Despite the evident limitations of the geography of the book as conceived by Febvre and Martin, 'it is only recently that the discipline of book history has consciously engaged with theories and methods in support of geographies of the book'.[19] Consequently, it has been suggested that 'If the history of the book is now an established discipline, the geography of the book is still making up its rules'.[20]

---

16   James A. Secord, 'Knowledge in Transit', *Isis* 95 (2004): 654–72.

17   Lucien Febvre and Henri-Jean Martin, *The Coming of the Book: The Impact of Print, 1450–1800* (London: Verso, 1976).

18   Miles Ogborn and Charles W.J. Withers, 'Introduction: Book Geography, Book History', in *Geographies of the Book*, ed. Miles Ogborn and Charles W.J. Withers (Farnham: Ashgate, 2010), 4–5.

19   Fiona A. Black, 'Constructing the Spaces of Print Culture: Book Historians' Visualization Preferences', in *Geographies of the Book*, ed. Ogborn and Withers, 79.

20   Leah Price, '*In Another Country: Colonialism, Culture, and the English Novel in India* (Review)', *Victorian Studies* 45 (2003): 334; Innes M. Keighren, 'Bringing

Such an assertion stems from the fact that scholars have (as in the case of book history) approached book geography from different disciplinary backgrounds and with different understandings of what constitutes geography in this context.

Within the geographical discipline itself, there has been a long running interest in the relationship between space, place and text. However, research in this area has tended to develop along two parallel lines of enquiry focusing either on spaces *in* texts (undertaken under the banner of 'literary geography') or spaces *of* texts (undertaken by what might be called textual geographers).[21] As Ogborn observes, 'On the one hand are those who use notions of cultural production and formal aesthetics to produce ever more complex readings of the meanings of texts, spaces and their conjunctions' and 'on the other hand are those whose concern with the geographies of production and dissemination, and with the embodied practices of reading and writing, serves to generate a material historical geography of texts'.[22] This chapter suggests, across the grain of this distinction, that attention to geographies of textual production, dissemination and consumption can provide more nuanced understandings of the way that textual meaning can be constructed and contested in specific historical and geographical circumstances. Thus whilst there are differences between the approaches adopted by literary and textual geographers, these distinctions should not be overemphasised.[23] Textual geographers need not eschew analysis of the spaces within texts and nor is imaginative and creative process the exclusive preserve of literary geographers.

Those scholars who utilise book history approaches to investigate textual or book geography examine the relationship between text and space at a variety of scales. For Mayhew, for instance, the geography of the book begins with the space of the printed page. He argues that historians of geography ought to pay more attention to the physicality of geographical works rather than simply concentrating upon 'the contexts and contents of geographical books'. As he explains, 'There has been a tendency to focus on the 'message' of historical texts, as if this can be meaningfully divided from the print medium in which it is expressed, and as if that medium is not in and of itself performing expressive functions'.[24] By contrast,

---

Geography to the Book: Charting the Reception of *Influences of Geographic Environment*', *Transactions of the Institute of British Geographers* 31 (2006): 528.

21   Angharad Saunders, 'Literary Geography: Reforging the Connections', *Progress in Human Geography* 34 (2010): 437.

22   Miles Ogborn, 'Mapping Worlds', *New Formations* 57 (2006): 149.

23   See the following for an overview of the development of literary geography as a field including its relationship with textual geography and literary studies: Saunders, 'Literary Geography'; Sheila Hones, 'Text as it Happens: Literary Geography', *Geography Compass* 2 (2008): 1301–17; Marc Brosseau, 'Geography's Literature', *Progress in Human Geography* 18 (1994): 333–53; Douglas C.D. Pocock, 'Geography and Literature', *Progress in Human Geography* 12 (1988): 87–102.

24   Robert J. Mayhew, 'Materialist Hermeneutics, Textuality and the History of Geography: Print Spaces in British Geography, c.1500–1900', *Journal of Historical Geography* 33 (2007): 487–8.

Mayhew himself deploys what he terms a 'materialist hermeneutics' approach in order to investigate 'what the history of the material forms of geography books can tell us of the social and intellectual status of geography' in the early-modern period.[25] Such work, though, invariably moves outwards from the printed page itself to consider a variety of individuals and practices implicated in the fashioning of particular printed objects. As Mayhew demonstrates elsewhere, for instance, an attention to the physical construction of the page can emphasise the role of editors and editing which in turn 'can deepen the move from a concern with the *history* of the book to the *historical geography* of the book by showing how spatially differentiated editorial practice was, this being a contributor to the geographically differentiated production and reception of knowledge'.[26]

However, if we accept that materiality can also influence how a work is received, then 'to speak of *the* reception of *a* book is problematic'.[27] Keighren argues that studies of book readership should explicitly identify which editions or volumes are under consideration and, whenever sources allow, address specific copies and study marginalia and provenance to demonstrate how real readers may have both responded to the materiality of a text and also how they might have physically altered it throughout its lifespan.

Others focus less resolutely on the spatial arrangement of the page and instead consider the various networks which were implicated in the production and circulation of printed products in and beyond the publishing and printing houses of specific locations. Raven, for instance, has attempted to reconstruct what he calls the 'bookscapes' of eighteenth-century London in order to 'offer a multivariate historical mapping of cultural production that evaluates connections between place, personnel, and product, and tracks changing processes and perceptions of literary and artistic endeavour'.[28] Drawing upon historical maps, land tax records, insurance documents and property records among other sources, Raven has described how we might come to appreciate London's multiple book trades using mapping techniques and historical sources which are less familiar to book historians. Doing so, he contends, opens up further avenues for research, as well as avoiding the tendency of the new cultural history to overlook the dynamics of change in specific sites and times. However, to prove effective such site-specific histories need to be brought into conversation with other histories and geographies so that we might begin to consider not only how specific works were produced and circulated within local contexts but also how they moved outwards, across

---

25   Mayhew, 'Materialist Hermeneutics, Textuality and the History of Geography', 471–2.

26   Robert J. Mayhew, 'Printing Posterity: Editing Varenius and the Construction of Geography's History', in *Geographies of the Book*, ed. Ogborn and Withers, 161.

27   Keighren, 'Bringing Geography to the book', 528.

28   James Raven, 'Constructing Bookscape: Experiments in Mapping the Sites and Activities of the London Book Trades in the Eighteenth Century', *Humanities Research Group Working Papers, University of Windsor*, 9 (2001): 1.

regional and national boundaries in particular material forms. When approached in this way, as Withers suggests, the geography of the book necessarily 'addresses the spatial, to include the displacement of texts, reading, and reviewing practices in different physical and social spaces and the questions of meaning and epistemic significance that arise from such matters of geography'.[29] Thus, as Keighren explains further,

> Attending to the spaces in which texts are composed, printed, distributed, sold, read and reviewed, the geography of the book attempts to situate ideas, practices and practitioners within geographical context, and to understand how knowledge and ideas are made mobile and circulate between these spaces.[30]

The more recent collaborative research of Withers and Keighren has focused on author–publisher relations in the context of early nineteenth-century travel and exploration narratives. Concerned with the publisher John Murray, in particular, they have built upon existing work highlighting the transformations that travellers' narratives could undergo prior to publication in order to problematise accepted categories such as author, editor and publisher as well as breaking down the traditional dichotomy between writing practices 'in the field' and those which take place 'at home' in preparation for publication. Withers and Keighren argue that is it fruitful to consider the fluid relationship between practices associated with authoring, editing and publishing rather than assigning discrete roles to different individuals. As they explain, in the context of travel and exploration narratives, 'editing was something that geographical authors did and ... the processes of geographical authoring and authorising was something publishers acting as editors did, sometimes in association with writers, on occasion not'.[31] This has important consequences for how we think about authorship and consequently authority. By demonstrating that publishers, editors and others not named on title pages had an important role in shaping the content of geographical works, this scholarship urges us look more closely at who we write in and out of the history of geography. However, in order to understand the impact that pre-publication decisions had upon geographical understandings, it is crucial to follow texts, in their various forms, into specific marketplaces. Such work, then, provides a strong rationale for re-examining even seemingly well-known travel and exploration texts, linking together questions of production, circulation and reception so that we may gain more nuanced understandings of how geographical ideas were constructed, communicated and contested in print in specific historical and geographical circumstances.

---

29   Withers, *Placing the Enlightenment*, 51.
30   Keighren, 'Bringing Geography to the Book', 528.
31   Charles W.J. Withers and Innes M. Keighren, 'Travels into Print: Authoring, Editing and Narratives of Travel and Exploration, c.1815–c.1857', *Transactions of the Institute of British Geographers* 36 (2011): 570.

Travelling Texts: Reserving the Right of Translation

In what follows, French and German renditions of *Missionary Travels* are considered, in order to explore how Livingstone's narrative was repackaged in different cultural contexts. Although the archive of publishing firm John Murray reveals relatively little about the negotiations which led to the production of authorised French- and German-language editions, examining the legal frameworks intended to govern the international trade in books in conjunction with the translated editions themselves can tell us much about the wider circulation of Livingstone's narrative.

The 1838 International Copyright Law proved effective in establishing bilateral agreements with several European states. It was designed to 'afford protection ... to the Authors of Books first published in Foreign Countries, and their Assigns, in cases where Protection shall be afforded in such Foreign Countries to the Authors of Books first published in her Majesty's Dominions ... '.[32] Together with the amendments made in 1844, which included the extension of protection to works produced in the whole of the British Empire, this legislation provided the basis for a series of agreements between Britain and the German states, the first of which was signed with Prussia in 1846.[33] A series of other agreements followed between different European powers until, as John Feather explains, 'there was gradually developing a network of reciprocal copyright protection which covered much of Europe'.[34] In 1851, an Anglo-French treaty was finally secured after a compromise was reached over translated works. This legislation protected authors against unauthorised translations providing strict conditions were met.[35] Firstly, the title page of the original work had to indicate, as *Missionary Travels* did, that the right of translation had been reserved. Additionally, to secure protection, works had to be registered both in Stationers' Hall in London and at the *Bureau de la Libraire* at the Ministry of the Interior in Paris within three months of publication.[36] There were restrictions too upon the time frame in which a protected translation could be published, with part of the translation having to appear within a year and the whole not more than three years since publication.[37] With its attention to translations, the 1851 Anglo-French agreement prompted an alteration to domestic law and in 1855 the Anglo-Prussian agreement was also amended to afford works produced

---

32 John Feather, *Publishing, Piracy and Politics: An Historical Study of Copyright in Britain* (London: Mansell, 1994), 157.

33 Feather, *Publishing, Piracy and Politics*, 160.

34 Feather, *Publishing, Piracy and Politics*, 160.

35 Catherine Seville, *The Internationalisation of Copyright Law: Books, Buccaneers and the Black Flag in the Nineteenth Century* (Cambridge: Cambridge University Press, 2006), 51–2.

36 Feather, *Publishing, Piracy and Politics*, 161.

37 Seville, *The Internationalisation of Copyright Law*, 51–2.

in the German state the same protection.[38] Nonetheless, while this legislation was intended to curb the trade in unauthorised translations there remained a band of enterprising foreign publishers who sought to provide volumes which, although not authorised or wholesale translations, adhered to the general principles and concepts expressed in the original editions.

Missionary Travels *in Germany*

There was an avid appetite for translated travelogues among German readers and publishers were keen to feed this demand. During the nineteenth century, as Tautz notes, translated accounts of travel were an important source of 'geographical, mercantile, and scientific knowledge of the world' not least because the 'German principalities … rarely participated in the colonial undertakings of their European neighbours'.[39] The demand for German-language travel literature remained high throughout the nineteenth century although the preferred style of travel writing changed over time. For as Sidorko's work on German accounts of Balkan travel highlights, nineteenth-century writers were influenced by Humboldt and thus 'combined natural scientific observations with ethnological, archaeological and linguistic insights'. More generally, as time went on, 'purely scientific reports were gradually replaced by journalistic accounts'.[40] As was also the case in Britain and America then, there was a growing market in Germany for narrative accounts of daring exploration.

Leipzig publisher Hermann Costenoble negotiated an agreement with Murray in April 1858 which gave him the 'exclusive right of translation into German' along with a set of clichés [printing blocks] for the 45 woodcuts which would appear in his edition of *Missionary Travels* in exchange for £100.[41] This sum was paid entirely to Livingstone as Murray had agreed to surrender all profits from foreign editions to the explorer. Indeed as Murray put it to Costenoble, he acted only as the 'Friend of Dr Livingstone' in brokering the deal.[42] However, Murray appeared less than friendly subsequently when a disagreement over the registration of copyright arose. Although the exact cause of the dispute is unclear, less than a month after

---

38   Seville, *The Internationalisation of Copyright Law*, 51–2.

39   Birgit Tautz, 'Cutting, Pasting, Fabricating: Late 18th-Century Travelogues and their German Translators between Legitimacy and Imaginary Nations', *German Quarterly* 79 (2006): 170.

40   Clemens P. Sidorko, 'Nineteenth-Century German Travelogues as Sources on the History of Daghestan and Chechnya', *Central Asia Survey* 31 (2002): 284; Susanne Stark, *Translation and Anglo-German Cultural Relations in the Nineteenth Century* (Clevedon: Multilingual Matters Ltd, 1999), 143–73. On Humboldt specifically, see Nicolaas Rupke, *Alexander von Humboldt: A Metabiography* (Chicago: University of Chicago Press, 2007).

41   J. Murray to H. Costenoble, April 27, 1858, John Murray Archive, MS 41912, National Library of Scotland.

42   J. Murray to H. Costenoble, April 27, 1858, John Murray Archive, MS 41912, National Library of Scotland.

Costenoble was granted the right of translation, Murray warned him that he had obtained 'from Leipzig, full evidence of the full & exact registration of Dr Livingstone's Copyright in Saxony' and complained that 'under the circumstances your proceedings have not been creditable & I warn you that in the event of any infringement by you of Dr Livingstone's rights, I shall take legal measurements to obtain justice'.[43] It is not surprising to find that Murray fiercely defended his author for, as he later explained, he believed that 'An author's book is like a weaver's web – if you let foreign pirates take his property with the same justice you might cut off so many inches from the weaver's broadcloth on its way to market' and 'To stop a pirated copy therefore is no more than to arrest a pickpocket'.[44]

Irrespective of this disagreement, however, Costenoble did go on to publish a two-volume translation of Livingstone's narrative titled *Missionsreisen und Forschungen in Süd-Afrika Während eines Sechzehnjährigen Aufenthalts im Innern des Continents.*[45] Philosopher and psychologist Hermann Lotze, who had already prepared an edition of Charles J. Andersson's *Lake Ngami or, Explorations and Discoveries During Four Years' Wanderings in the Wilds of Southwestern Africa* for the German market, undertook the translation.[46] Echoing Rupke's earlier assertion that it is not only the text that is transformed in the process of translation, several paratextual features were altered, repositioned or omitted in the preparation of this new edition. The title page for instance dispensed with Livingstone's affiliations and the image of the tsetse fly which was used to emphasise the scientific quality of the work. Similarly, the dedication which highlighted Livingstone's personal connection to Roderick Murchison was removed. In John Murray's English edition, the preface had allowed Livingstone to directly address questions of authorship and authenticity in ways entirely conventional in English exploration narratives, apologising for his failings as an author but at the same time indirectly claiming that the unpolished character of his narrative was testament to its veracity. In his role as translator, though, Lotze assumed responsibility for the literary value of the new text, having been employed to render Livingstone's narrative accessible to a particular audience. While Lotze's translation retained the introduction in which the explorer reminded readers of his humble background and famously claimed that he would rather walk across Africa again than write another book, Livingstone's more personal preface was relocated to the second volume. While translation offered Lotze the opportunity to stamp his own authority upon Livingstone's text, he also had to ensure that it

---

43 J. Murray to H. Costenoble, May 25, 1858, John Murray Archive, MS 41913, National Library of Scotland.

44 J. Murray to W.E. Gladstone, February 28, 1860, John Murray Archive, MS 41913, National Library of Scotland.

45 David Livingstone, *Missionsreisen und Forschungen in Süd-Afrika Während eines Sechzehnjährigen Aufenthalts im Innern des Continents*, Hermann Lotze (trans.), 2 vols. (Leipzig: H. Costenoble, 1858).

46 Charles J. Andersson, *Reisen in Südwest-Afrika bis zum See Ngami in den Jahren 1850–1854*, Hermann Lotze (trans.) 2 vols (Leipzig: H. Costenoble, 1858).

remained convincing as an 'authentic' account. Hence the importance of the prominent declaration on the title page that this was an 'Autorisierte, vollständig Ausgabe für Deutschland [Authoritative and complete edition for Germany]', a fact which Costenoble's advertisements also always emphasised.

However, holding the exclusive right of German translation did not mean that Costenoble's edition circulated without competition. In 1859, a fellow Leipzig publisher Otto Spamer produced a volume dedicated to *Livingstone, the Missionary.* Spamer's text was not an attempt at direct or wholesale translation, but instead offered readers an overview of African exploration to date, before moving on to consider Livingstone's adventures in particular.[47] Costenoble's edition was clearly a superior product in terms of reproducing the highly detailed narrative from the English edition. However, by dispensing with much of the scientific information and situating Livingstone's travels amidst a wider tradition of exploration, Spamer created a work which was more accessible to the majority of readers. This book ran to only 296 pages making it considerably shorter than Costenoble's two-volume tome and, while there was less text, the work was heavily illustrated with ninety-two in-text illustrations, seven full-page illustrations and a map of Africa. Spamer's volume also contained Livingstone's portrait, an essential element which was included in so many texts dedicated to relaying the explorer's travels. While the illustrations were clearly based on those that had appeared in Costenoble's edition (and Murray's previously), those responsible for the images appearing in Spamer's edition, as well being arguably less skilled than the original craftsmen, altered some of the images in significant ways. This is clear in the case of the image that originally appeared under the title 'Lake Ngami, discovered by Oswell, Murray, and Livingstone'. Rather than depicting the moment the three men 'discovered' the lake in 1849, as the title suggests it ought to, the engraving actually shows Livingstone admiring the lake with his wife and children while an African servant brews tea in the foreground. The image was constructed by combining portraits of Livingstone and his family with the artist Alfred Ryder's drawing of the lake, made in 1850. While Livingstone had indeed returned to the lake that year, the scene that he described in his letters and journals differed markedly from that shown in the book. In reality both the pregnant Mary Livingstone and the explorer's children were constantly ill. Yet, as Tim Barringer notes, 'In *"Lake Ngami"* the archetypal roles of the Victorian middle-class household are enacted: Livingstone, the paterfamilias, is at the top of a pyramid comprising his wife, children, servants and possessions. Mary Livingstone is seen enacting the stereotypical role of "Angel in the house", caring for the young and the sick'.[48] However, in the version of this

---

47　*Erforschungsreisen im Innern Afrikas. In Schilderungen der bekanntesten älteren und neueren Reisen insbesondere der großen Entdeckungen im südlichen Afrika während der Jahre 1840–1856 durch Dr David Livingstone* (Leipzig: O. Spamer, 1859).

48　Tim Barringer, 'Fabricating Africa: Livingstone and the Visual Image 1850–1874', in *David Livingstone and the Victorian Encounter with Africa*, ed. John M. MacKenzie (London: National Portrait Gallery, 1996), 179.

scene in Spamer's book, Livingstone is depicted by his children's side beckoning instructively towards the Lake, rather than standing protectively over his wife's shoulder. Mary Livingstone meanwhile now stood watching over the black servant brewing tea in the centre of the picture. While such changes were relatively subtle they point toward differing expectations among British and German readers as to the construction of the family unit and more specifically about the role of women therein. However, while this example highlights that elements of Livingstone's text were altered to ensure their fit with German cultural norms, the fact that the majority of images contained within the shorter account of Livingstone's travels provided by Otto Spamer remained relatively similar to those contained with the Costenoble edition (and that of Murray previously) may well have served to convince readers that they were not missing out on significant detail by opting for a less expensive alternative. Certainly several reviewers pointed out that while the book might be specifically aimed at younger readers there was also much to recommend it to the public at large.[49]

Costenoble, though, claimed that Spamer's book was the product of plagiarism, prompting a public dispute, not dissimilar to that which erupted in England and then again in America when publishers sought to produce rival volumes which interfered with the official, authorised accounts. Costenoble's accusations of plagiarism prompted Spamer to produce a pamphlet and advertisement defending his actions. In the advertisement, Spamer claimed that he would not personally engage in a debate over the value of either work but would rather let two geographical authorities settle the matter.[50] A review in Petermann's *Mittheilungen*, the most influential German geographical periodical of the period, was quoted as a means of demonstrating the worth of his account for public education and entertainment. What Spamer failed to reveal, however, was that these comments appeared within a longer review which concentrated to a larger degree on Costenoble's edition.[51] The reviewer had much praise for the authorised translation, commending its high standard in the light of Livingstone's own peculiar style of writing and its value as the only complete German version of Livingstone's narrative. In his own comments on Costenoble, Spamer quoted the *Berliner Zeitschrift für Erdkunde* which was rather less positive about the

---

49 'Livingstone's Reisen in Deutschem Gewände', *Magazin für die Literatur des Auslandes* 56 (1859): 52; 'Literatur. *Erforschungsreisen in Innern Afrikas* [Review]', *Berliner Revue* 16 (1859): 172.

50 Otto Spamer, *Vehme oder Justiz? Appellation an die öffentliche Meinung im Betreff eines Gutachtens des Leipziger Sachverständigen-Vereins I. Section vom 10. November 1858, wodurch dieselbe den Inhalt des angeklagten Buchs: 'Livingstone, der Missionär' etc. für gesetzmässig erklärt und dennoch dessen Verleger auf's empfindlichste an der Ehre kränkt, eine Streitschrift zu seiner Rechtfertigung und im Interesse des gesammten deutschen Buchhandels geschrieben von Otto Spamer.* (Leipzig, 1859) [Not seen].

51 'Literatur', *Mittheilungen aus Justus Perthes' Geographischer Anstalt über wichtige neue erforschungen auf des Gesammtgebiete der Geographie von Dr A. Petermann*, (1858), 562.

worth of the authorised translation. This reviewer suggested that, since English was the common language of geographers, it was not so much a direct translation that was required but more a volume which reorganised and clarified the contents of the original for the wider reading public, something which Lotze's volume failed to do. Although Spamer did not reproduce them in the advertisement, the publisher also alluded to a list of 'ridiculous and incomprehensible' translations which the reviewer had highlighted in the full article.[52] Through the selective use of periodical reviews, Spamer thus attempted to convince potential readers that Costenoble's 'authorised' version was not necessarily more instructive than his own 'independent' work. This episode highlights that Livingstone's authority over his narrative was challenged in the German context both because of issues inherent in translation and because of the failure of copyright legislation to deter the publication of rival versions of the same narrative.

Missionary Travels *in France*

The circulation of *Missionary Travels* in French-language editions further testifies to the significance of the transnational market for narratives of travel and exploration. Publisher Louis Hachette, one of France's largest publishing houses at this time, began negotiating terms for the production of a French translation of *Missionary Travels* before the English edition had appeared.[53] It would, however, be 1859 before *Explorations dans l'intérieur de l'Afrique australe* would appear in print.[54] Hachette's authorised translation was published as a single volume and, in fulfilment of copyright regulations, indicated on its title page that it had been *'ouvrage traduit de l'anglais avec l'autorisation de l'auteur'* [translated from English with the author's permission].[55] The translator was identified as Henriette Loreau, who was responsible for translating a number of well-known works into French for Hachette, including travel books. In 1862, for instance, she was responsible for Richard F. Burton's *Voyage aux grand lacs de l'Afrique orientale*, and in 1866 David and Charles Livingstone's *Explorations du Zambèse et de*

---

52   'Livingstone. Erforschungsreisen in Afrika, unser "Buch der Reisen", II Band', in *Erforschungreisen im Innern Afrika's in Schilderungen der bekanntesten älteren und neueren Reisen insbesondere der großen Entdeckungen im südlichen Afrika während der Jahre 1840–1856 durch Dr David Livingstone* (Leipzig: O. Spamer, 1860), n.p.

53   Hachette was a principal supplier of school textbooks and also owned an extensive network of railway bookstalls which followed the example set by the English company W.H. Smith. On Hachette, see Christine Haynes, *Lost Illusions: The Politics of Publishing in Nineteenth-Century France* (Cambridge, MA: Harvard University Press, 2010), 154–86; Eileen S. DeMarco, *Reading and Riding: Hachette's Railroad Bookstore Network in Nineteenth-Century France* (Cranbury NJ: Associated University Presses, 2010).

54   Louis Hachette to John Murray, October 5, 1857, John Murray Archive, MS40500, National Library of Scotland.

55   David Livingstone, *Explorations dans l'intérieur de l'Afrique australe*, Henriette Loreau (trans.) (Paris: Hachettte, 1859).

ses affluents.[56] Periodical reviewers of Hachette's edition of *Missionary Travels* regularly applauded Loreau's talent as a translator before going on to discuss Livingstone's merits as an explorer. Charles-Louis Chassin told readers of the *Revue de l'Instruction Publique de la Literature et des Sciences*, for instance, that the French edition was beautifully printed and illustrated and would be a work of lasting value, being consulted often by those interested in geography, natural and human history, while the fidelity and clarity of Loreau's translation was specifically praised.[57] Similarly, the *Bibliotheque Universelle Revue Suisse et Etrangère* claimed that Loreau had successfully conveyed Livingstone's unique style of writing, combining perfect simplicity with a remarkable talent for description. The result, the reviewer claimed, was a popular book which would be sought after by all ages and classes of readers.[58] *La Tour du Monde's* review, though, did not discuss the merits of the translation in relation to the Murray volume, instead providing bibliographic details for both English and French editions before noting that Livingstone's narrative had now sold 50,000 copies across both countries, further emphasising that the interest in Livingstone's narrative went far beyond his native country.[59]

As elsewhere, this interest would give rise to rival accounts often aimed at children or other readers who might find the authorised volumes inaccessible. Author Henry Paumier, for instance, took advantage of the time between the appearance of the English and French editions to construct *L'Afrique ouverte ou une esquisse des decouvertes du Dr Livingstone*. The book was published in 1858 by Meyrueis and included within the *Nouvelle bibliotheque des familles*. At only 132 pages long and heavily illustrated it was aimed primarily at younger readers, and at 1 franc it was more accessible than Hachette's volume. Although not an attempt at direct translation, Paumier's book clearly drew heavily upon John Murray's edition. It began by recounting Livingstone's humble upbringing before summarising his African adventures, which were illustrated by a fold-out map and several images based around those in the authorised English edition. Like Spamer's text, that produced by Paumier dispensed with much scientific detail. Eye-catching illustrations which were originally accompanied by lengthy textual descriptions were now included with little contextual information, instead

---

56   Richard F. Burton, *Voyage aux grand lacs de l'Afrique orientale*, H. Loreau (trans.) (Paris: Hachette, 1862); David Livingstone and Charles Livingstone, *Explorations du Zambèse et de ses affluents*, H. Loreau (trans.) (Paris: Hachette, 1866).

57   Charles-Louis Chassin, 'Explorations dans l'Interior de l'Afrique australe [Review]', *Revue de l'Instruction publique de la littérature et des sciences* 43 (1869): 677.

58   'Bulletin littéraire et bibliographique. France', *Bibliotheque universelle revue suisse et étrangère* 5 (1859): 148.

59   'Nouvelle découvertes du docteur Livingstone', *Le Tour du monde nouvelle journal des voyages*, 1 (1860): 62.

being used to excite the imagination rather than aid understanding.[60] Although not reviewed as widely as Hachette's edition of *Missionary Travels*, Paumier's book was received favourably in a number of periodicals, precisely because it summarised what were perceived to be the most interesting (that is, adventurous) elements of *Missionary Travels*.[61] Although there is no evidence to suggest that Paumier's volume sparked the same controversy that similar English and German works did, this case reinforces the earlier claim that, in practice, Livingstone and Murray could not retain absolute control over either the text or the illustrations within *Missionary Travels* or its authorised translations once they were in the wider marketplace.

Conclusion

This chapter began by suggesting that while *Missionary Travels* was a well-documented British publishing success the extent to which it might be considered an international or even global literary event was a question which was yet to be fully explored. By following Livingstone's book into French and German marketplaces, this chapter has improved our understanding of the impact of Livingstone's travels, particularly among non-specialist audiences. While foreign-language piracies and abridgements of travel narratives have received scant scholarly attention to date, the case of *Missionary Travels* suggests that they could play an important role in engaging readerships which may not have had access to authorised editions. This chapter has considered the strategies used by a range of publishers and authors as they attempted to convince readers that their particular publications were authoritative sources. The French and German volumes considered above remind us that while there were mechanisms in place to regulate the circulation of texts across national and linguistic boundaries, such mechanisms were routinely challenged by foreign publishers, authors, translators and reviewers. It provides a rationale for further studies of translations of *Missionary Travels* in particular and of the international circulation of travel and exploration literature more widely. *Missionary Travels* and the texts that it inspired constitute at the very least an international publishing success, the scale of which will only be appreciated after further analyses of this type.

By exploring the French and German circulation of the most famous mid-Victorian travel narrative, *Missionary Travels and Researches in South Africa*, this chapter has raised a number of general points which have significance beyond

---

60   Compare, for example, the use of illustrations of Loandan hairstyles in Henry Paumier, *L'Afrique ouverte ou une Esquisse des decouvertes du Dr Livingstone*, 108–9 and Livingstone, *Missionary Travels*, 450–51.

61   'L'Afrique Ouverte', *Revue critique des livres nouveaux* 27 (1859): 118–19; 'Bulletin littéraire et bibliographique. France', *Bibliotheque universelle revue suisse et étrangère* 7 (1860): 326–7.

this particular case. The chapter began by surveying recent work in the history and geography of science and history and geography of the book and suggested that this work provides a rationale for a critical re-examination of nineteenth-century literatures of travel and exploration. Recent conceptualisations of the geography of the book in particular encourage us to link questions of production, circulation and reception together in a way that is attentive both to physical form and content. This necessarily leads us to consider how texts circulated beyond the publishing house in various forms and to question how texts were received in particular times and spaces. Adopting such an approach, this chapter suggests, can help us to better understand the conditions surrounding the construction and circulation of geographical ideas in specific times and spaces.

Beyond this, recent work in the history of science examining a 'global Darwin' who shaped and was shaped by a variety of cultural contexts and encounters might prompt us to consider the extent to which we can identify a 'global Livingstone' constructed through printed text and image.[62] Whether this is a feasible enterprise or not is, of course, open to question. We should absolutely be attentive to ways in which in Livingstone's life and legacy was constructed beyond Britain as a result of textual circulation and translation. Indeed, the examples explored above have been offered as a means of opening up further debate about Livingstone, encouraging scholars to look more closely at the impact he and his writings had in a variety of widely distributed locations. Yet to speak of *a* 'global Livingstone' may risk closing down the conversation by underplaying the nuances of reception and the stubborn specificity of historical and geographical circumstance.

---

62   For critical reflection on a 'global Darwin', see James A. Secord, 'Global Darwin', in *Darwin*, ed. William Brown and Andrew C. Fabian (Cambridge: Cambridge University Press, 2010), 31–57 and Elshakry, *Reading Darwin in Arabic*, 6–9.

Afterword
Connections, Institutions, Languages

Charles W.J. Withers

The issues addressed in this book were encountered, witnessed, and lived at first hand. Writing in October 1858 on the nature and the consequences of travel and exploration in the nineteenth century, the British social theorist and political commentator Harriet Martineau noted how

> One of the discontents of our saucy modern day is at the smallness of the globe we live on: Between the recent discoveries in astronomy, on the one hand, and the prodigious achievements in geographical exploration on the other, together with the saving of time from steam-travelling, we seem to have obtained a command over the spaces of the globe which considerably diminishes the popular reverence for the mysteries of our planet.[1]

Martineau was herself a traveller, spending two years in the United States in the mid-1830s; her *How to Observe: Morals and Manners* (1838) may even be seen as an early example of a how-to travel guide. While the world had become both smaller and larger by the mid-nineteenth century for the reasons Diarmid A. Finnegan and Jonathan Jeffrey Wright disclose in their introduction, Martineau knew from personal experience that hers was a world of difference and inequality: the anti-slavery views expressed in her *Society in America* (1837) and her anti-colonial stance on the Indian Mutiny of 1857 show Martineau's authorship to be rooted in a close understanding by herself and her contemporaries – she counted Thomas Malthus, Charles Lyell, Charles Darwin, and John Stuart Mill amongst her many acquaintances – of exactly those issues of empire, encounter, and exchange that are the subject of enquiry here.[2]

Scholars of the nineteenth century face different challenges. Even when past actions and events are retrievable, their meanings and consequences are often elusive. There is always the danger of attributing agency incorrectly. Local circumstances may not readily speak to global ones, and vice versa. Archival traces are uneven, the

---

1 [Harriet Martineau], 'Travel During the Last Half Century', *Westminster Review* 70 (October 1858), 426–65, quote from page 426.

2 Ella Dzelainis and Cora Kaplan, eds, *Harriet Martineau: Authorship, Society, and Empire* (Manchester: Manchester University Press, 2011). On Martineau's *How to Observe* as a proto-methodological instructional text in travel and exploration, see Felix Driver, *Geography Militant: Cultures of Exploration and Empire* (Oxford: Blackwell, 2001), 51, 61–2.

product of institutional exigency, individual diligence, and sheer good fortune. What do such circumstances mean, then, when studying things so literally all-encompassing as 'global knowledge' and for so long a period as the nineteenth century?

The chapters assembled here illuminate these issues in different ways. By place, the subjects addressed include global, sub-continental and national geographies: 'Canada' as it was understood in the early nineteenth century, the United States, Germany and France, the island nation of Haiti, India and Burma in the final years of that century. More prescribed local sites are highlighted: civic museums in Belfast, the Royal Botanic Garden at Kew near London. What we may think of as exemplary 'micro-sites' of global knowledge making are brought into view: ships at sea as they became, briefly and messily, mobile spaces of discernment and depiction; archaeological digs; display cabinets in the drawing rooms of the well-to-do; the tents and studios of female botanists; the wards of medical missionaries; the bartering rooms of fur traders; or, in the exhibition cases of leading scientific institutions, scale models of imperial labour. By topic, emphasis is given to processes at once revelatory and exhibitive: voyaging (in the Pacific), picturing (paintings of plants, sketches of specimens), collecting (orchids, coins, potsherds), tending (garden plots, ill patients), displaying (in museums and at home), authoring and translating (in different editions and in foreign-language works of others' accounts in English).

Individually and severally, the authors and editors affirm the importance of thinking geographically about the cognitive content and the reach of 'global knowledge'. We are invited to think of this as, broadly, knowledge that helped build and sustain empires yet as something commonly rooted in a close-grained empiricism and prompted by sponsoring institutions and shown to receptive audiences, both institutions and audiences being located at some distance from the sites of knowledge production. There is an insistence, albeit one variously expressed and differently illustrated, upon what in recent years have become concepts central to the spatial and temporal analyses of science, religion, and empire. These include (the list is far from exhaustive): place (as location and grounded meaning); space (as physical territory and social construction); site (as institutional setting), sight (the representational forms employed to depict knowledge); objects (with particular meaning such as instruments of science and medicine, or, more broadly, history's 'material turn'); and print history (as a changing technology, or, in different forms, the means by which people in one place were informed about people and things in another).

There is not the space in a brief afterword to expand upon these points. But it is noteworthy that these and other categories used to address knowledge about the world in, broadly, the 'long nineteenth century', are not fixed, epistemologically speaking, any more than is the object under scrutiny, global knowledge, or the period in question.[3] There may be advantage in such 'definitional indeterminacy'

---

3   On place, space and institutions in nineteenth-century science, for example, see the essays in David N. Livingstone and Charles W.J. Withers, eds, *Geographies of Nineteenth-Century Science* (Chicago: University of Chicago Press, 2011). On print history, local

as the editors note also in the Introduction. There are certainly attendant dangers. By its very nature the category 'global knowledge' may require such qualification by place, theme, or timing that it loses cogency. Can we, in the face of archival contingencies, limitations of time, linguistic incompetencies, ever really do justice at a global scale to (say) the making of oceanographic science, tropical botanising, or the writing and reception of books of travel? How do we realise 'the global' through 'the local', or make small stories speak to grand narratives?

These comments are made not in criticism but from empathy. We are all cognisant of the difficulties intrinsic to writing geographically sensitive work where the subject under review – whose dimensions were differently encountered, exhibited, and exchanged on a global scale – may only be revealed and explained in local context. I would argue, too, that there is an important difference within the term 'global knowledge' between elements of knowledge which can be shown to be global in significance and reach, and that more particular knowledge of the globe which derives from being collected, measured or in other ways produced in one place and assessed, explained, and made known to different audiences in yet other places. The first is about the balance between scale and scope. Because a genuinely global geography is difficult, we resort either to national studies in comparative context or to thematic studies whose purpose is to reveal the global interconnectedness of places and processes whose uneven consequences at that scale are not of primary concern. It may be for these reasons that much of what, commonly, is taken to be 'global history' (global geography) is thematically specific and conceptually framed as 'networks' – commodity-focused geo-histories of coffee, salt, sugar, slavery, even of science, for example.[4] The second is about the presumption implicit in works of fine-grained analysis over how knowledge somewhere (there and not elsewhere) is and was produced and consumed locally and how its categories and consequences were then transmitted – between places,

studies and grand narratives, see James A. Secord, 'Knowledge in Transit', *Isis* 95 (2004): 654–72. On the idea of 'the global' emerging as an historical and geographical category as the result of exploration and observation, see Denis E. Cosgrove, *Apollo's Eye: A Cartographic Genealogy of the Earth in the Western Imagination* (Baltimore: Johns Hopkins University Press, 2001) and Joyce E. Chaplin, *Round About the Earth: Circumnavigation from Magellan to Orbit* (New York: Simon and Schuster, 2012). The category 'exploration', and its constituent elements and variant expressions, is the subject of Dane Kennedy, ed., *Reinterpreting Exploration: The West in the World* (Oxford and New York: Oxford University Press, 2014). On the idea of world history (and world geography) as 'commodity history' – global knowledge through trade – see for example Kenneth Pomeranz and Steven Topik, eds, *The World That Trade Created: Society, Culture and the World Economy, 1400 to the Present* (London: M. E. Sharpe, 1999). On sites and sights – the interconnections between places and venues, and the practices of observation and representation – see Lorraine Daston and Elizabeth Lunbeck, eds, *Histories of Scientific Observation* (Chicago: University of Chicago Press, 2011).

4   Bruce Mazlish, *The New Global History* (London: Routledge, 2006).

over time, and at different scales – such that the knowledge may be presumed 'global' (the 'somewhere' becomes an 'everywhere'). In this sense, it is not enough simply to show that things and people locally had connections with things and people more distant without demonstrating how both shaped, and were shaped by, the other.[5] By the same token, we must be cautious about uncritically assuming equivalence between 'modernity' and what C.A. Bayly has considered those 'global uniformities in the state, religion, political ideologies, and economic life' characteristic of the nineteenth century. This was an age, after all, that lacked agreement over common metrological standards, the yard or the metre, was without an agreed standard as to civic time, and, until 1884, had no single prime meridian as a global base point from which to measure time and space. The world was encountered, exchanged, and interconnected in the nineteenth century (as it had been in earlier periods albeit perhaps in different ways), but the modernity of the 1800s did not beget uniformity.[6]

Let me move from the authors' claims and my own necessarily perfunctory remarks in introduction to highlight some possibilities concerning the wider project of global histories and geographies in the nineteenth century. These are indicative not prescriptive, and deliberately raise more questions than it is possible to answer (as the many question marks will reveal). I offer three sub-headings using the terms flagged in my sub-title and hope, too, that the relationships between them will be apparent.

Connections

One of the principal ways by which knowledge about the globe became global in the early modern period (before 1800) was through letter writing and networks of correspondence.[7] Nineteenth-century communities of knowledge and the globalisation of that knowledge may similarly be explored and explained with

---

5   On these issues, see for example Doreen Massey, *For Space* (London: Sage, 2005); Philip E. Ethington, 'Placing the Past: "Groundwork" for a Spatial Theory of History', *Rethinking History* 11 (2007), 465–93; and the essays in Peter Meusburger, David N. Livingstone and Heike Jöns, eds, *Geographies of Science* (Heidelberg: Springer, 2010) and other volumes in the 'Knowledge and Space' series under the editorship of Peter Meusburger at Heidelberg.

6   Christopher A. Bayly, *The Birth of the Modern World, 1780–1914: Global Connections and Comparisons* (Oxford: Blackwell, 2004), 1. On the temporal and spatial differences involved in the making of nineteenth-century 'modernity', see Stephen Kern, *The Culture of Time and Space 1880–1918* (Cambridge, MA: Harvard University Press, 2003 edition), and, from a literary perspective, Adam Barrows, *The Cosmic Time of Empire: Modern Britain and World Literature* (Berkeley: University of California Press, 2011).

7   On this, see Dan Edelstein and Paula Findlen, *Mapping the Republic of Letters: Exploring Correspondence and Intellectual Community in the Early Modern Period*

reference to epistolary cultures – many of the 'big' names of 'big' science such as Charles Darwin have been shown to be so from the letters they wrote and received. So, too, knowledge was made by turning letters and correspondence into print. But it was the nineteenth century that produced the administrative technologies of the Post Office to regulate the geographies of correspondence. If there is one key difference, however, between the early modern period and the nineteenth century it is not of type but in scale: in the greatly increased number of periodicals, specialist journals, journals of review, and popular periodicals produced, in-part, through new 'steam' and other technologies, and so on. The general implications of this are understood.[8] But there is still much to know about the geographies of nineteenth-century print technologies: how and where they were produced and read, the labour markets of the skilled and the unskilled, the cultures of review in popular context and of refereeing in specialist context, to name only a few. Were the global and local print geographies of the metropolitan north reproduced in the nineteenth-century 'global south'? If diffusionist models are blunt instruments, how are we to explain the globalising of geographies and technologies of print culture from non-western, even non-elite, perspectives?[9]

As Harriet Martineau knew at first hand, steam-powered print was one thing, but steam power in the form of ocean-going ships was similarly transformative, technically and conceptually. New geographies of globalisation were, at heart, technically driven: global space 'shrank' because sailing distances between places were reduced in time; cargoes could be carried in greater bulk at greater speed; new networks of imperial commercial transactions wove the world together. In a profound sense, nineteenth-century empires and their attendant global knowledge were engineered accomplishments. This claim is borne out in the growth of railway networks and their concomitant geographies. Perhaps most clearly of all nineteenth-century technological developments in terms of the impact upon global space and communicative networks, it is apparent in the laying of trans-Atlantic

(1500–1800), http://republicofletters.stanford.edu and the Stanford Spatial History Project http://www.stanford.edu/group/spatialhistory/cgi-bin/site/index.php.

8 In the British context, see John Barnes, Bill Bell, Rimi J. Chatterjee, Wallace Kirsop and Michael Winship, 'A Place in the World', in *The Cambridge History of the Book in Britain. Volume 6: 1830–1914*, ed. David McKitterick (Cambridge: Cambridge University Press, 595–634); Leslie Howsam, 'The History of the Book in Britain, 1801–1914', in *The Oxford Companion to the Book*, ed. Michael Suarez, S.J. and Henry R. Woudhuysen (2 vols, Oxford: Oxford University Press), I: 180–87; Simon Eliot, 'Some Trends in British Book Publishing, 1800–1919', in *Literature in the Marketplace: Nineteenth-Century British Publishing and Reading Practices*, ed. John O. Jordan and Robert L. Patten (Cambridge: Cambridge University Press, 1994, 19–43); Aileen Fyfe, *Steam-Powered Knowledge: William Chambers and the Business of Publishing, 1820–1860* (Chicago: University of Chicago Press, 2012).

9 On this, see for example the essays in Ulrike Freitag and Achim von Oppen, eds, *Translocality: The Study of Globalising Processes from a Southern Perspective* (Amsterdam: Brill, 2010).

telegraph cables from the late 1860s, and, later in the century, in global sub-marine telegraphy. Telegraphy transformed ideas about what 'the global' meant, in commerce, in the running of empires, in the doing of science.[10] Technology has, always, an uneven global history.[11] That history was produced and shaped by different geographies – places with access to the telegraph; communicative 'blind spots' of almost continental dimension until well after 1900 (much of mainland Asia for example); commercial networks were not the same as governmental ones. How were conceptions of the empire altered by such informational spaces? How were 'main office' transactional places – in for example banking and insurance, or in maritime trade – transformed by ease of communication with 'branch' offices? How did being 'wired' into the world change being in the world?

Institutions

The chapters gathered here succeed in demonstrating the nature of global geographies and the experience of empire in the nineteenth century because they look at the lives, works, and experiences of individuals. The idea of 'global' or colonial lives – of entangled biographies as illustrative of global processes otherwise too complex and impersonal to chart in detail – is an established theme.[12] But as scholars such as Zoë Laidlaw have shown – and to connect with the foregoing theme – the 'careering' of innumerable colonial individuals across the globe and the networking of the world as empire was, often, made possible because of the institutions of government. In its more detailed geographies, Laidlaw's *Colonial Connections* is a study of New South Wales and the Cape Colony and the mechanisms of British control there in the thirty years from 1815.[13] In its wider implications, it is a case study in changing metropole–colony relations, in charting the shift from government to governance in the early Colonial Office and, vitally, in demonstrating the impersonal nature of imperial exchange and encounter: as she notes, 'Imperial administrators continued to assert metropolitan centrality, but their reliance on affective, nuanced, networks of personal connection and

---

10   Ben Marsden and Crosbie Smith, *Engineering Empires: A Cultural History of Technology in Nineteenth-Century Britain* (Basingstoke: Palgrave Macmillan, 2005). On telegraphy, see Roland Wenzlhuemer, *Connecting the Nineteenth-Century World: The Telegraph and Globalization* (Cambridge: Cambridge University Press, 2013).

11   Daniel R. Headrick, *Technology: A World History* (Oxford: Oxford University Press, 2009).

12   In Britain, for the early modern period, see Miles Ogborn, *Global Lives: Britain and the World, 1550–1800* (Cambridge: Cambridge University Press, 2008), and, for the nineteenth century, David Lambert and Alan Lester, eds, *Colonial Lives across the British Empire: Imperial Careering in the Long Nineteenth Century* (Cambridge: Cambridge University Press, 2006).

13   Zoë Laidlaw, *Colonial Connections 1815–45: Patronage, The Information Revolution and Colonial Government* (Manchester: Manchester University Press, 2005).

critical colonial rule up to the mid-1830s changed as more emphasis was placed on impersonal bureaucracy and system'.[14]

This is not to propose the study for their own sake of formal institutions that either facilitated or reported empire, and it is certainly not to call for a typology of institutions of strictly different sort – scientific organisations, commercial bodies, government bureaus, and such like – when the evidence points to complexity and hybridity, not to essential distinctions. It is, rather, to suggest that we might consider how institutions undertook the roles they did and to show, following Jon Klancher's discussion of the arts and the sciences in late eighteenth- and early nineteenth-century Britain, how institutional administrators of places like art galleries, lecture halls, and public theatres, were also cultural producers and scientific 'brokers', 'go-betweens' mediating between imperial knowledge's makers and empire's diverse publics. In nineteenth-century urban Britain, and in the principal population centres of her colonies, numerous civic natural history museums collectively exhibited the fruits and tools of empire as emporium and as the products of unfettered empiricism. More than any other institution perhaps, nineteenth-century museums displayed and dismayed the world. So, too, could libraries or, rather, what they facilitated might do so. In Britain after the 1850 Public Libraries Act, public libraries established a new means of public access to knowledge in ways that, in their criteria of selection, physical organisation and organisation of readers across different social classes and places, were broadly the same across the country. In that sense, the same institution looked and ran much the same wherever you were, *contra* Laidlaw's analysis of Britain's Colonial Office at work overseas in the early nineteenth century. Looking 'internally', as it were, at how different sorts of institutions worked, at what was possible to do and say and write within any given setting and what not, and not simply at their external consequences as if organisations acted straightforwardly and of their own volition, may cast further light on how more exactly knowledge about the globe was made and received. In short, where, how, and why did imperial bodies work and become trained – as individuals, as institutional members, as paid employees, as social beings, as empire's subjects?[15]

---

14   Laidlaw, *Colonial Connections 1815–45*, 5.

15   These thoughts are prompted, amongst others, by Jon Klancher, *Transfiguring the Arts and Sciences: Knowledge and Cultural Institutions in the Romantic Age* (Cambridge: Cambridge University Press, 2013); the essays in Simon Schaffer, Lissa Roberts, Kapil Raj and James Delbourgo, eds, *The Brokered World: Go-Betweens and Global Intelligence, 1770–1820* (Science History Publications: Sagamore Beach, 2009); John M. Mackenzie, *Museums and Empire: Natural History, Human Cultures and Colonial Identities* (Manchester: Manchester University Press, 2009); David McKitterick, 'Libraries, Knowledge and Public Identity' in *The Organisation of Knowledge in Victorian Britain*, ed. Martin Daunton (Oxford: Oxford University Press, 2005), 287–312; and, with regard to science, the essays in Aileen Fyfe and Bernard Lightman, eds, *Science in the Marketplace: Nineteenth-Century Sites and Experiences* (Chicago: University of Chicago Press, 2007) and in Livingstone and Withers, eds, *Geographies of Nineteenth-Century Science*.

Languages

If definitional indeterminacy in studying the spaces of global knowledge is useful yet risky, even a cursory glance across disciplines discloses many other 'slippery' labels in current use: 'trans-national', 'cosmopolitanism', 'international'; 'internationalism' and 'internationalisation'; 'globalisation' and the portmanteau neologism 'glocalisation'. At root, these terms, no less perhaps than 'place' and 'space', their counterparts in terminological imprecision, speak to just those difficulties of making 'the local' and 'the global' relate one to another that the contributors engage with and that I have noted above and elsewhere.[16] There is not the space here to pursue the genealogies of these several terms, but let me briefly consider one implication of working with them in addressing global matters at local scales, and vice versa. The implication – simply stated, but far from simple in its resolution – is this: what about the national? National frames of reference are, often, taken for granted: the British Empire, Britain, German science, French imperialism, and so on. One argument would suggest that the consequence of working either with 'the local' or 'the global' and certainly with their combination has been the relative elision of 'the national' as a focus of study: we are left either with longer-distance networks, cycles of exchange and accumulative regimes or with their local, grounded, manifestations. Are we witnessing, as one historian of science sees it, 'an end to national science'.[17]

Reports of the death of the nation(al) are, of course, exaggerated, but it is interesting, nevertheless, how such terms as 'transnational' and 'translocal', 'internationalism' and 'internationalisation' (as with 'global history'), look above and beyond the nation. A counter argument would have it that the effect of using such terms and of thinking in this way has, paradoxically, been to 'produce' the nation in ever stronger ways as the necessary frame or scale of reference. In one discussion of the idea of a transnational history of science, the authors make the following observation:

> Other approaches such as world history and new global history share with transnational history the wish to abandon Euro- and US-centric viewpoints, explaining the role of historical actors and agencies in international networks, and focusing on the circulation of people, objects and ideas. Transnational history approaches appear to differ from these perspectives because of their

---

16   Charles W.J. Withers, 'Place and the "Spatial Turn" in Geography and History', *Journal of the History of Ideas* 70 (2009): 637–58. On the 'spatial turn' more widely, see Barney Warf and Santa Arias, eds, *The Spatial Turn: Interdisciplinary Perspectives* (New York: Routledge, 2009).

17   Lewis Pyenson, 'An End to National Science: The Meaning and the Extension of Local Knowledge', *History of Science* 40 (2002): 251–90. On the idea of circuits and networks over and above nations, see Alan Lester, 'Imperial Circuits and Networks: Geographies of the British Empire', *History Compass* 4 (2006): 124–41.

focus on the modern and contemporary periods, and the reappraisal of nations' role in shaping the past. Indeed, because of these distinctive chronological and theoretical features, some scholars have argued that transnational history pays more attention to nation states than do its alternatives.[18]

As 'global geography' might substitute here for 'global history', so 'internationalism' (heuristically, if not semantically, a close relative of 'transnational') has been used to consider the shared development of issues over and above the national while keeping the nation in full view.[19] Transnational history has been seen by one interlocutor as 'a way of seeing', not as a specific set of instruments for historical (or geographical) work.[20] But both for that term and for others, we do need to know what ends we have in view quite apart from the practical difficulties of doing a more-than-one-theme transnational or global history.

Matters of language are more than terminological. 'Translation' is more than linguistic equivalence since the term embraces movement over space and elements of an almost ontological transformation as well as of linguistic change: even books in one language, Secord reminds us, effectively change their meaning according to their different readers.[21] But what would a Francophone global knowledge look like? A German one? Is it necessary to have an empire to have global knowledge in the ways disclosed here? For the Spanish (who had had an empire) but not the Portuguese (who still had one), global knowledge in the nineteenth century

---

18   Simone Turchetti, Néstor Herran and Soraya Boudia, 'Introduction: Have We Ever Been 'Transnational'? Towards a History of Science Across and Beyond Borders', *British Journal for the History of Science* 45 (2012): 319–36, quote at page 322. As the title suggests, this is an introduction to a theme issue on the transnational history of science. For the perspective of historians on this category, see Christopher A. Bayly, Sven Beckert, Matthew Connelly, Isabel Hofmeyr, Wendy Kozol and Patricia Seed, 'AHR Conversation: On Transnational History', *American Historical Review* 111 (2006): 1441–64.

19   Elisabeth Crawford, Terry Shinn and Sverker Sörlin, 'The Nationalisation and Denationalization of the Sciences: An Introductory Essay', in *Denationalizing Science: The Contexts of International Scientific Practice*, ed. Elisabeth Crawford, Terry Shinn and Sverker Sörlin (Kluwer: Dordrecht, 1993), 1–42; Martin H. Geyer and Johannes Paulman, eds, *The Mechanics of Internationalism: Culture, Society, and Politics from the 1840s to the First World War* (Oxford: Oxford University Press, 2001).

20   Beckert in Bayley et al., 'AHR Conversation: On Transnational History', 1454.

21   James A. Secord, *Victorian Sensation: The Extraordinary Publication, Reception, and Secret Authorship of* The Natural History of Creation (Chicago: University of Chicago Press, 2001). On the connections between translation and my remarks above on the terminology of 'global history', see Victor Roudometof, 'Transnationalism, Cosmopolitanism and Glocalisation', *Current Sociology* 53 (2005): 113–35, and Sundar Sarukkai, 'Translation as Method: Implications for History of Science', in *The Circulation of Knowledge Between Britain, India and China: The Early-Modern World to the Twentieth Century*, ed. Bernard Lightman, Gordon McOuat and Larry Stewart (Leiden: Brill, 2013): 311–29.

might read as narratives of loss and restitution, of colonial dispossession rather than of imperial display. For the Americans, such notions are hardly applicable in the nineteenth century: empire and global knowledge borne of military and commercial hegemony would come only after 1900.[22]

It does matter what you call things. Imperial exchange, in the nineteenth century as before and after, was not always conducted on equal terms. What was on display at one end of a network of imperial encounter and exchange may, for others elsewhere, be interpreted as the result of an act of dispossession. Looked at from one point of view, Cypriot archaeology, tropical plant hunting, and fur trading in Arctic Canada are instances of global knowledge; from another, this is plunder, depredation of nature – rapacious encounter and unequal exchange as 'global pillage'.[23] This is to offer a stark dichotomy (I do so for effect although my point about terminology and interpretative significance holds). If this is to caution that the terms and interpretations we offer, the themes we study, are not symmetrical – that is to say, the relationships between 'metropole' and 'colony' were not equal and it is our responsibility to show how and why global knowledge was unevenly made, distributed, and disseminated – then it is also to note that we as researchers produce asymmetries in the languages, methods and approaches we adopt. Consider the case of botany as a global practice in the long nineteenth century, and its connected forms: plant hunting, economic botany, systems of classification. Looked at from Kew, or Paris, or Cambridge, the botanical sciences and centres of botanical calculation were about exactly these things – collecting, classifying, utilising. Looked at from Calcutta, or western China and Tibet, the practices used in collecting, classification and in according plants utility may have been the same, but the structures of meaning they produced were often not. In Erik Mueggler's work, the botanical archive lay not in metropolitan imperial repositories but in indigenous knowledge systems 'in the field'; for Kapil Raj and Khyati Nagar, the making of botanical knowledge depended upon its circulation between Calcutta and Kew, a circulation that was also a translation, a mediation between classificatory systems and, even, alternative epistemologies.[24]

---

22   On these matters, see the essays in Morag Bell, Robin Butlin and Michael Heffernan, eds, *Geography and Imperialism* (Manchester: Manchester University Press, 1995); Mark Bassin, *Imperial Visions: Nationalist Imagination and Geographical Expansion in the Russian Far East, 1840–1865* (Cambridge: Cambridge University Press, 1999); Karen Morin, *Civic Discipline: Geography in America, 1860–1890* (Farnham: Ashgate, 2011); Neil Smith, *American Empire: Roosevelt's Geographer and the Prelude to Globalization* (Berkeley and Los Angeles: University of California Press, 2003).

23   I take the term from Larry Stewart who sees this being visited by the west on non-Western natures and cultures from the age of James Cook onwards: Larry Stewart, 'Global Pillage: Science, Commerce, and Empire', in *The Cambridge History of Science. Volume 4: Eighteenth-Century Science*, ed. Roy Porter (Cambridge: Cambridge University Press, 2003), 825–44.

24   Arguably, a metropolitan global view of botany is offered by, for example, Richard Drayton, *Nature's Government: Science, Imperial Britain, and the 'Improvement' of the World*

* * *

As Harriet Martineau knew, the world in the nineteenth century was a far from equal place and knowledge about that world was far from equally accessible. Her experiences were not typical. For most women, people of colour and many in the lower social classes, the opportunity to encounter a world beyond their immediate locale, to exchange news about it gleaned either from print or picture, or to visit an exhibition and marvel at the world's wonders were limited. Admitting this to be so, the places, spaces, sites and so on differently encountered by contemporaries constituted, for them, a lived reality albeit an unequal one.

For historians and geographers looking at the spaces (and places, and sites) of global knowledge, the categories we use, the connections we hope to elucidate and the languages we employ to do so are not so evidently realities as they are what Nicolaas Rupke has called 'assignments':

> Studying spatial distribution does not merely mean determining natural realms of occurrence. The categories of spatiality in which we examine science are themselves invariably constitutively influenced by our own locations as historians of science – whether these categories be as concrete as national territories and museums of natural history or as abstract as denominational discourses and rhetorical spaces, and whether we think of them as material 'container spaces' or as social co-productions of the scientific endeavour itself.[25]

There is much to be excited about in considering global history and geography, in the potential of 'transnational' approaches to global knowledge, in the consideration from different perspectives of the articulation of the local and the global, and in showing what exchange and encounter meant and what it has produced, now and in the past.[26]

(New Haven: Yale University Press, 2000), Jim Endersby, *Imperial Nature: Joseph Hooker and the Practice of Victorian Science* (Chicago: University of Chicago Press, 2008), and Nuala C. Johnson, *Nature Displaced, Nature Displayed: Order and Beauty in Botanical Gardens* (London: I. B. Tauris, 2011). Works that stress the making of botany as a global science from a more evidently 'colonial' perspective are Kapil Raj, *Relocating Modern Science: Circulation and the Construction of Knowledge in South Asia and Europe, 1650–1900* (Basingstoke: Palgrave Macmillan, 2007); Khyati Nagar, 'Between Calcutta and Kew: The Divergent Circulation and Production of *Hortus Bengalensis* and *Flora Indica*', in *The Circulation of Knowledge Between Britain, India and China*, ed. Lightman, McOuat and Stewart, 153–78; Erik Mueggler, *The Paper Road: Archive and Experience in the Botanical Exploration of West China and Tibet* (Berkeley: University of California Press, 2011).

25 Nicolaas Rupke, 'Afterword: Putting the Geography of Science in its Place', in *Geographies of Nineteenth-Century Science*, ed. Livingstone and Withers, 439–54, quote from page 450.

26 Harry Liebersohn, 'A Half Century of Shifting Narrative Perspectives on Encounters', in *Reinterpreting Exploration*, ed. Kennedy, 38–53.

Bibliography

Abir-Am, Pnina G. and Outram, Dorinda, eds. *Uneasy Careers and Intimate Lives: Women in Science 1789–1979*. New Brunswick, NJ: Rutgers University Press, 1987.

Adams, Ruth. 'The V&A: Empire to Multiculturalism?' *Museum and Society* 8 (2010): 63–79.

Alberti, Samuel J.M.M. 'Natural History Collections and Their Owners in Nineteenth Century Provincial England'. *British Journal for the History of Science* 35 (2002): 291–311.

Allen, David E. *The Naturalist in Britain: A Social History*. London: Allen Lane, 1976.

Ambirajan, S. 'Malthusian Population Thesory and Indian Famine Policy in the Nineteenth Century'. *Population Studies* 30 (1976): 5–14.

Anderson, Warwick. 'From Subjugated Knowledge to Conjugated Subjects: Science and Globalisation, or Postcolonial Studies of Science?' *Postcolonial Studies* 12 (2009): 389–400.

Anderson, Warwick. 'Making Global Health History: The Postcolonial Worldliness of Biomedicine'. *Social History of Medicine* 27 (2014): 372–84.

Andrew Graham's Observations on Hudson's Bay, 1767–91, ed. Glyndwr Williams (London: Hudson's Bay Record Society, 1969), xxx–xxxvi, 388–96.

Appadurai, Arjun, ed. *The Social Life of Things: Commodities in Cultural Perspective*. Cambridge: Cambridge University Press, 1986.

Arnold, David. '"Illusory Riches": Representations of the Tropical World, 1850–1940'. *Singapore Journal of Tropical Geography* 21 (2000): 6–18.

Auerbach, Jeffrey. *The Great Exhibition of 1851: A Nation on Display*. New Haven: Yale University Press, 1999.

Auerbach, Jeffrey and Hoffenberg, Peter H., eds. *Britain, the Empire, and the World at the Great Exhibition of 1851*. Aldershot: Ashgate, 2008.

Aylmer, G.E. 'Navy, State, Trade, and Empire'. In *The Oxford History of the British Empire: Volume I: The Origins of Empire: British Overseas Enterprise to the Close of the Seventeenth Century*, edited by Nicholas Canny, 467–79. Oxford: Oxford University Press, 1998.

Baigent, Elizabeth. 'Travelling Bodies, Texts and Reputations: The Gendered Life and Afterlife of Kate Marsden and Her Mission to Siberian Lepers in the 1890s'. *Studies in Travel Writing* 18 (2014): 34–56.

Baker, Michael J.N. *Our Three Selves: The Life of Radclyffe Hall*. New York: William Morrow, 1985.

Balfour-Paul, Jenny. *Indigo*. London: British Museum Press, 1998.

Ballantyne, Tony and Burton, Antoinette, eds. *Moving Subjects: Gender, Mobility and Intimacy in an Age of Global Empire*. Urbana: University of Illinois Press, 2009.

Barnes, Irene H. *Between Life and Death: The Story of C.E.Z.M.S. Medical Missions in India, China, and Ceylon*. London: Marshal Brothers and Church of England Zenana Missionary Society, 1901.

Barnes, John, Bell, Bill, Chatterjee, Rimi J., Kirsop, Wallace and Winship, Michael, 'A Place in the World'. In *The Cambridge History of the Book in Britain. Volume 6: 1830–1914*, edited by David McKitterick, 595–634. Cambridge: Cambridge University Press.

Barringer, Tim. 'Fabricating Africa: Livingstone and the Visual Image 1850–1874'. In *David Livingstone and the Victorian Encounter with Africa*, edited by John M. MacKenzie, 169–200. London: National Portrait Gallery, 1996.

Barrow, John. *Voyages of Discovery and Research within the Arctic Regions, from the Year 1818 to the Present Time*. London: John Murray, 1846.

Barrows, Adam. *The Cosmic Time of Empire: Modern Britain and World Literature*. Berkeley: University of California Press, 2011.

Basalla, George. 'The Spread of Western Science'. *Science* 156 (1967): 611–22.

Bashford, Alison. 'Malthus and Colonial History'. *Journal of Australian Studies* 36 (2012): 99–110.

Bashford, Alison. *Global Population: History, Geopolitics, and Life on Earth*. New York: Columbia University Press, 2014.

Bashford, Alison and Chaplin, Joyce. *The New Worlds of Thomas Robert Malthus*. Princeton: Princeton University Press, forthcoming.

Bassin, Mark. *Imperial Visions: Nationalist Imagination and Geographical Expansion in the Russian Far East, 1840–1865*. Cambridge: Cambridge University Press, 1999.

Batten, Charles. *Pleasurable Instruction: Form and Convention in Eighteenth Century Travel Literature*. Berkeley: University of California Press, 1978.

Baur, John E. 'Faustin Soulouque, Emperor of Haiti His Character and His Reign'. *The Americas* 6 (1949): 131–66.

Bayles, Ruth Margaret Bowman. 'Understanding Local Science: The Belfast Natural History Society in the Mid-Nineteenth Century'. In *Science and Irish Culture: Why the History of Science Matters in Ireland*, edited by David Attis and Charles Mollan, 139–69. Dublin: Royal Dublin Society, 2004.

Bayles, Ruth Margaret Bowman. 'Science in its Local Context: The Belfast Natural History and Philosophical Society'. PhD Diss., Queen's University Belfast, 2005.

Bayles, Ruth Margaret Bowman. 'The Belfast Natural History Society in the Nineteenth Century: A Communication Hub'. In *Belfast: The Emerging City, 1850–1914*, edited by Olwen Purdue, 105–24. Dublin: Irish Academic Press, 2013.

Bayly, Christopher A. *Imperial Meridian: The British Empire and the World, 1780–1830*. London: Longmans, 1989.

Bayly, Christopher A. *The Birth of the Modern World, 1780–1914: Global Connections and Comparisons*. Oxford: Blackwell, 2004.

Bayly, Christopher A., Beckert, Sven, Connelly, Matthew, Hofmeyr, Isabel, Kozol, Wendy and Seed, Patricia. 'AHR Conversation: On Transnational History'. *American Historical Review* 111 (2006): 1441–64.

Beaglehole, J.C. *The Life of Captain James Cook*. Stanford: Stanford University Press, 1974.

Beattie, Judith Hudson. '"My Best Friend": Evidence of the Fur Trade Libraries Located in the Hudson's Bay Company Archives'. *Épilogue* 8 (1993): 1–32.

Bell, Morag, Butlin, Robin and Heffernan, Michael, eds. *Geography and Imperialism*. Manchester: Manchester University Press, 1995.

Belyea, Barbara. *Dark Storm Moving West*. Calgary: University of Calgary Press, 2007.

Bennett, Tony. *The Birth of the Museum: History, Theory, Politics*. London and New York: Routledge, 1995.

Binnema, Ted. *Common and Contested Ground: A Human and Environmental History of the Northwestern Plains*. Norman: University of Oklahoma Press, 2001.

Binnema, Ted. 'How Does a Map Mean? Old Swan's Map of 1801 and the Blackfoot World'. In *From Rupert's Land to Canada*, edited by T. Binnema, G.J. Ens and R.C. Macleod, 201–24. Edmonton, University of Alberta Press, 2001.

Binnema, Ted. 'Theory and Experience: Peter Fidler and the Transatlantic Indian'. In *Native Americans and Anglo-American Culture, 1750–1850: The Indian Atlantic*, edited by Tim Fulford and Kevin Hutchings, 155–70. Cambridge: Cambridge University Press, 2009.

Binnema, Ted. *Enlightened Zeal: The Hudson's Bay Company and Scientific Networks, 1670–1870*. Toronto: University of Toronto Press, 2014.

Blunt, Alison. *Travel, Gender and Imperialism: Mary Kingsley and West Africa*. New York: Guilford, 1994.

Bose, Pradip Kumar. *Health and Society in Bengal: A Selection from Late 19th-Century Bengali Periodicals*. Calcutta: Sage, 2006.

Bourguet, Marie Noëlle, Licoppe, Christian and Sibum, H. Otto, eds. *Instruments, Travel and Science: Itineraries of Precision from the Seventeenth to the Twentieth Century*. London: Routledge, 2002.

Bourke, Marie. *The Story of Irish Museums 1790–2000: Culture, Identity and Education*. Cork: Cork University Press, 2011.

Brandon, Jennie J. 'Faustin Soulouque – President and Emperor of Haiti'. *Negro History Bulletin* 15 (1951): 34–7.

Brett, Charles E.B. *Buildings of Belfast, 1700–1914*, rev edn. Belfast: Friar's Bush Press, 1985.

Brett, Charles E.B. *Buildings of North County Down*. Belfast: Ulster Architectural Heritage Society, 2002.

Brock, Michael. *The Great Reform Act*. London: Hutchinson, 1973.

Brosseau, Marc. 'Geography's Literature'. *Progress in Human Geography* 18 (1994): 333–53.

Broughton, P. 'Astronomical Observations by Peter Fidler and Others in "Canada" 1790–1820'. *Journal of the Royal Astronomical Society of Canada* 103 (2009): 141–51.

Browne, Janet. 'Biogeography and Empire'. In *Cultures of Natural History*, edited by N. Jardine, J.A. Secord and E.C. Spary, 305–21. Cambridge: Cambridge University Press, 1996.

Browne, Janet. 'Corresponding Naturalists'. In *The Age of Scientific Naturalism*, edited by Bernard Lightman and Michael S. Reidy, 157–70. London: Pickering and Chatto, 2014.

Buchenau, Jürgen. 'Global Darwin: Multicultural Mergers'. *Nature* 462 (19 November 2009): 284–5.

Burnett, D. Graham. 'Hydrographic Discipline Among the Navigators: Charting an 'Empire of Commerce and Science' in the Nineteenth-Century Pacific'. In *The Imperial Map: Cartography and the Mastery of Empire*, edited by James R. Ackerman, 185–260. Chicago: University of Chicago Press, 2009.

Burrow, John. *Evolution and Society: A Study in Victorian Social Theory*. Cambridge: Cambridge University Press, 1966.

Butler, Patricia. *Irish Botanical Illustrators and Flower Painters*. Suffolk: Antique Collectors Club, 2000.

Byrne, Angela. *Geographies of the Romantic North: Science, Antiquarianism, and Travel, 1790–1830*. New York: Palgrave Macmillan, 2013.

Camerini, Jane R. 'Wallace in the Field'. *Osiris* 2nd ser. 11 (1996): 44–65.

Campbell, Albert A. *Belfast Naturalists' Field Club: Its Origin and Progress*. Belfast: Hugh Greer, 1938.

Carter, Paul. *The Road to Botany Bay: An Essay in Spatial History*. London: Faber and Faber, 1987.

Casid, Jill H. *Sowing Empire: Landscape and Colonization.* London: University of Minnnesota Press, 2005.

Catanach, Ian. 'Plague and the Tensions of Empire in India, 1896–1916'. In *Imperial Medicine and Indigenous Societies*, edited by David Arnold, 149–71. Manchester: Manchester University Press, 1988.

Cesnola, Luigi Palma di. *Cyprus: Its Ancient Cities, Tombs and Temples*. London: John Murray, 1877.

Cesnola, Luigi Palma di. *A Descriptive Atlas of the Cesnola Collection of Cypriote Antiquities in the Metropolitan Museum of Art, New York*. Boston: James R. Osgood and Company, 1885.

Chakrabarti, Pratik. *Western Science in Modern India: Metropolitan Methods, Colonial Practices*. Delhi: Permanent Black, 2004.

Challis, Debbie. *From the Harpy Tomb to the Wonders of Ephesus: British Archaeologists in the Ottoman Empire 1840–1880*. London: G. Duckworth and Co, 2008.

Chambers, David Wade and Gillespie, Richard. 'Locality in the History of Science: Colonial Science, Technoscience, and Indigenous Knowledge'. *Osiris* 2nd ser. 15 (2000): 221–40.

Chaplin, Joyce E. *Round About the Earth: Circumnavigation from Magellan to Orbit.* New York: Simon and Schuster, 2012.

Childs, Elizabeth C. 'Big Trouble: Daumier, Gargantua, and the Censorship of Political Caricature'. *Art Journal* 51 (1992): 26–37.

Clayton, Daniel. *Islands of Truth: The Imperial Fashioning of Vancouver Island.* Vancouver: UBC Press, 2000.

Cleaver, Anne Hoffman and Stann, E.J., eds. *Voyage to the Southern Ocean – Letters of Lieutenant William Reynolds of the United States Exploring Expedition 1838–1842.* Annapolis: Naval Institute Press, 1988.

Cock, Randolph. 'Scientific Servicemen in the Royal Navy and the Professionalism of Science, 1816–55'. In *Science and Beliefs: From Natural Philosophy to Natural Science, 1700–1900*, edited by David M. Knight and Matthew D. Eddy, 95–111. Aldershot: Ashgate, 2005.

Cohn, Bernard. *Colonialism and its Forms of Knowledge: The British in India.* Princeton: Princeton University Press, 1996.

Colley, Linda. *Captives: Britain, Empire and the World, 1600–1850.* London: Pimlico, 2003.

Coole, Diane and Frost, Samantha. 'Introducing the New Materialisms'. In *New Materialisms: Ontology, Agency, and Politics*, edited by Diane Coole and Samantha Frost, 1–46. Durham, NC: Duke University Press, 2010.

Corfield, Richard. *The Silent Landscape: Discovering the World of the Oceans in the Wake of HMS Challenger's Epic 1872 Mission to Explore the Sea Bed.* London: John Murray, 2005.

Cosgrove, Denis. *Apollo's Eye: A Cartographic Genealogy of the Earth in the Western Imagination.* Baltimore: Johns Hopkins University Press, 2001.

Crawford, Elisabeth, Shinn, Terry and Sörlin, Sverker. 'The Nationalisation and Denationalization of the Sciences: An Introductory Essay'. In *Denationalizing Science: The Contexts of International Scientific Practice*, edited by Elisabeth Crawford, Terry Shinn and Sverker Sörlin, 1–42. Kluwer: Dordrecht, 1993.

Cronin, William. *Changes in the Land: Indians, Colonists and the Ecology of New England.* New York: Hill & Wang, 1983.

Crosbie, Barry. *Irish Imperial Networks: Migration, Social Communication and Exchange in Nineteenth-Century India.* Cambridge: Cambridge University Press, 2012.

Crosby, Alfred. *Ecological Imperialism: The Biological Expansion of Europe, 900–1900.* Cambridge: Cambridge University Press, 1986.

Cundall, Frank. *Reminiscences of the Colonial and Indian Exhibition.* London: W. Clowes, 1886.

Dain, Bruce. *A Hideous Monster of the Mind: American Race Theory in the Early Republic.* Cambridge, MA: Harvard University Press, 2002.

Daniels, Stephen and Nash, Catherine. 'Lifepaths: Geography and Biography'. *Journal of Historical Geography* 30 (2004): 449–58.

Dash, J. Michael. 'Nineteenth-Century Haiti and the Archipelago of the Americas: Anténor Firmin's Letters from St. Thomas'. *Research in African Literatures* 35 (2004): 44–53.

Daston, Lorraine. 'On Scientific Observation'. *Isis* 99 (2008): 97–110.

Daston, Lorraine and Lunbeck, Elizabeth, eds. *Histories of Scientific Observation*. Chicago: University of Chicago Press, 2011.

Davis, Mike. *Late Victorian Holocausts: El Niño Famines and the Making of the Third World*. London: Verso, 2002.

Deacon, Margaret. *Scientists and the Sea, 1650–1900: A Study of Marine Science*. Aldershot: Ashgate, 1971.

Deane, Arthur, ed. *Belfast Natural History and Philosophical Society, Centenary Volume, 1821–1921: A Review of the Activities of the Society for 100 years with Historical Notes, and Memoirs of Many Distinguished Members*. Belfast: Belfast Natural History and Philosophical Society, 1924.

DeMarco, Eileen S. *Reading and Riding: Hachette's Railroad Bookstore Network in Nineteenth-Century France*. Cranbury, NJ: Associated University Presses, 2010.

Dening, Greg. *Mr Bligh's Bad Language: Passion, Power and Theatre on the Bounty*. Cambridge: Cambridge University Press, 1992.

Desmond, Ray. *The History of the Royal Botanic Gardens, Kew*. London: Kew Publishing, 2007.

Dobbs, Arthur. *An Account of the Countries Adjoining to Hudson's Bay*. London: J. Robinson, 1749.

Dodge, Ernest S. *The Polar Rosses: John and James Clark Ross and their Explorations*. London: Faber and Faber, 1973.

Dolan, Brian. *Exploring European Frontiers: British Travellers in the Age of Enlightenment*. Basingstoke: Palgrave Macmillan, 2000.

Drayton, Richard. *Nature's Government: Science, Imperial Britain, and the 'Improvement' of the World*. New Haven: Yale University Press, 2000.

Dritsas, Lawrence. 'From Lake Nyassa to Philadelphia: A Geography of the Zambesi Expedition, 1858–1864'. *British Journal for the History of Science* 38 (2005): 35–52.

Dritsas, Lawrence. *Zambesi: David Livingstone and Expeditionary Science in Africa*. London and New York: I.B. Tauris, 2010.

Driver, Felix. 'Editorial: Field-Work in Geography'. *Transactions of the Institute of British Geographers* 25 (2000): 267–8.

Driver, Felix. *Geography Militant*. Oxford: Blackwell, 2001.

Driver, Felix. 'Hidden Histories Made Visible? Reflections on a Geographical Exhibition'. *Transactions of the Institute of British Geographers* 38 (2013): 420–35.

Driver, Felix. '*Missionary Travels:* Livingstone, Africa and the Book'. *Scottish Geographical Journal* 129 (2013): 164–78.

Driver, Felix and Ashmore, Sonia. 'The Mobile Museum: Collecting and Circulating Indian Textiles in Britain'. *Victorian Studies* 52 (2010): 353–85.

Driver, Felix and Jones, Lowri. *Hidden Histories of Exploration: Researching the RGS-IBG Collections*. London: Royal Holloway, University of London, 2009.

Drummond, James Lawson. *Thoughts on the Study of Natural History; and on the Importance of Attaching Museums of the Productions of Nature, to National Seminaries of Education, Addressed to the Proprietors of the Belfast Institution*. Belfast: F.D. Finlay, 1820.

Drummond, James Lawson. *Letters to a Young Naturalist on the Study of Nature and Natural Theology*. London: Longman, Rees, Orme, Brown, and Green, 1831.

Dubois, Laurent. *Haiti: The Aftershocks of History*. New York: Metropolitan Books, 2011.

Du Tertre, Jean-Baptiste. *Histoire Générale des Antilles Habitées par les François, Vol. 2*. Paris: T. Jolly, 1667.

Dzelainis, Ella and Kaplan, Cora, eds. *Harriet Martineau: Authorship, Society, and Empire* Manchester: Manchester University Press, 2011.

Edney, Matthew. *Mapping an Empire: The Geographical Construction of British India 1765–1843*. Chicago: University of Chicago Press, 1997.

Edwards, Elizabeth. 'Photographic Uncertainties: Between Evidence and Reassurance'. *History and Anthropology* 25 (2014): 171–88.

Eliot, Simon. 'Some Trends in British Book Publishing, 1800–1919'. In *Literature in the Marketplace: Nineteenth-Century British Publishing and Reading Practices*, edited by John O. Jordan and Robert L. Patten, 19–43. Cambridge: Cambridge University Press, 1994.

Elshakry, Marwa. 'Global Darwin: Eastern Enchantment'. *Nature* 461 (29 October 2009): 1200–201.

Elshakry, Marwa. 'When Science Became Western: Historiographical Reflections'. *Isis* 101 (2010): 98–109.

Elshakry, Marwa. *Reading Darwin in Arabic, 1860–1950*. Chicago: University of Chicago Press, 2013.

Endersby, Jim. *Imperial Nature: Joseph Hooker and the Practice of Victorian Science*. Chicago: University of Chicago Press, 2008.

Ethington, Philip E. 'Placing the Past: 'Groundwork' for a Spatial Theory of History'. *Rethinking History* 11 (2007): 465–93.

Evans, Christopher. '"Delineating Objects": Nineteenth-Century Antiquarian Culture and the Project of Archaeology'. In *Visions of Antiquity: The Society of Antiquaries of London, 1707–2007*, edited by Susan Pearce, 267–305. London: Society of Antiquaries of London, 2007.

Fan, Fa Ti. 'The Global Turn in the History of Science'. *East Asian Science Technology and Society* 6 (2012): 249–58.

Feather, John. *Publishing, Piracy and Politics: An Historical Study of Copyright in Britain*. London: Mansell, 1994.

Febvre, Lucien and Martin, Lucien. *The Coming of the Book: The Impact of Print, 1450–1800*. London: Verso, 1976.

Finnegan, Diarmid A. 'The Spatial Turn: Geographical Approaches in the History of Science'. *Journal of the History of Biology* 41 (2008): 369–88.

Finnegan, Diarmid A. *Natural History Societies and Civic Culture in Victorian Scotland*. London: Pickering and Chatto, 2009.

Fischer, Sibylle. *Modernity Disavowed: Haiti and the Cultures of Slavery in the Age of Revolution*. Durham, NC: Duke University Press, 2004.

Flint, Kate. *The Transatlantic Indian, 1776–1930*. Princeton: Princeton University Press, 2009.

Forgan, Sophie. 'Building the Museum: Knowledge, Conflict, and the Power of Place'. *Isis* 96 (2005): 572–85.

Forsyth, Isla. 'The More-Than-Human Geographies of Field Science'. *Geography Compass* 7 (2013): 527–39.

Fox, Christopher, Porter, Roy and Wokler, Robert, eds. *Inventing Human Science: Eighteenth-Century Domains*. Berkeley: University of California Press, 1995.

Franklin, John. *Narrative of a Second Expedition to the Shores of the Polar Sea, in the Years 1825, 1826, and 1827*. London: John Murray, 1828.

Fraser, Antonia. *Perilous Question: The Drama of the Great Reform Bill 1832*. London: Weidenfeld and Nicolson, 2013.

Fraser, Ian 'Father and Son – A Tale of Two Cities'. *Ulster Medical Journal* 37 (1968): 1–39.

Freitag Ulrike and von Oppen, Achim, eds. *Translocality: The Study of Globalising Processes from a Southern Perspective*. Amsterdam: Brill, 2010.

Fyfe, Aileen. *Steam-Powered Knowledge: William Chambers and the Business of Publishing, 1820–1860*. Chicago: University of Chicago Press, 2012.

Fyfe, Aileen and Lightman, Bernard, eds. *Science in the Marketplace: Nineteenth-Century Sites and Experiences*. Chicago: University of Chicago Press, 2007.

Garner, Margaret A.K. *Robert Workman of Newtownbreda, 1835–1921*. Belfast: William Mullan & Son Ltd, 1969.

Gates, Barbara T. *Kindred Nature: Victorian and Edwardian Women Embrace the Living World*. Chicago: University of Chicago Press, 1998.

Geyer, Martin H. and Paulman, Johannes, eds. *The Mechanics of Internationalism: Culture, Society, and Politics from the 1840s to the First World War*. Oxford: Oxford University Press, 2001.

Gieryn, Thomas F. 'City as Truth-Spot: Laboratories and Field-Sites in Urban Studies'. *Social Studies of Science* 36 (2006): 5–38.

Glover, Winifred. 'In the Wake of Captain Cook: The Travels of Gordon Augustus Thomson (1799–1886), Principal Donor of Ethnographic Objects to the Ulster Museum, Belfast'. *Familia* 9 (1993): 46–61.

Glover, Winifred. 'The Folks Back Home: Connections Between Ethnography and Folk Life'. *Journal of Museum Ethnography* 9 (1997): 21–32.

Glover, Winifred. 'Power and Collecting: Big Men Talking'. *Journal of Museum Ethnography* 15 (2003): 19–24.

Golinski, Jan. 'American Climate and the Civilization of Nature'. In *Science and Empire in the Atlantic World*, edited by James Delbourgo and Nicholas Dew, 153–74. New York: Routledge, 2008.

Golinski, Jan. 'Is It Time to Forget Science? Reflections on Singular Science and Its History'. *Osiris* 2nd ser. 27 (2012): 19–36.

Goodman, Jordan. *The Rattlesnake: A Voyage of Discovery to the Coral Sea.* London: Faber and Faber, 2005.

Goring, Elizabeth. *A Mischievous Pastime. Digging in Cyprus in the Nineteenth Century.* Edinburgh: National Museums of Scotland, 1988.

Gosden, Chris and Knowles, Chantal. *Collecting Colonialism: Material Culture and Colonial Change.* Oxford: Berg, 2001.

Gosden, Chris and Marshall, Yvonne. 'The Cultural Biography of Objects'. *World Archaeology* 31 (1999): 169–78.

Greeley, Horace. *Art and Industry: As Represented in the Exhibition at the Crystal Palace, New York, 1853–4.* New York: J.S. Redfield, 1853.

Greenblatt, Stephen. 'Resonance and Wonder'. In *Exhibiting Cultures: The Poetics and Politics of Museum Display*, edited by Ivan Karp and Steven D. Lavine, 42–56. Washington, DC and London: Smithsonian Institution, 1991.

Gregory, Derek and Duncan, James S. eds. *Writes of Passage: Reading Travel Writing.* London: Routledge, 1999.

Guelke, Jeanne K. and Morin, Karen M. 'Gender, Nature, Empire: Women Naturalists in Nineteenth Century British Travel Literature'. *Transactions of the Institute of British Geographers* 26 (2001): 306–26.

Gunning, Lucia P. *The British Consular Service in the Aegean and the Collection of Antiquities for the British Museum.* Farnham: Ashgate, 2009.

Gurney, Alan. *The Race to the White Continent: Voyages to the Antarctic.* London: Norton, 2000.

Hale, Piers. *Political Descent: Malthus, Mutualism and the Politics of Evolution in Victorian England.* Chicago: University of Chicago Press, 2014.

Hall, Catherine and Rose, Sonya O. 'Introduction: Being at Home with the Empire'. In *At Home with the Empire: Metropolitan Culture and the Imperial World*, edited by Catherine Hall and Sonya O. Rose, 1–31. Cambridge: Cambridge University Press, 2006.

Hall, Radclyffe. *Adam's Breed.* London: Virago Modern Classics, 1985.

Hallett, Robin, ed. *The Niger Journal of Richard and John Lander.* London: Routledge and Kegan Paul, 1965.

Hamilakis, Yannis. 'From Ethics to Politics'. In *Archaeology and Capitalism, From Ethics to Politics*, edited by Yannis Hamilakis and Philip Duke, 15–40. Walnut Creek: Left Coast Press, 2007.

Harvey, David. *The New Imperialism.* Oxford: Oxford University Press, 2003.

Harvey, David C. 'Broad Down Devon: Archaeological and Other Stories'. *Journal of Material Culture* 15 (2010): 345–67.

Hasty, William and Peters, Kimberly. 'The Ship in Geography and the Geographies of Ships'. *Geography Compass* 6 (2012): 660–76.

Haynes, Christine S. *Lost Illusions: The Politics of Publishing in Nineteenth-Century France.* Cambridge, MA: Harvard University Press, 2010.

Headrick, Daniel R. *The Tools of Empire, Technology and European Imperialism in the Nineteenth Century*. Oxford: Oxford University Press, 1981.

Headrick, Daniel R. *Technology: A World History*. Oxford: Oxford University Press, 2009.

Henderson, Louise C. '"Everyone will die laughing": John Murray and the Publication of David Livingstone's *Missionary Travels'*. Livingstone Online, Welcome Trust Centre for the History of Medicine at UCL. http://www.livingtsoneonline.ucl.ac.uk/companion.php?id=HIST2.

Henderson, Louise C. 'Geography, Travel and Publishing in Mid-Nineteenth Century Britain'. PhD Diss., Royal Holloway University of London, 2006.

Henderson, Louise C. 'David Livingstone's *Missionary Travels* in Britain and America: Exploring the Wider Circulation of a Victorian Travel Narrative'. *Scottish Geographical Journal* 129 (2013): 179–93.

Heringman, Noah. *Sciences of Antiquity: Romantic Antiquarianism, Natural History, and Knowledge Work*. Oxford: Oxford University Press, 2013.

Hill, Jude M. 'Travelling Objects: The Wellcome Collection in Los Angeles, London and Beyond'. *Cultural Geographies* 13 (2006): 340–66.

Hill, Kate. *Culture and Class in English Public Museums, 1850–1914*. Aldershot: Ashgate, 2005.

Hirschfeld, Charles. 'America on Exhibition: The New York Crystal Palace'. *American Quarterly* 9 (1957): 101–16.

Hodge, Joseph M. 'Science and Empire: An Overview of the Historical Scholarship'. In *Science and Empire: Knowledge and Networks of Science Across the British Empire, 1800–1970*, edited by Brett M. Bennett and Joseph M. Hodge, 3–29. Basingstoke: Palgrave Macmillan, 2011.

Hogarth, David G. *A Wandering Scholar in the Levant*. London: MacMillan, 1896.

Hollerbach, Anne Larsen. 'Of Sangfroid and Sphinx Moths: Cruelty, Public Relations, and the Growth of Entomology in England, 1800–1840'. *Osiris* 2nd ser. 11 (1996): 201–20.

Holmes, Richard. *The Age of Wonder: How the Romantic Generation Discovered the Beauty and Terror of Science*. London: Harper Press, 2009.

Hones, Sheila. 'Text as it Happens: Literary Geography'. *Geography Compass* 2 (2008): 1301–17.

Hooker, Joseph Dalton. *Flora Antarctica*. London: Reeve Brothers, 1844.

Hooker, William Jackson. *Museum of Economic Botany: Or, A Popular Guide to the Useful and Remarkable Vegetable Products of the Museum of the Royal Gardens of Kew*. London: Longman, Brown, Green, and Longman, 1855.

Houston, Stuart, Ball, Tim and Houston, Mary, eds. *Eighteenth-Century Naturalists of Hudson Bay*. Montreal and Kingston: McGill-Queen's University Press, 2003.

Howe, Stephen. 'Questioning the (Bad) Question: "Was Ireland a Colony?"'. *Irish Historical Studies* 36 (2008): 138–52.

Howsam, Leslie. 'The History of the Book in Britain, 1801–1914'. In *The Oxford Companion to the Book*, edited by Michael Suarez, S.J. and Henry R. Woudhuysen. 2 volumes. I: 180–87. Oxford: Oxford University Press.

Hulme, Mike. 'Problems with Making and Governing Global Kinds of Knowledge'. *Global Environment Change* 20 (2010): 558–64.

Igler, David. *The Great Ocean: Pacific Worlds from Captain Cook to the Gold Rush*. New York: Oxford University Press, 2013.

Irving, Sarah. *Natural Science and the Origins of the British Empire*. London: Pickering & Chatto, 2008.

James, Patricia. *Population Malthus: His Life and Times*. London: Routledge, 1979.

James, Patricia, ed. *The Travel Diaries of T.R. Malthus*. Cambridge: Cambridge University Press, 1966.

Jamieson, John. *The History of the Royal Belfast Academical Institution, 1810–1960*. Belfast: W. Mullan, 1959.

Jasanoff, Maya. *Edge of Empire: Lives, Culture, and Conquest in the East, 1750–1850*. New York: Alfred A. Knopf, 2005.

Jeffery, Keith, ed. *An Irish Empire? Aspects of Ireland and the British Empire*. Manchester: Manchester University Press, 1996.

Johnson, Nuala C. *Nature Displaced, Nature Displayed: Order and Beauty in Botanical Gardens*. London: I.B. Tauris, 2011.

Keighren, Innes M. 'Bringing Geography to the Book: Charting the Reception of *Influences of Geographic Environment*'. *Transactions of the Institute of British Geographers* 31 (2006): 525–40.

Keighren, Innes M. *Bringing Geography to Book: Ellen Semple and the Reception of Geographical Knowledge*. London: I.B. Tauris, 2010.

Keighren, Innes M. and Withers, Charles W.J. 'Questions of Inscription and Epistemology in British Travelers' Accounts of Early Nineteenth-Century South America'. *Annals of the Association of American Geographers* 101 (2011): 1331–46.

Kennedy, Dane, ed. *Reinterpreting Exploration: The West in the World*. Oxford and New York: Oxford University Press, 2014.

Kennedy, David. 'The Perils of the Midday Sun: Climatic Anxieties in the Colonial Tropics'. In *Imperialism and the Natural World*, edited by John M. McKenzie, 118–40. Manchester: Manchester University Press, 1990.

Kenny, Judith. 'Climate, Race and Imperial Authority: The Symbolic Landscape of the British Hill Station in India'. *Annals of the Association of American Geographers* 85 (1995): 694–714.

Kenny, Kevin, ed. *Ireland and the British Empire*. Oxford: Oxford University Press, 2004.

Kern, Stephen. *The Culture of Time and Space 1880–1918*. Cambridge, MA: Harvard University Press, 2003 edition.

Kiely, Thomas. 'Charles Newton and the Archaeology of Cyprus'. *Cahiers du Centre d'Etudes Chypriotes* 40 (2010): 231–51.

Killingray, David, Lincoln, Margarette and Rigby, Nigel, eds. *Maritime Empires: British Imperial Maritime Trade in the Nineteenth Century*. Woodbridge: Boydell Press, 2004.

Klancher, Jon. *Transfiguring the Arts and Sciences: Knowledge and Cultural Institutions in the Romantic Age*. Cambridge: Cambridge University Press, 2013.

Knapp, Sandra, Sanders, Lynn and Baker, William. 'Alfred Russel Wallace and the Palms of the Amazon'. *Palms* 46 (2002): 109–19.

Kohler, Robert E. 'A Generalist's Vision'. *Isis* 96 (2005): 224–9.

Kohler, Robert E. 'History of Science: Trends and Prospects'. In *Knowing Global Environments: New Historical Perspectives on the Field Sciences*, edited by Jeremy Vetter, 212–40. New Brunswick, NJ: Rutgers University Press, 2011.

Kriegel, Lara. *Grand Designs: Labor, Empire, and the Museum in Victorian Culture*. Durham, NC: Duke University Press, 2007.

Kuklick, Henrika and Kohler, Robert E. 'Introduction'. *Osiris* 2nd ser. 11 (1996): 1–14.

Kumar, Deepak and Basu, Raj Sekhar, eds. *Medical Encounters in British India*. New Delhi: Oxford University Press, 2013.

Kumar, Prakash. *Indigo Plantations and Science in Colonial India*. Cambridge: Cambridge University Press, 2012.

Laidlaw, Zoe. *Colonial Connections 1815–1845: Patronage, the Information Revolution and Colonial Government*. Manchester: Manchester University Press, 2005.

Lambert, David and Lester, Alan, eds. *Colonial Lives Across the British Empire: Imperial Careering in the Long Nineteenth Century*. Cambridge: Cambridge University Press, 2006.

Lambert, David, Martins, Luciana and Ogborn, Miles. 'Currents, Visions and Voyages: Historical Geographies of the Sea'. *Journal of Historical Geography* 32 (2006): 479–93.

Lang, Robert H. *Cyprus: Its History, its Present Resources and Future Prospects*. London: MacMillan and Co., 1878.

Lang, Robert H. 'On Archaic Survivors in Cyprus'. *Journal of the Anthropological Institute of Great Britain and Ireland* 16 (1887): 186–8.

Lang, Robert H. 'Reminiscences – Archaeological Research in Cyprus'. *Blackwood's Magazine* 177 (1905): 622–39.

Lang, Robert H. and Poole, Stuart. 'Narrative of Excavations in a Temple at Dali (Idalium) in Cyprus'. *Transactions of the Royal Society of Literature XI*, 2nd ser. (1878): 30–79.

Latour, Bruno. *Reassembling the Social: An Introduction to Actor-Network-Theory*. Oxford: Oxford University Press, 2005.

Law, John. 'And if the Global were Small and Noncoherent? Method, Complexity and the Baroque'. *Environment and Planning D: Society and Space* 22 (2004): 13–26.

Leitch, Maurice. *Burning Bridges*. London: Hutchinson, 1989.

Lester, Alan. *Imperial Networks: Creating Identities in Nineteenth-Century South Africa and Britain*. London: Routledge, 2001.

Lester, Alan. 'Imperial Circuits and Networks: Geographies of the British Empire'. *History Compass* 4 (2006): 124–41.

Levere, T.H. *Science and the Canadian Arctic: A Century of Exploration, 1818–1918.* Cambridge: Cambridge University Press, 1993.

Levine, Philippa. *The Amateur and the Professional: Antiquarians, Historians and Archaeologists in Victorian England 1838–1886.* Cambridge: Cambridge University Press, 1996.

Lightman, Bernard, McOuat, Gordon and Stewart, Larry, eds. *The Circulation of Knowledge Between Britain, India and China: The Early-Modern World to the Twentieth Century.* Leiden: Brill, 2013.

Livingstone, David. *Missionary Travels and Researches in South Africa.* London: John Murray, 1857.

Livingstone, David. *A Popular Account of Missionary Travels and Researches in South Africa.* London: John Murray, 1861.

Livingstone, David N. 'The Spaces of Knowledge: Contributions Towards a Historical Geography of Science'. *Environment and Planning D: Society and Space* 13 (1995): 5–35.

Livingstone, David N. 'Tropical Climate and Moral Hygiene: The Anatomy of a Victorian Debate'. *British Journal for the History of Science* 32 (1999): 93–110.

Livingstone, David N. *Science, Space and Hermeneutics: Hettner Lecture 2001.* Heidelberg: University of Heidelberg, 2001.

Livingstone, David N. *Putting Science in its Place: Geographies of Scientific Knowledge.* Chicago: University of Chicago Press, 2003.

Livingstone, David N. 'Text, Talk and Testimony: Geographical Reflections on Scientific Habits. An Afterword'. *British Journal for the History of Science* 38 (2005): 93–100.

Livingstone, David N. 'Science, Text and Space: Thoughts on the Geography of Reading'. *Transactions of the Institute of British Geographers* 30 (2005): 391–401.

Livingstone, David N. *Dealing with Darwin: Place, Politics and Rhetoric in Religious Engagements with Evolution.* Baltimore: Johns Hopkins University Press, 2014.

Livingstone, David N. and Withers, Charles W.J., eds. *Geographies of Nineteenth-Century Science.* Chicago: University of Chicago Press, 2011.

Livingstone, Justin D. 'The Meaning and Making of *Missionary Travels:* The Sedentary and Itinerant Discourses of a Victorian Bestseller'. *Studies in Travel Writing* 15 (2011): 267–92.

Livingstone, Justin D. '*Missionary Travels*, Missionary Travails: David Livingstone and the Victorian Publishing Industry'. In *David Livingstone: Man, Myth and Legacy*, edited by S. Worden, 32–51. Edinburgh: National Museums Scotland, 2012.

Lloyd, Christopher. *The British Seaman 1200–1860: A Social Survey.* London: Paladin, 1970.

270 Spaces of Global Knowledge

Logan, Rayford. *Haiti and the Dominican Republic*. New York: Oxford University Press, 1968.

Longair, Sarah and McAleer, John, eds. *Curating Empire: Museums and the British Imperial Experience*. Manchester: Manchester University Press, 2012.

Lorimer, Hayden. 'Telling Small Stories: Spaces of Knowledge and the Practice of Geography'. *Transactions of the Institute of British Geographers* 28 (2003): 197–217.

Lorimer, Hayden and Spedding, Nick. 'Excavating Geography's Hidden Spaces'. *Area* 34 (2002): 294–302.

Lorimer, Hayden and Spedding, Nick. 'Locating Field Science: A Geographical Family Expedition to Glen Roy, Scotland'. *British Journal for the History of Science* 38 (2005): 13–33.

Lorimer, Hayden and Withers, Charles W.J. 'Introduction'. *Geographers Biobibliographical Studies* 32 (2013): 1–5.

Losano, Antonia. 'A Preference for Vegetables: The Travel Writings and Botanical Art of Marianne North'. *Women's Studies* 26 (1997): 423–48.

Loughney, Claire. 'Colonialism and the Development of the English Provincial Museum, 1823–1914'. PhD Diss., Newcastle University, 2006.

Lyte, Charles. *Frank Kingdon Ward: The Last of the Great Plant Hunters*. London: John Murray, 1989.

McConnell, Anita. *No Sea Too Deep: The History of Oceanographic Instruments*. Bristol: Adam Hilger Ltd., 1982.

McCook, Stuart. '"It May Be Truth, But It Is not Evidence": Paul du Chaillu and the Legitimisation of Evidence in the Field Sciences'. *Osiris* 2nd ser. 11 (1996): 177–97.

McDonough, Terence, ed. *Was Ireland a Colony? Economics, Politics, and Culture in Nineteenth-Century Ireland*. Dublin: Irish Academic Press, 2005.

McEwan, Cheryl. 'Gender, Science and Physical Geography in Nineteenth Century Britain'. *Area* 30 (1998): 215–23.

McFadden, Elizabeth. *The Glitter and the Gold: A Spirited Account of the Metropolitan Museum of Art's First Director, the Audacious and High-handed Luigi Palma di Cesnola*. New York: Dial Press, 1971.

McKechnie, Charlotte. 'Spiders, Horror and Animal Others in Late-Victorian Empire Fiction'. *Journal of Victorian Culture* 17 (2012): 505–16.

McKeich, Cherie. 'Botanical fortunes. T N Mukharji, international exhibitions, and Trade between India and Australia'. *reCollections* 3 (2008): 1–12.

MacKenzie, John M. *Museums and Empire: Natural History, Human Cultures and Colonial Identities*. Manchester: Manchester University Press, 2009.

McKitterick, David. 'Libraries, Knowledge and Public Identity'. In *The Organisation of Knowledge in Victorian Britain*, edited by Martin Daunton, 287–312. Oxford: Oxford University Press, 2005.

Maddrell, Avril. *Complex Locations: Women's Geographical Work in the UK 1850–1970*. Oxford: Wiley-Blackwell, 2009.

Magee, Gary B. and Thompson, Andrew S. *Empire and Globalisation: Networks of People, Goods and Capital in the British world, c.1850–1914*. Cambridge: Cambridge University Press, 2010.

Malthus, Thomas R. *An Essay on the Principle of Population*. Edited by Patricia James. Cambridge: Cambridge University Press, 1989.

Marouby, Christian. 'Adam Smith and the Anthropology of the Enlightenment: The 'Ethnographic' Sources of Economic Progress'. In *The Anthropology of the Enlightenment*, edited by Larry Wolff and Marco Cipolloni, 85–102. Stanford: Stanford University Press, 2007.

Marsden, Ben and Smith, Crosbie. *Engineering Empires: A Cultural History of Technology in Nineteenth-Century Britain*. Basingstoke: Palgrave Macmillan, 2005.

Marshall, P.J. and Williams, Glyndwr. *The Great Map of Mankind: Perceptions of New Worlds in the Age of Enlightenment*. Cambridge, MA: Harvard University Press, 1982.

Massey, Doreen. *For Space*. London: Sage, 2005.

Mathur, Saoni. *India by Design: Colonial History and Cultural Display*. Berkeley: University of California Press, 2007.

Mayhew, Robert, J. 'Mapping Science's Imagined Community: Geography as a Republic of Letters, 1600–1800'. *British Journal for the History of Science* 38 (2005): 73–92.

Mayhew, Robert J. 'Materialist Hermeneutics, Textuality and the History of Geography: Print Spaces in British Geography, c.1500–1900'. *Journal of Historical Geography* 33 (2007): 466–88.

Mayhew, Robert J. *Malthus: The Life and Legacies of an Untimely Prophet*. Cambridge, MA: Harvard University Press, 2014.

Mazlish, Bruce. *The New Global History*. London: Routledge, 2006.

Meek, Ronald. *Social Science and the Ignoble Savage*. Cambridge: Cambridge University Press, 1976.

Merriman, Nick. *Beyond the Glass Case: The Past, the Heritage and the Public in Britain*. Leicester: Leicester University Press, 1991.

Meusburger, Peter, Livingstone, David N. and Jöns, Heike, eds. *Geographies of Science*. Heidelberg: Springer, 2010.

Michalski, Katarzyna and Michalski, Sergiusz. *Spider*. London: Reaktion Books, 2010.

Middleton, Dorothy. *Victorian Lady Travellers*. London: Routledge and Kegan Paul, 1965.

Millar, Sarah Louise. 'Science at Sea: Soundings and Instrumental Knowledge in British Polar Expedition Narratives, c. 1818–1848'. *Journal of Historical Geography* 42 (2013): 77–87.

Miller, Daniel Philip and Reill, Peter Hanns, eds. *Visions of Empire: Voyages, Botany, and Representations of Nature*. Cambridge: Cambridge University Press, 1996.

Moodie, D. Wayne and Kaye, Barry. 'Indian Agriculture in the Fur Trade Northwest'. *Prairie Forum* 11 (1986): 171–83.

Moon, Paul. *A Savage Country: The Untold Story of New Zealand in the 1820s*. Auckland, N.Z: Penguin, 2012.

Morin, Karen M. 'Peak Practices: Englishwomen's "Heroic" Adventures in the Nineteenth Century American West'. *Annals of the Association of American Geographers* 89 (1999): 489–514.

Morin, Karen M. *Civic Discipline: Geography in America, 1860–1890*. Farnham: Ashgate, 2011.

Moser, Stephanie. *Wondrous Curiosities: Ancient Egypt at the British Museum*. Chicago: University of Chicago Press, 2006.

Moser, Stephanie. 'On Disciplinary Culture: Archaeology as Fieldwork and Its Gendered Associations'. *Journal of Archaeological Method and Theory* 14 (2007): 235–63.

Moyn, Samuel and Sartori, Andrew. 'Approaches to Global Intellectual History'. In *Global Intellectual History* edited by Samuel Moyn and Andrew Sartori. New York: Columbia University Press, 2013.

Mueggler, Erik. *The Paper Road: Archive and Experience in the Botanical Exploration of West China and Tibet*. Berkeley: University of California Press, 2011.

Mukharji, Trailokya Nath. *Art-Manufactures of India (Specially Compiled for the Glasgow International Exhibition 1888)*. Calcutta: Superintendent of Government Printing, 1888.

Mukharji, Trailokya Nath. *A Visit to Europe*. Calcutta: W. Newman & Co, 1889.

Myres, John Linton. *Handbook of the Cesnola Collection of Antiquities from Cyprus*. New York: Metropolitan Museum of New York, 1914.

Myres, John Linton. 'The Amathus Bowl: A Long-Lost Masterpiece of Oriental Engraving'. *Journal of Hellenic Studies* 53 (1933): 25–39.

Nakazawa, Nobuhiko. 'Malthus's Political Views in 1798: A "Foxite" Whig?' *History of Economics Review* 56 (2012): 14–28.

Nappi, Carla. 'The Global and Beyond: Adventures in Local Historiographies of Science'. *Isis* 104 (2013): 102–10.

National Botanic Gardens. *The Art of Flowers: National Botanic Gardens, Glasnevin Bicentenary Exhibition 1995 Catalogue*. Dublin: National Botanic Gardens, 1995.

National Botanic Gardens. *Plant Treasures: Two Hundred Years of Botanical Illustration from the National Botanic Gardens, Glasnevin*. Dublin: National Botanic Gardens, 2002.

Naylor, Simon. 'The Field, the Museum and the Lecture Hall: The Spaces of Natural History in Victorian Cornwall'. *Transactions of the Institute of British Geographers* 27 (2002): 494–513.

Naylor, Simon. 'Introduction: Historical Geographies of Science – Places, Contexts, Cartographies'. *British Journal for the History of Science* 38 (2005): 1–12.

Nead, Lynda. 'Gender and the City'. In *The Victorian World*, edited by Martin Hewitt, 291–307. London: Routledge, 2012.

Nelson, Charles E. 'The Lady of the Rhododendrons – Charlotte Wheeler Cuffe 1867–1967'. *Rhododendrons* (1982): 33–41.

Nesbitt, Noel. *A Museum in Belfast: A History of the Ulster Museum and its Predecessors*. Belfast: Ulster Museum, 1979.

Newton, Charles Thomas. *Essays on Art and Archaeology*. New York: MacMillan, 1880.

Nicholls, David. *From Dessalines to Duvalier: Race, Colour and National Independence in Haiti*. New Brunswick, NJ: Rutgers University Press, 1996.

Norwood, Vera. *Made From This Earth: American Women and Nature*. Chapel Hill: University of North Carolina Press, 1993.

O'Brien, Conor Cruise. *The Great Melody: A Thematic Biography of Edmund Burke*. London: Sinclair Stevenson, 1992.

O'Brien, Gerard. 'State Intervention and the Medical Relief of the Irish Poor, 1787–1850'. In *Medicine, Disease and the State in Ireland, 1650–1940*, edited by Greta Jones and Elizabeth Malcolm, 195–207. Cork: Cork University Press, 1999.

O'Brien, Karen. *Women and Enlightenment in Eighteenth Century Britain*. Cambridge: Cambridge University Press, 2009.

O'Brien, Patrick. 'Historiographical Traditions and Modern Imperatives in the Restoration of Global History'. *Journal of Global History* 1 (2006): 3–39.

O'Connor, Daniel, ed. *Three Centuries of Mission: The United Society for the Propagation of the Gospel 1701–2000*. London: Continuum, 2000.

O'Hara, Glen. '"The Sea is Swinging into View": Modern British Maritime History in a Globalised World'. *English Historical Review* 124 (2009): 1109–34.

Ogborn, Miles. 'Writing Travels: Power, Knowledge and Ritual on the English East Indian Company's Early Voyages'. *Transactions of the Institute of British Geographers* 27 (2002): 155–71.

Ogborn, Miles. 'Mapping Worlds'. *New Formations* 57 (2006): 145–9.

Ogborn, Miles. *Indian Ink: Script and Ink in the Making of the English East India Company*. Chicago: University of Chicago Press, 2007.

Ogborn, Miles. *Global Lives: Britain and the World, 1550–1800*. Cambridge: Cambridge University Press, 2008.

Ogborn, Miles and Withers, Charles W.J., eds. *Geographies of the Book*. Farnham: Ashgate, 2010.

Osterhammel, Jürgen and Petersson, Niels P. *Globalization: A Short History*. Princeton: Princeton University Press, 2005.

Outram, Dorinda. 'New Spaces in Natural History'. In *Cultures of Natural History*, edited by Nicholas Jardine, James A. Secord and Emma C. Spary, 249–65. Cambridge: Cambridge University Press, 1996.

Parry, W.E. *Voyage for the Discovery of a North-West Passage from the Atlantic to the Pacific*. London: J. Murray, 1821.

Pearce, Edward. *Reform! The Fight for the 1832 Reform Act*. London, Jonathan Cape, 2003.

Pettitt, Claire. *Dr Livingstone, I Presume? Missionaries, Journalists, Explorers and Empire.* London: Profile Books, 2007.

Philbrick, Nathaniel. *Sea of Glory*. London: Harper Perennial, 2005.

Phillips, Patricia. *The Scientific Lady: A Social History of Women's Scientific Interests 1520–1918*. London: Weidenfeld and Nicolson, 1990.

Pickstone, John V. 'Museological Science? The Place of the Analytical/ Comparative in Nineteenth-century Science, Technology and Medicine'. *History of Science* 32 (1994): 111–38.

Pietsch, Tamson. 'A British Sea: Making Sense of Global Space in the Late Nineteenth Century'. *Journal of Global History* 5 (2010): 423–46.

Ploszajska, Teresa. 'Constructing the Subject: Geographical Models in English Schools, 1870–1944'. *Journal of Historical Geography* 22 (1996): 388–98.

Pocock, Douglas C.D. 'Geography and Literature'. *Progress in Human Geography* 12 (1988): 87–102.

Pocock, J.G.A. *Barbarism and Religion: Volume 4: Barbarians, Savages and Empires*. Cambridge: Cambridge University Press, 2005.

Podruchny, Carolyn. *Making the Voyageur World: Travelers and Traders in the North American Fur Trade*. Lincoln: University of Nebraska Press, 2006.

Pomeranz, Kenneth and Topik, Steven, eds. *The World That Trade Created: Society, Culture and the World Economy, 1400 to the Present*. London: M.E. Sharpe, 1999.

Post, Robert C. 'Reflections of American Science and Technology at the New York Crystal Palace Exhibition'. *Journal of American Studies* 17 (1983): 337–56.

Price, Leah. '*In Another Country: Colonialism, Culture, and the English Novel in India* (Review)'. *Victorian Studies* 45 (2003): 333–4.

Pullen, J.M. 'Notes from Malthus: The Inverarity Manuscript'. *History of Political Economy* 13 (1981): 794–811.

Pullen, J.M. and Parry, Trevor Hughes, eds. *T. R. Malthus: The Unpublished Papers in the Collection of Kanto Gakuen University*. 2 volumes. Cambridge: Cambridge University Press, 1997–2004.

Pyenson, Lewis. 'An End to National Science: The Meaning and the Extension of Local Knowledge'. *History of Science* 40 (2002): 251–90.

Qureshi, Sadiah. *Peoples on Parade: Exhibitions, Empire, and Anthropology in Nineteenth-Century Britain*. Chicago: University of Chicago Press, 2011.

Raj, Kapil. *Relocating Modern Science: Circulation and the Construction of Scientific Knowledge in South Asia and Europe, Seventeenth to Nineteenth Centuries*. Delhi: Permanent Black, 2006.

Raven, James. 'Constructing Bookscape: Experiments in Mapping the Sites and Activities of the London Book Trades in the Eighteenth Century'. In *Mappa Mundi: Mapping Culture/Mapping the World*, edited by J. Murray, 35–59. University of Windsor, Working Papers in the Humanities, 2001.

Reidy, Michael. *Tides of History: Ocean Science and Her Majesty's Navy.* Chicago: University of Chicago Press, 2008.

Reidy, Michael and Rozwadowski, Helen. 'The Spaces in Between: Science, Ocean, Empire'. *Isis* 105 (2014): 338–51.

Richards, William C. *A Day in the New York Crystal Palace.* New York: G.P. Putnam & Co., 1853.

Richardson, John. *Fauna Boreali-Americana; or the Zoology of the Northern Parts of British America.* London: John Murray, 1829.

Rico, Monica. *Nature's Noblemen: Transatlantic Masculinities and the Nineteenth-Century American West.* New Haven: Yale University Press, 2013.

Rivière, Marc Serge. 'From Belfast to Mauritius: Charles Telfair (1778–1833), Naturalist and a Product of the Irish Enlightenment'. *Eighteenth-Century Ireland* 21 (2006): 125–44.

Rodger, N.A.M. 'Sea-Power and Empire, 1688–1793'. In *The Oxford History of the British Empire: Volume II: The Eighteenth Century*, edited by P.J. Marshall, 169–83. Oxford: Oxford University Press, 1998.

Rodger, N.A.M. *The Command of the Ocean: A Naval History of Britain, 1649–1815.* London: Penguin, 2006.

Rodrick, Anne B. *Self-Help and Civic Culture: Citizenship in Victorian Birmingham.* Aldershot: Ashgate, 2004.

Rosenberg, Emily. *A World Connecting, 1870–1945.* Cambridge, MA: Harvard University Press, 2012.

Rosenman, Helen. *Two Voyages to the South Seas.* Melbourne: Melbourne University Press, 1992.

Ross, James Clark. *A Voyage of Discovery and Research in the Southern and Antarctic Regions During the Years 1839–1843.* 2 volumes. London: John Murray, 1847.

Roudometof, Victor. 'Transnationalism, Cosmopolitanism and Glocalisation'. *Current Sociology* 53 (2005): 113–35.

Royal Botanic Garden. *A Popular Guide to the Royal Botanic Garden of Belfast: Published for the Garden.* Belfast: W. &. G. Agnew, 1851.

Rozwadowski, Helen. *Fathoming the Ocean.* Cambridge, MA: Harvard University Press, 2005.

Rupke, Nicolaas. *Richard Owen: Victorian Naturalist.* New Haven: Yale University Press, 1994.

Rupke, Nicolaas. 'Translation Studies in the History of Science: The Example of *Vestiges*'. *British Journal for the History of Science* 33 (2000): 209–22.

Rupke, Nicolaas. *Alexander von Humboldt: A Metabiography.* Chicago: University of Chicago Press, 2007.

Russell, Peter A. *How Agriculture Made Canada: Farming in the Nineteenth Century.* Montreal and Kingston: McGill-Queen's University Press, 2012.

Ryan, James R. 'Visualizing Imperial Geography: Halford Mackinder and the Colonial Office Visual Instruction Committee, 1902–11'. *Cultural Geographies* 1 (1994): 157–76.

Ryan, James R. 'Photography, Visual Revolutions, and Victorian Geography'. In *Geography and Revolution*, edited by David N. Livingstone and Charles W.J. Withers, 199–238. Chicago: University of Chicago Press, 2005.

Safier, Neil. 'Global Knowledge on the Move: Itineraries, Amerindian Narratives, and Deep Histories of Science'. *Isis* 101 (2010): 133–45.

Said, Edward W. *Culture and Imperialism*. London: Chatto and Windus, 1993.

Said, Edward W. *Orientalism*. New York: Vintage Books, 1994.

Salmond, Anne. *The Trial of the Cannibal Dog: Captain Cook in the South Seas*. London: Penguin, 2003.

Samson, Jane. *Imperial Benevolence: Making British Authority in the Pacific Islands*. Honolulu: University of Hawaii Press, 1998.

Saunders, Angharad. 'Literary Geography: Reforging the Connections'. *Progress in Human Geography* 34 (2010): 436–52.

Schaffer, Simon. 'Newton on the Beach: The Information Order of the *Principia Mathematica*'. *History of Science* 47 (2009): 243–76.

Schaffer, Simon, Roberts, Lissa, Raj, Kapil and Delbourgo, James, eds. *The Brokered World: Go-Betweens and Global Intelligence, 1780–1820*. Sagamore Beach, MA: Science History Publications, 2009.

Schlee, Susan. *The Edge of an Unfamiliar World: A History of Oceanography*. New York: E.P. Dutton & Co. Inc., 1973.

Secord, Anne. 'Corresponding Interests: Artisans and Gentlemen in Nineteenth-Century Natural History'. *The British Journal for the History of Science* 27 (1994): 383–408.

Secord, James. 'Knowledge in Transit'. *Isis* 95 (2004): 654–72.

Secord, James. 'How Scientific Conversation Became Shop Talk'. *Transactions of the Royal Historical Society* 17 (2007): 129–56.

Secord, James. 'Global Darwin'. In *Darwin*, edited by William Brown and Andrew C. Fabian, 31–57. Cambridge: Cambridge University Press, 2010.

Seville, Catherine. *The Internationalisation of Copyright Law: Books, Buccaneers and the Black Flag in the Nineteenth Century*. Cambridge: Cambridge University Press, 2006.

Shapin, Steven. 'Placing the View from Nowhere: Historical and Sociological Problems in the Location of Science'. *Transactions of the Institute of British Geographers* 23 (1998): 5–12.

Shaw, Wendy M.K. *Possessors and Possessed: Museums, Archaeology, and the Visualization of History in the Late Ottoman Empire*. Berkeley: University of California Press, 2003.

Shermer, Michael. *In Darwin's Shadow*. Oxford: Oxford University Press, 2011.

Shiebinger, Londa and Swan, Claudia, eds. *Colonial Botany: Science, Commerce and Politics in the Early Modern World*. Philadelphia: University of Pennsylvania Press, 2005.

Short, J.R. *Cartographic Encounters: Indigenous Peoples and the Exploration of the New World*. London: Reaktion, 2009.

Shteir, Ann B. *Cultivating Women, Cultivating Science: Flora's Daughters and Botany in England, 1760–1860*. Baltimore: Johns Hopkins University Press, 1996.

Sidorko, Clemens P. 'Nineteenth-Century German Travelogues as Sources on the History of Daghestan and Chechnya'. *Central Asia Survey* 31 (2002): 283–99.

Sivasundaram, Sujit. 'Introduction: Global Histories of Science'. *Isis* 101 (2010): 95–7.

Sivasundaram, Sujit. 'Science and the Global: On Methods, Questions and Theory'. *Isis* 101 (2010): 146–58.

Smiles, Sam. 'Record and Reverie: Representing British Antiquity in the Eighteenth Century'. In *Enlightening the British: Knowledge, Discovery and the Museum in the Eighteenth Century*, edited by R.G.W. Anderson, Marjorie L. Caygill, A.G. MacGregor and L. Syson, 176–84. London: British Museum Press, 2003.

Smith, Bernard. *European Vision and the South Pacific*. New Haven: Yale University Press, 1985.

Smith, Charlotte F.H. and Stevenson, Michelle. 'Modelling Cultures: 19th Century Indian Clay Figures'. *Museum Anthropology* 33 (2010): 37–48.

Smith, Mark M. *How Race is Made: Slavery, Segregation, and the Senses*. Chapel Hill: University of North Carolina Press, 2006.

Smith, Neil. *American Empire: Roosevelt's Geographer and the Prelude to Globalization*. Berkeley: University of California Press, 2003.

Soloway, Beverley. 'The Fur Traders' Garden: Horticultural Imperialism in Rupert's Land, 1670–1770'. In *Irish and Scottish Encounters with Indigenous Peoples: Canada, the United States, New Zealand, and Australia*, edited by Graeme Morton and D.A. Wilson, 287–303. Montreal and Kingston: McGill-Queen's University Press, 2013.

Sörlin, Sverker. 'Ordering the World for Europe: Science as Intelligence and Information as Seen from the Northern Periphery'. *Osiris* 2nd ser. 15 (2000): 51–69.

Sorrenson, Richard. 'The Ship as a Scientific Instrument in the Eighteenth Century'. *Osiris* 2nd ser. 11 (1996): 221–36.

Stark, Susanne. *Translation and Anglo-German Cultural Relations in the Nineteenth Century*. Clevedon: Multilingual Matters Ltd, 1999.

Stein, Claudia and Cooter, Roger. 'Visual Objects and Universal Meanings: AIDS Posters and the Politics of Globalisation and History'. *Medical History* 55 (2011): 85–108.

Stendall, J.A.S. 'New Municipal Museum and Art Gallery in Belfast', *North Western Naturalist* 4 (1929): 170–72.

Stepan, Nancy Leys. *Picturing Tropical Nature*. London: Reaktion Books, 2001.

Stewart, Larry. 'Global Pillage: Science, Commerce, and Empire'. In *The Cambridge History of Science. Volume 4: Eighteenth-Century Science*, edited by Roy Porter, 825–44. Cambridge: Cambridge University Press, 2003.

Stillman, William J. *Report of W. J. Stillman on the Cesnola Collection*. New York: Thompson and Moreau, 1885.

Stocking, George. *Victorian Anthropology*. New York: Free Press, 1987.

Tatton-Brown, Veronica, ed. *Cyprus in the 19th Century AD: Fact, Fancy and Fiction*. Oxford: Oxbow Books, 2001.

Tautz, Birgit. 'Cutting, Pasting, Fabricating: Late 18th-Century Travelogues and their German Translators Between Legitimacy and Imaginary Nations'. *German Quarterly* 79 (2006): 155–74.

Thomas, Nicholas. *Entangled Objects: Exchange, Material Culture, and Colonialism in the Pacific*. Cambridge, MA: Harvard University Press, 1991.

Thomas, Nicola J. 'Embodying Imperial Spectacle: Dressing Lady Curzon, Vicereine of India 1899–1905'. *Cultural Geographies* 14 (2007): 369–400.

Thompson, William. *The Natural History of Ireland*. 4 volumes. London: Reeve, Benham and Reeve, 1849–56.

Thrift, Nigel, Driver, Felix and Livingstone, David N. 'Editorial'. *Environment and Planning D: Society and Space* 13 (1995): 1.

Todes, Daniel. 'Global Darwin: Contempt for Competition'. *Nature* 462 (5 November 2009): 36–7.

Trouillot, Michel-Rolph. *Silencing the Past: Power and the Production of History*. Boston: Beacon Press, 1995.

Turchetti, Simone, Herran, Néstor and Boudia, Soraya. 'Introduction: Have We Ever Been 'Transnational'? Towards a History of Science Across and Beyond Borders'. *British Journal for the History of Science* 45 (2012): 319–36.

Turnbull, David. 'Local Knowledge and Comparative Scientific Traditions'. *Knowledge and Policy* 6 (1993–4): 29–54.

Vetter, Jeremy, ed. *Knowing Global Environments: New Historical Perspectives on the Field Sciences*. New Brunswick, NJ: Rutgers University Press, 2011.

Wallace, Alfred Russel. *Palm Trees of the Amazon and their Uses*. London: J. Van Voorst, 1853.

Wallerstein, Immanuel. *The Modern World System I: Capitalist Agriculture and the Origins of the European World-Economy in the Sixteenth Century*. New York: Academic Press, 1974.

Walsh, Oonagh. 'The Dublin University Mission Society, 1890–1905'. *History of Education* 24 (1995): 61–72.

Warf, Barney and Arias, Santa, eds. *The Spatial Turn: Interdisciplinary Perspectives*. New York: Routledge, 2009.

Wenzlhuemer, Roland. *Connecting the Nineteenth-Century World: The Telegraph and Globalization*. Cambridge: Cambridge University Press, 2013.

Whatmore, Sarah. *Hybrid Geographies*. London: Sage, 2002.

Whelan, Frederick G. *Enlightenment Political Thought and Non-Western Societies: Sultans and Savages*. London: Routledge, 2009.

Wilkes, Charles. *Narrative of the United States Exploring Expedition During the Years 1838, 1839, 1840, 1841, 1842. 5 vols. and Atlas*. Philadelphia: Lea and Blanchard, 1845.

Williams, Glyn. *The Death of Captain Cook: A Hero Made and Unmade*. London: Profile, 2008.

Williams, Glyn. *Naturalists at Sea: From Dampier to Darwin*. New Haven: Yale University Press, 2013.

Williams, Glyndwr. 'The Hudson's Bay Company and its Critics in the Eighteenth Century'. *Transactions of the Royal Historical Society* 5th ser. 20 (1970): 149–71.

Winch, Donald. *Riches and Poverty: An Intellectual History of Political Economy in Britain, 1750–1834*. Cambridge: Cambridge University Press, 1996.

Winks, Robin W. ed. *The Oxford History of the British Empire: Volume V: Historiography*. Oxford: Oxford University Press, 1999.

Wintle, Claire. 'Model Subjects: Representations of the Andaman Islands at the Colonial and Indian Exhibition, 1886'. *History Workshop Journal* 67 (2009): 194–207.

Withers, Charles W.J. *Geography, Science and National Identity: Scotland Since 1520*. Cambridge: Cambridge University Press, 2001.

Withers, Charles W.J. *Placing the Enlightenment: Thinking Geographically about the Age of Reason*. Chicago: University of Chicago Press, 2007.

Withers, Charles W.J. 'Place and the "Spatial Turn" in Geography and History'. *Journal of the History of Ideas* 70 (2009): 637–58.

Withers, Charles W.J. and Finnegan, Diarmid A. 'Natural History Societies, Fieldwork and Local Knowledge in Nineteenth-Century Scotland: Towards a Historical Geography of Civic Science'. *Cultural Geographies* 10 (2003): 334–53.

Withers, Charles W.J. and Keighren, Innes M. 'Travels into Print: Authoring, Editing and Narratives of Travel and Exploration, c.1815–c.1857'. *Transactions of the Institute of British Geographers* 36 (2011): 560–73.

Wolff, Larry. *Inventing Eastern Europe: The Map of Civilization on the Mind of the Enlightenment*. Stanford: Stanford University Press, 1994.

Wright, Jonathan Jeffrey. *The 'Natural Leaders' and their World: Politics, Culture and Society in Belfast, c. 1801–1832*. Liverpool: Liverpool University Press, 2012.

Wright, Jonathan Jeffrey. '"The Belfast Chameleon": Ulster, Ceylon and the Imperial Life of Sir James Emerson Tennent'. *Britain and the World* 6 (2013): 192–219.

Wright, Jonathan Jeffrey. '"The Perverted Graduates of Oxford": Priestcraft, "Political Popery" and the Transnational Anti-Catholicism of Sir James Emerson Tennent'. In *Transnational Perspectives on Modern Irish History*, edited by Niall Whelehan, 127–48. London: Routledge, 2014.

Wrigley, E.A. and Souden, David, eds. *The Works of Thomas Robert Malthus*. 8 volumes. London: Pickering, 1986.

Wrobel, David M. *Global West, American Frontier*. Albuquerque: University of New Mexico Press, 2013.

Yusoff, Kathryn. 'Climates of Sight: Mistaking Visibilities, Mirages and "Seeing Beyond" in Antarctica'. In *High Places: Cultural Geographies of Mountains and Ice*, edited by Denis Cosgrove and Veronica Della Dora, 64–86. London: I.B. Tauris, 2008.

Zeller, Suzanne. 'The Spirit of Bacon: Science and Self-Perception in the Hudson's Bay Company, 1830–1870'. *Scientia Canadensis* 13 (1989): 79–101.

Zeller, Suzanne. *Inventing Canada: Early Victorian Science and the Idea of a Transcontinental Nation*. Montreal and Kingston: McGill-Queen's University Press, 2009.

Index

Note: page references in italics refer to figures.